GREEN CHEMISTRY

FUNDAMENTALS OF SUSTAINABLE

CHEMICAL SCIENCE

AND TECHNOLOGY

Stanley E. Manahan

2004

ChemChar Research, Inc
Publishers
Columbia, Missouri U.S.A.

ChemChar Research, Inc.
2005 Woodlea Drive
Columbia, MO 65201 U.S.A.
http://www.chemchar.com
chemchar@headgap.com

International Standard Book Number: 0-9749522-0-6

Printed in the United States of America

TABLE OF CONTENTS

THE AUTHOR

Stanley E. Manahan is Professor of Chemistry at the University of Missouri-Columbia, where he has been on the faculty since 1965. He received his A.B. in chemistry from Emporia State University in 1960 and his Ph.D. in analytical chemistry from the University of Kansas in 1965. Since 1968 his primary research and professional activities have been in environmental chemistry, toxicological chemistry, and waste treatment. His classic textbook, *Environmental Chemistry*, 8th ed (CRC Press, Boca Raton, Florida, 2004) has been in print continuously in various editions since 1972 and is the longest standing title on this subject in the world. Other books that he has written are *Toxicological Chemistry and Biochemistry*, 3rd ed. (CRC Press/Lewis Publishers, 2001), *Fundamentals of Environmental Chemistry,* 2nd ed. (CRC Press/Lewis Publishers, 1997), *Industrial Ecology: Environmental Chemistry and Hazardous Waste* (CRC Press/Lewis Publishers, 1999), *Environmental Science and Technology* (CRC Press/Lewis Publishers, 1997), *Hazardous Waste Chemistry, Toxicology and Treatment* (Lewis Publishers, 1992), *Quantitative Chemical Analysis*, (Brooks/Cole, 1986), and *General Applied Chemistry*, 2nd ed. (Willard Grant Press, 1982). He has lectured on the topics of environmental chemistry, toxicological chemistry, waste treatment, and green chemistry throughout the U.S. as an American Chemical Society Local Section Tour Speaker, and has presented plenary lectures on these topics in international meetings in Puerto Rico; the University of the Andes in Mérida, Venezuela; Hokkaido University in Japan; the National Autonomous University in Mexico City; and Italy. He was the recipient of the Year 2000 Award of the Environmental Chemistry Division of the Italian Chemical Society. His research specialty is gasification of hazardous wastes.

PREFACE

Green Chemistry: Fundamentals of Sustainable Chemical Science and Technology was written to provide an overview of the emerging discipline of green chemistry along with the fundamental chemical principles needed to understand this science. It goes without saying that sustainability must be achieved if humankind is to survive with any sort of reasonable living standard on Planet Earth. Chemists and chemical science have an essential role to play in achieving sustainability. In the chemical sciences, green chemistry has developed since the 1990s as a key to sustainability. And it is crucial that nonchemists have an understanding of green chemistry and how it can be used to achieve sustainability, not just for humans, but for all life forms as well, on our fragile planet. Therefore, this book includes a basic introduction to the principles of chemistry for those readers who may have little or no prior knowledge of this subject.

Laudable as its goals and those who work to achieve them are, green chemistry has developed a somewhat narrow focus. For the most part, it has concentrated largely on chemical synthesis, more specifically organic synthesis. It has not been sufficiently inclusive of other areas pertinent to the achievement of sustainability, such as environmental chemistry and the science of industrial ecology. This book attempts to integrate these and other pertinent disciplines into green chemistry. In so doing, it recognizes five overlapping and interacting environmental spheres. Four of these have long been recognized by practitioners of environmental science. They are (1) the biosphere, (2) the hydrosphere, (3) the geosphere, and (4) the atmosphere. But, to be realistic, a fifth sphere must be recognized and studied. This is the anthrosphere, which consists of all of the things that humans have made and the systems that they operate throughout the environment. Highways, buildings, airports, factories, cultivated land, and a huge variety of structures and systems produced by human activities are part of Earth as we know it and must be dealt with in any comprehensive view of the environment. A basic aspect of this book is to deal with the five environmental spheres and to discuss how — for better or worse — the anthrosphere is an integral part of this Earth system.

Chapters 1–4 of this book introduce the basic concepts of chemistry and green chemistry. Chapter 1, "Chemistry, Green Chemistry and Environmental Chemistry," includes a brief "minicourse" in chemistry that introduces the reader to fundamental ideas of atoms, elements, compounds, chemical formulas, and chemical equations so that the reader can have the background to understand these aspects in later chapters. Chapter 2, "The Elements: Basic Building Blocks of Green Chemicals," introduces the elements and fundamentals of atomic structure. It develops an abbreviated version of the periodic table consisting of the first 20 elements to give the reader an understanding of this important foundation of chemistry. It also points out the green aspects of these elements, such as elemental hydrogen as a means of energy storage and transport and fuel for nonpolluting fuel cells. Chapter 3, "Compounds: Safer Materials for a Safer World," explains chemical bonding, chemical formulas, and the concept of the mole. It points out how some chemical compounds are greener than others, for example, those that are relatively more biodegradable compared to ones that tend to persist in the environment. With an understanding of chemical compounds, Chapter 4. "Chemical Reactions: Making Materials Safely Without Damaging The Environment," discusses how compounds are made and changed and introduces the idea of stoichiometry. It develops some key ideas of green chemistry such as atom economy and illustrates what makes some chemical reactions more green than others.

It is impossible to consider green chemistry in a meaningful manner without consideration of organic chemistry. Furthermore, given the importance of biosynthesis and the biological effects of toxic substances, it is essential to have a basic understanding of biochemistry. These subjects are covered in Chapter 5, "The Wonderful World of Carbon: Organic Chemistry and Biochemistry."

Chapter 6, "Energy Relationships," discusses the crucial importance of energy in green chemistry. It explains how abundant, sustainable, environmentally friendly energy sources are a fundamental requirement in maintaining modern societies in a sustainable manner. Chapter 7, "Green Water," discusses water resources and the environmental chemistry of water. The environmental chemistry of the atmosphere is covered in Chapter 8, "Air and The Atmosphere." This chapter also explains how the atmosphere is a sustainable source of some important raw materials, such as nitrogen used to make nitrogen fertilizers. The biosphere is discussed in Chapter 9, "The Biosphere: How The Revolution in Biology Relates to Green Chemistry." Obviously, protection of the biosphere is one of the most important goals of green chemistry. This chapter explains how the biosphere is a renewable source of some key raw materials. The geosphere is introduced in Chapter 10. "The Geosphere, Soil, And Food Production: The Second Green Revolution In Agriculture." Soil and its role in producing food and raw materials are discussed in this chapter. The concepts of the anthrosphere and industrial ecology are covered in Chapter 11, "The Anthrosphere and Industrial Ecology." Feedstocks, which are required to support the chemical industry are discussed in Chapter 12, "Feedstocks: Maximum Utilization of Renewable and Biological Materials." Emphasis is placed on renewable feedstocks from biological sources in place of depletable petroleum feedstocks.

Terrorism has become a central problem of our time. A unique feature of this book is its coverage of this topic in Chapter 13, "Terrorism, Toxicity, and Vulnerability: Chemistry in Defense of Human Welfare." Included are agents of terrorism such as military poisons, means of detecting terrorist threats, and measures that may be taken to reduce such threats. Because of the threats posed by toxic agents, toxicological chemistry is introduced and discussed in this chapter.

Reader feedback is eagerly solicited. Questions and suggestions may be forwarded to the author at manahans@missouri.edu.

1. CHEMISTRY, GREEN CHEMISTRY, AND ENVIRONMENTAL CHEMISTRY

1.1. CHEMISTRY IS GOOD

Chemistry is the science of matter. Are you afraid of chemistry? Many people are and try to avoid it. But avoiding chemistry is impossible. That is because all matter, all things, the air around us, the water we must drink, and all living organisms are made of chemicals. People who try to avoid all things that they regard as chemical may fail to realize that chemical processes are continuously being carried out in their own bodies. These are processes that far surpass in complexity and variety those that occur in chemical manufacturing operations. So, even those people who want to do so cannot avoid chemistry. The best course of action with anything that cannot be avoided and that might have an important influence on our lives (one's chemistry professor may come to mind) is to try to understand it, to deal with it. To gain an understanding of chemistry is probably why you are reading this book.

Fundamentals of Green Chemistry is written for a reader like you. It seeks to present a body of chemical knowledge from the most fundamental level within a framework of the relationship of chemical science to human beings, their surroundings, and their environment. Face it, the study of chemistry based upon facts about elements, atoms, compounds, molecules, chemical reactions, and other basic concepts needed to understand this science is found by many to be less than exciting. However, these concepts and many more are essential to a meaningful understanding of chemistry. Anyone interested in green chemistry clearly wants to know how chemistry influences people in the world around us. So this book discusses the real world aspects of chemistry, introducing chemical principles as needed.

During the approximately two centuries that chemical science has been practiced on an ever-increasing scale, it has enabled the production of a wide variety of goods that are valued by humans. These include such things as plastics, fertilizers, paints, and life-saving pharmaceuticals that humans use and value. Without the persistent efforts of chemists and the enormous productivity of the chemical industry, nothing approaching the high standard of living enjoyed in modern industrialized societies would be possible.

But there can be no denying that in years past, and even at present, chemistry has been misused in many respects, resulting in harm to the environment and living things, including humans. It is now obvious that chemical science must be turned away from emphasis upon the exploitation of limited resources and the production of increasing amounts of products that ultimately end up as waste and toward the application of chemistry in ways that provide for human needs without damaging the Earth support system upon which all living things depend. Fortunately, the practice of chemical science and industry is moving steadily in the direction of environmental friendliness and resource sustainability.

The practice of chemistry in a manner that maximizes its benefits while eliminating or at least greatly reducing its adverse impacts has come to be known as **green chemistry**, the topic of this book.

As will be seen in later chapters of this book, the practice of chemistry is divided into several major categories. Most elements other than carbon are involved with **inorganic chemistry**. Common examples of inorganic chemicals are water, salt (sodium chloride), air pollutant sulfur dioxide, and lime. Carbon occupies a special place in chemistry because it is so versatile in the kinds of chemical species (compounds) that it forms. Most of the more than 20 million known chemicals are substances based on carbon known as organic chemicals and addressed by the subject of **organic chemistry**. The unique chemistry of carbon is addressed specifically in Chapter 5, "The Wonderful World of Carbon: Organic Chemistry and Biochemicals." The underlying theory and physical phenomena that explain chemical processes are explained by **physical chemistry**. Living organisms carry out a vast variety of chemical processes that are important in green chemistry and environmental chemistry. The chemistry that living organisms perform is addressed under the topic of **biochemistry**, which is also covered in Chapter 5. It is always important to know the identities and quantities of various chemical species present in a system, including various environmental systems. Often, significant quantities of chemical species are very low, so sophisticated means must be available to detect and quantify such species. The branch of chemistry dealing with the determination of kinds and quantities of chemical species is **analytical chemistry**.

As the chemical industry developed and grew during the early and mid 1900s, most practitioners of chemistry remained unconcerned with and largely ignorant of the potential for harm — particularly damage to the outside environment — of their products and processes. Environmental chemistry was essentially unknown and certainly not practiced by most chemists. Incidents of pollution and environmental damage, which were many and severe, were commonly accepted as a cost of doing business or blamed upon the industrial or commercial sectors. However, nobody is more qualified to accept responsibility for environmental damage from chemical products or processes than are chemists who have the knowledge to understand how such harmful effects came about. As the detrimental effects of chemical manufacture and use became more obvious and severe, chemists were forced, often reluctantly, to deal with them. At present, enlightened chemists and chemical engineers do not view the practice of environmentally beneficial chemistry and manufacturing as a burden, but rather as an opportunity that challenges human imagination and ingenuity.

1.2. THE ENVIRONMENT AND THE FIVE ENVIRONMENTAL SPHERES

Compared to the generally well defined processes that chemists study in the laboratory, those that occur in the environment are rather complex and must be viewed in terms of simplified models. A large part of this complexity is due to the fact that environmental chemistry must take into account five interacting and overlapping compartments or spheres of the environment, which affect each other and which undergo continual interchanges of matter and energy. Traditionally, environmental science has considered water, air, earth, and life — that is, the **hydrosphere**, the **atmosphere**, the **geosphere**, and the **biosphere**. When considered at all, human activities were generally viewed as undesirable perturbations on these other spheres, causing pollution and generally adverse effects. Such a view is too narrow, and we must include a fifth sphere, the anthrosphere, consisting of the things humans make and do. By regarding the anthrosphere as an integral part of the environment, humans can modify their anthrospheric activities to do minimal harm to the environment, or to even improve it.

Figure 1.1 shows the five spheres of the environment, including the anthrosphere, and some of the exchanges of material between them. Each of these spheres is described briefly below.

The atmosphere is a very thin layer compared to the size of Earth, with most atmospheric gases lying within a few kilometers of sea level. In addition to providing oxygen for living organisms, the atmosphere provides carbon dioxide required for plant photosynthesis, and nitrogen that organisms use to make proteins. The atmosphere serves a vital protective function in that it absorbs highly energetic ultraviolet radiation from the sun that would kill living organisms exposed to it. A particularly important part of the atmosphere in this respect is the stratospheric layer of ozone, an ultraviolet-absorbing form of elemental oxygen. Because of its ability to absorb infrared radiation by which Earth loses the energy that it absorbs from the sun, the atmosphere stabilizes Earth's surface temperature. The atmosphere also serves as the medium by which the solar energy that falls with greatest intensity in equatorial regions is redistributed away from the Equator. It is the medium in which water vapor evaporated from oceans as the first step in the hydrologic cycle is transported over land masses to fall as rain over land.

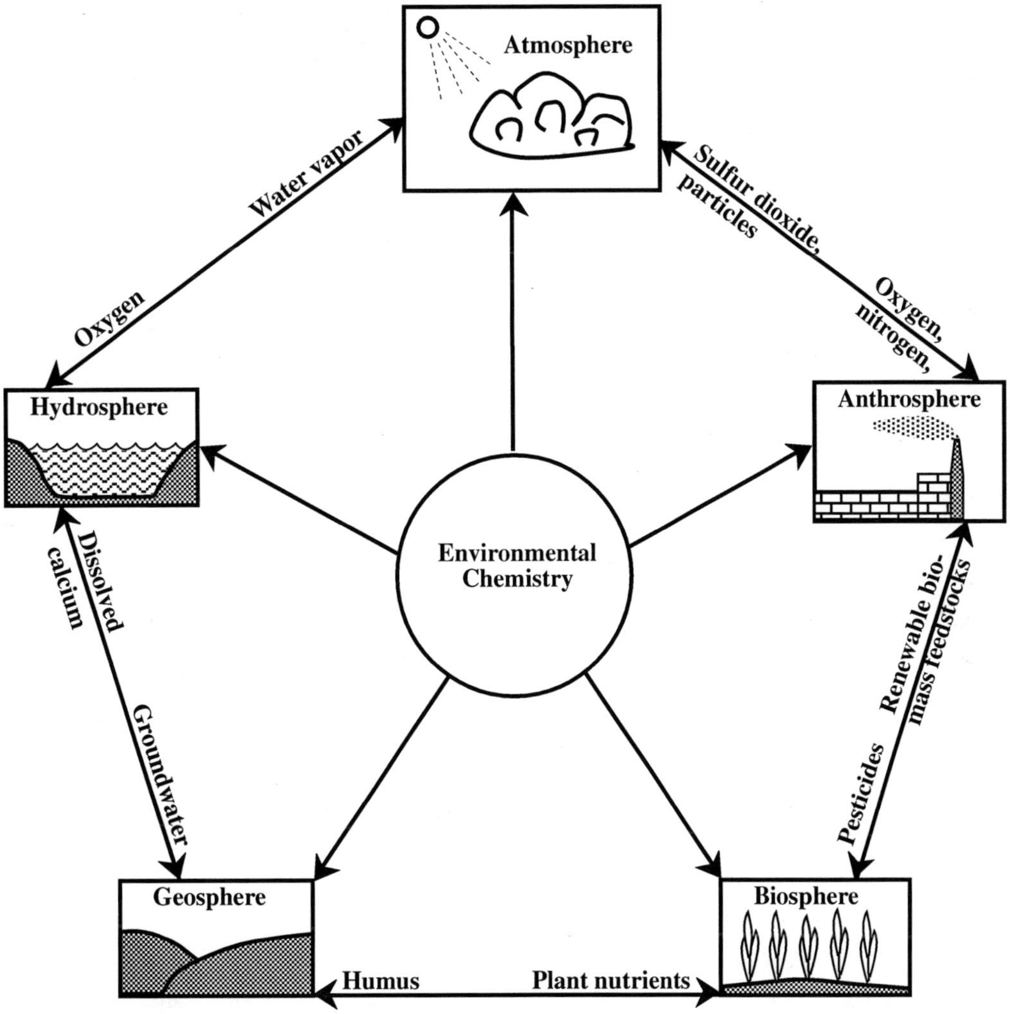

Figure 1.1. Illustration of the five major spheres of the environment. These spheres are closely tied together, interact with each other, and exchange materials and energy. A meaningful examination of environmental sciences must include all five of these spheres, including the anthrosphere.

Earth's water is contained in the hydrosphere. Although frequent reports of torrential rainstorms and flooded rivers produced by massive storms might give the impression that a large fraction of Earth's water is fresh water, more than 97% of it is seawater in the oceans. Most of the remaining fresh water is present as ice in polar ice caps and glaciers. A small fraction of the total water is present as vapor in the atmosphere. The remaining liquid fresh water is that available for growing plants and other organisms and for industrial uses. This water may be present on the surface as lakes, reservoirs, and streams, or it may be underground as groundwater.

The solid part of earth, the geosphere, includes all rocks and minerals. A particularly important part of the geosphere is soil, which supports plant growth, the basis of food for all living organisms. The **lithosphere** is a relatively thin solid layer extending from Earth's surface to depths of 50–100 km. The even thinner outer skin of the lithosphere known as the **crust** is composed of relatively lighter silicate-based minerals. It is the part of the geosphere that is available to interact with the other environmental spheres and that is accessible to humans.

The biosphere is composed of all living organisms. For the most part, these organisms live on the surface of the geosphere on soil, or just below the soil surface. The oceans and other bodies of water support high populations of organisms. Some life forms exist at considerable depths on ocean floors. In general, though, the biosphere is a very thin layer at the interface of the geosphere with the atmosphere. The biosphere is involved with the geosphere, hydrosphere, and atmosphere in **biogeochemical cycles** through which materials such as nitrogen and carbon are circulated.

Through human activities, the anthrosphere has developed strong interactions with the other environmental spheres. Many examples of these interactions could be cited. By cultivating large areas of soil with domestic crops, humans modify the geosphere and influence the kinds of organisms in the biosphere. Humans divert water from its natural flow, use it, sometimes contaminate it, then return it to the hydrosphere. Emissions of particles to the atmosphere by human activities affect visibility and other characteristics of the atmosphere. The emission of large quantities of carbon dioxide to the atmosphere by combustion of fossil fuels may be modifying the heat-absorbing characteristics of the atmosphere to the extent that global warming is almost certainly taking place. The anthrosphere perturbs various biogeochemical cycles.

The effect of the anthrosphere over the last two centuries in areas such as burning large quantities of fossil fuels is especially pronounced upon the atmosphere and has the potential to change the nature of Earth significantly. According to Nobel Laureate Paul J. Crutzen of the Max Planck Institute for Chemistry, Mainz, Germany, this impact is so great that it will lead to a new global epoch to replace the halocene epoch that has been in effect for the last 10,000 years since the last Ice Age. Dr. Crutzen has coined the term **anthropocene**, a term derived from anthropogenic, to describe the new epoch that may be upon us.

1.3. WHAT IS ENVIRONMENTAL CHEMISTRY?

The practice of green chemistry must be based upon **environmental chemistry** defined as *the study of the sources, reactions, transport, effects, and fates of chemical species in water, soil, air, and living environments and the effects of technology thereon.*[1] Figure 1.2 illustrates this definition of environmental chemistry with an important type of environmental chemical species. In this example, two of the ingredients required for the formation of photochemical smog — nitric oxide and hydrocarbons — are emitted to the atmosphere from vehicles and transported through the atmosphere by wind and air currents. In the atmosphere, energy from sunlight brings about photochemical reactions that convert nitric oxide and hydrocarbons to ozone, noxious organic compounds, and particulate matter, all characteristic of photochemical smog. Various harmful effects are manifested, such as visibility-obscuring particles in the atmosphere, or ozone, which is unhealthy when inhaled by humans, or toxic to plants. Finally, the smog products end up on soil, deposited on plant surfaces, or in bodies of water.

Figure 1.1 showing the five environmental spheres may provide an idea of the complexity of environmental chemistry as a discipline. Enormous quantities of materials are exchanged among the five environmental spheres. In addition to variable flows of materials, there are variations in temperature, intensity of solar radiation, mixing, and other factors that strongly influence chemical conditions and behavior.

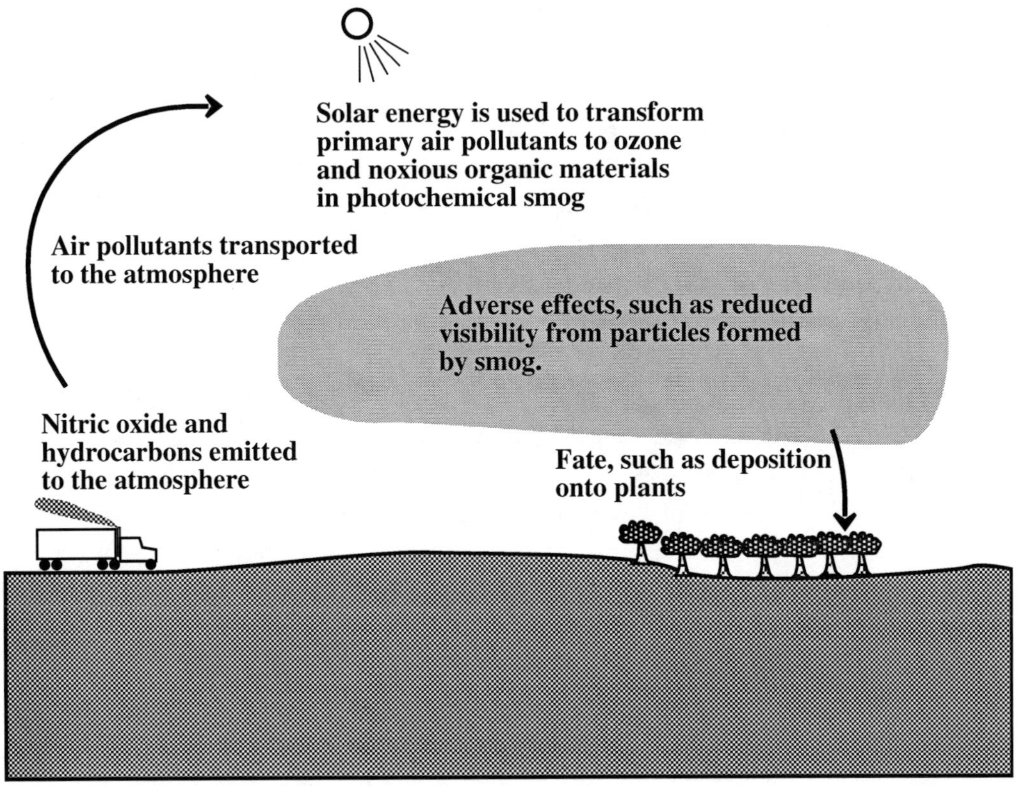

Figure 1.2. Illustration of the definition of environmental chemistry with a common environmental contaminant.

Throughout this book the role of environmental chemistry in the practice of green chemistry is emphasized. Green chemistry is practiced to minimize the impact of chemicals and chemical processes upon humans, other living organisms, and the environment as a whole. It is only within the framework of a knowledge of environmental chemistry that green chemistry can be successfully practiced.

There are several highly interconnected and overlapping categories of environmental chemistry. **Aquatic chemistry** deals with chemical phenomena and processes in water. Aquatic chemical processes are very strongly influenced by microorganisms in the water, so there is a strong connection between the hydrosphere and biosphere insofar as such processes are concerned. Aquatic chemical processes occur largely in "natural waters" consisting of water in oceans, bodies of fresh water, streams, and underground aquifers. These are places in which the hydrosphere can interact with the geosphere, biosphere, and atmosphere and is often subjected to anthrospheric influences. Aspects of aquatic chemistry are considered in various parts of this book and are addressed specifically in Chapter 7, "Green Water."

Atmospheric chemistry is the branch of environmental chemistry that considers chemical phenomena in the atmosphere. Two things that make this chemistry unique are the extreme dilution of important atmospheric chemicals and the influence of photochemistry. Photochemistry occurs when molecules absorb photons of high-energy visible light or ultraviolet radiation, become energized ("excited"), and undergo reactions that lead to a variety of products, such as photochemical smog. In addition to reactions that occur in the gas phase, many important atmospheric chemical phenomena take place on the surfaces of very small solid particles suspended in the atmosphere and in droplets of liquid in the atmosphere. Although no significant atmospheric chemical reactions are mediated by organisms in the atmosphere, microorganisms play a strong role in determining species that get into the atmosphere. As examples, bacteria growing in the absence of oxygen, such as in cow's stomachs and under water in rice paddies, are the single greatest source of hydrocarbon in the atmosphere because of the large amounts of methane that they emit. The greatest source of organic sulfur compounds in the atmosphere consists of microorganisms in the oceans that emit dimethyl sulfide. Atmospheric chemistry is addressed specifically in Chapter 8, "Air and the Atmosphere."

Chemical processes that occur in the geosphere involving minerals and their interactions with water, air, and living organisms are addressed by the topic of geochemistry. A special branch of geochemistry, soil chemistry, deals with the chemical and biochemical processes that occur in soil. Aspects of geochemistry and soil chemistry are covered in Chapter 10 of this book, "The Geosphere, Soil, and Food Production: The Second Green Revolution in Agriculture."

Environmental biochemistry addresses biologically mediated processes that occur in the environment. Such processes include, as examples, the biodegradation of organic waste materials in soil or water and processes within biogeochemical cycles, such as denitrification, which returns chemically bound nitrogen to the atmosphere as nitrogen gas. The basics of biochemistry are presented in Chapter 5, "The Wonderful World of Carbon: Organic Chemistry and Biochemicals," and in Chapter 9, "The Biosphere: How the Revolution in Biology Relates to Green Chemistry." Chapter 12, "Feedstocks: Maximum Utilization of Renewable and Biological Materials," discusses how chemical processes carried out by organisms can produce material feedstocks needed for the practice of green chemistry. The toxic effects of chemicals are of utmost concern to chemists and the public. Chapter 13, "Terroism, Toxicity, and Vulnerability: Chemistry in Defense of Human Welfare," deals with aspects of these toxic effects and discusses **toxicological chemistry**.

Although there is not a formally recognized area of chemistry known as "anthrospheric chemistry," most of chemical science and engineering developed to date deals with chemistry carried out in the anthrosphere. Included is industrial chemistry, which is very closely tied to the practice of green chemistry. A good way to view "anthrospheric chemistry" from a green chemistry perspective is within the context of **industrial ecology**. Industrial ecology considers industrial systems in a manner analogous to natural ecosystems. In a system of industrial ecology, various manufacturing and processing operations carry out "industrial metabolism" on materials. A successful industrial ecosystem is well balanced and diverse, with various enterprises that generate products for each other and use each other's products and potential wastes. A well-functioning industrial ecosystem recycles materials to the maximum extent possible and produces little — ideally no — wastes. Therefore, a good industrial ecosystem is a green chemical system.

1.4. ENVIRONMENTAL POLLUTION

Environmental chemistry has developed in response to problems and concerns regarding environmental pollution. Although awareness of chemical pollution had increased significantly in the two decades following World War II, the modern environmental movement dates from the 1962 publication of Rachel Carson's classic book *Silent Spring*. The main theme of this book was the concentration of DDT and other mostly pesticidal chemicals through the food chain, which caused birds at the end of the chain to produce eggs with soft shells that failed to produce viable baby birds. The implication was that substances harming bird populations might harm humans as well.

Around the time of the publication of Silent Spring another tragedy caused great concern regarding the potential effects of chemicals. This was the occurrence of approximately 10,000 births of children with badly deformed or missing limbs as a result of their mothers having taken the pharmaceutical thalidomide to alleviate the effects of morning sickness at an early stage of pregnancy.

The 1960s were a decade of high concern and significant legislative action in the environmental arena aimed particularly at the control of water and air pollutants. By around 1970, it had become evident that the improper disposal of chemicals to the geosphere was also a matter of significant concern. Although many incidents of such disposal were revealed, the one that really brought the problem into sharp focus was the Love Canal site in Niagara Falls, New York. This waste dump was constructed in an old abandoned canal in which large quantities of approximately 80 waste chemicals had been placed for about two decades starting in the 1930s. It had been sealed with a clay cap and given to the city. A school had been built on the site and housing constructed around it. By 1971 it became obvious that the discarded chemicals were leaking through the cap. This problem led eventually to the expenditure of many millions of dollars to remediate the site and to buy out and relocate approximately one thousand households. More than any other single incident the Love Canal problem was responsible for the passage of legislation, including Superfund, to clean up hazardous waste sites and to prevent their production in the future.

By about 1970 it was generally recognized that air, water, and land pollution was reaching intolerable levels. As a result, various countries passed and implemented laws designed to reduce pollutants and to clean up waste chemical sites at a cost that has easily exceeded one trillion dollars globally. In many respects, this investment has been strikingly successful. Streams that had deteriorated to little more than stinking waste drainage ditches (the Cuyahoga River in Cleveland, Ohio, once caught on fire from petroleum waste floating on its surface) have been restored to a healthy and productive condition. Despite a much increased population, the air quality in smog-prone Southern California has improved markedly. A number of dangerous waste disposal sites have been cleaned up. Human exposure to toxic substances in the workplace, in the environment, and in consumer products has been greatly reduced. The measures taken and regulations put in place have prevented devastating environmental problems from occurring.

Initially, serious efforts to control pollution were based on a **command and control** approach, which specifies maximum concentration guideline levels of substances that can be allowed in the atmosphere or water and places limits on the amounts or concentrations of pollutants that can be discharged in waste streams. Command and control efforts to diminish pollution have resulted in implementation of various technologies to remove or neutralize pollutants in potential waste streams and stack gases. These are so-called end-of-pipe measures. As a result, numerous techniques, such as chemical precipitation of water pollutants, neutralization of acidic pollutants, stack gas scrubbing, and waste immobilization have been developed and refined to deal with pollutants after they are produced.

Release of chemicals to the environment is now tracked in the U.S. through the Toxics Release Inventory TRI, under requirements of the Emergency Planning and Community Right to Know Act, which requires that information be provided regarding the release of more than 300 chemicals. The release of approximately one billion kilograms of these chemicals is reported in U.S. each year. Not surprisingly, the chemical industry produces the most such substances, followed by primary metals and paper manufacture. Significant amounts are emitted from transportation equipment, plastics, and fabricated metals, with smaller quantities from a variety of other enterprises. Although the quantities of chemicals released are high, they are decreasing, and the publicity resulting from the required publication of these releases has been a major factor in decreasing the amounts of chemicals released.

Although much maligned, various pollution control measures implemented in response to command and control regulation have reduced wastes and improved environmental quality. Regulation-based pollution control has clearly been a success and well worth the expense and effort. However, it is much better to prevent the production of pollutants rather than having to deal with them after they are made. This was recognized in United States with the passage of the 1990 Pollution Prevention Act. This act explicitly states that, wherever possible, wastes are not to be generated and their quantities are to be minimized. The means for accomplishing this objective can range from very simple measures, such as careful inventory control and reduction of solvent losses due to evaporation, to much more sophisticated and drastic approaches, such as complete redesign of manufacturing processes with waste minimization as a top priority. The means for preventing pollution are best implemented through the practice of green chemistry, which is discussed in detail in the following section.

1.5. WHAT IS GREEN CHEMISTRY?

The limitations of a command and control system for environmental protection have become more obvious even as the system has become more successful. In industrialized societies with good, well-enforced regulations, most of the easy and inexpensive measures that can be taken to reduce environmental pollution and exposure to harmful chemicals have been implemented. Therefore, small increases in environmental protection now require relatively large investments in money and effort. Is there a better way? There is, indeed. The better way is through the practice of green chemistry.

Green chemistry can be defined as the practice of chemical science and manufacturing in a manner that is sustainable, safe, and non-polluting and that consumes minimum amounts of materials and energy while producing little or no waste material. The practice of green chemistry begins with recognition that the production, processing, use, and eventual disposal of chemical products may cause harm when performed incorrectly. In accomplishing its objectives green chemistry and green chemical engineering may modify or totally redesign chemical products and processes with the objective of minimizing wastes and the use or generation of particularly dangerous materials.

To a degree, we are still finding out what green chemistry is. That is because it is a rapidly evolving and developing subdiscipline in the field of chemistry. And it is a very exciting time for those who are practitioners of this developing science. Basically, green chemistry harnesses a vast body of chemical knowledge and applies it to the production, use, and ultimate disposal of chemicals in a way that minimizes consumption of materials, exposure of living organisms, including humans, to toxic substances, and damage to the environment. And it does so in a manner that is economically feasible and cost effective. In one sense, green chemistry is the most efficient possible practice of chemistry and the least costly when all of the costs of the practice of chemistry, including hazards and potential environmental damage are taken into account.

Green chemistry is sustainable chemistry. There are several important respects in which green chemistry is sustainable:

- Economic: At a high level of sophistication green chemistry normally costs less in strictly economic terms (to say nothing of environmental costs) than chemistry as it is normally practiced.
- Materials: By efficiently using materials, maximum recycling, and minimum use of virgin raw materials, green chemistry is sustainable with respect to materials.
- Waste: By reducing insofar as possible, or even totally eliminating their production, green chemistry is sustainable with respect to wastes.

1.6. GREEN CHEMISTRY AND SYNTHETIC CHEMISTRY

Synthetic chemistry is the branch of chemical science involved with developing means of making new chemicals and developing improved ways of synthesizing existing chemicals. A key aspect of green chemistry is the involvement of synthetic chemists in the practice of environmental chemistry. Synthetic chemists, whose major objective has always been to make new substances and to make them cheaper and better, have come relatively late to the practice of environmental chemistry. Other areas of chemistry have been involved much longer in pollution prevention and environmental protection. From the beginning, analytical chemistry has been a key to discovering and monitoring the severity of pollution problems. Physical chemistry has played a strong role in explaining and modeling environmental chemical phenomena. The application of physical chemistry to atmospheric photochemical reactions has been especially useful in explaining and preventing harmful atmospheric chemical effects including photochemical smog formation and stratospheric ozone depletion. Other branches of chemistry have been instrumental in studying various environmental chemical phenomena. Now the time has arrived for the synthetic chemists, those who make chemicals and whose activities drive chemical processes, to become intimately involved in making the manufacture, use, and ultimate disposal of chemicals as environmentally friendly as possible.

Before environmental and health and safety issues gained their current prominence, the economic aspects of chemical manufacture and distribution were relatively simple. The economic factors involved included costs of feedstock, energy requirements, and marketability of product. Now, however, costs must include those arising from regulatory compliance, liability, end-of-pipe waste treatment, and costs of waste disposal. By eliminating or greatly reducing the use of toxic or hazardous feedstocks and catalysts and the generation of dangerous intermediates and byproducts, green chemistry eliminates or greatly reduces the additional costs that have come to be associated with meeting environmental and safety requirements of conventional chemical manufacture.

As illustrated in Figure 1.3, there are two general and often complemetary approaches to the implementation of green chemistry in chemical synthesis, both of which challenge the imaginations of chemists and chemical engineers. The first of these is to use existing chemicals but make them by more environmentally benign processes. The second approach is to substitute other chemicals that are made by environmentally benign approaches. In some cases, a combination of the two approaches is used.

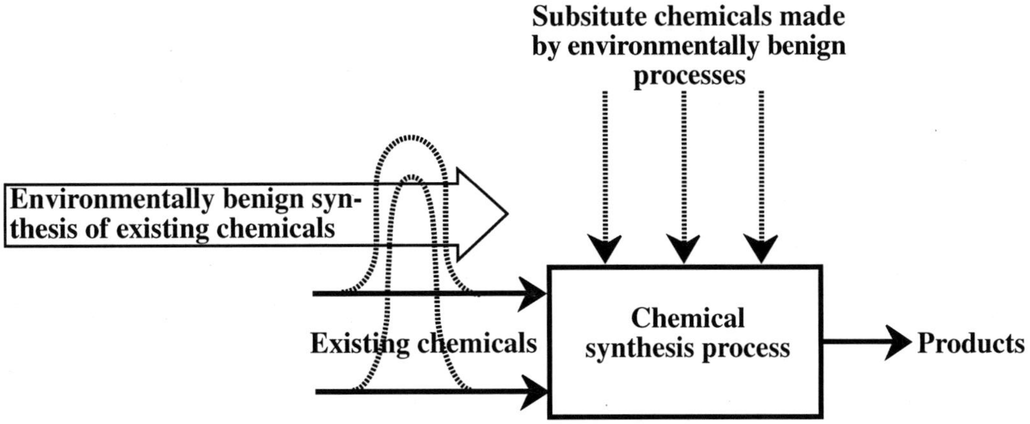

Figure 1.3. Two general approaches to the implementation of green chemistry. The dashed loops on the left represent alternative approaches to environmentally benign means of providing chemicals already used for chemical synthesis. A second approach, where applicable, is to substitute entirely different, environmentally safer raw materials.

Yield and Atom Economy

Traditionally, synthetic chemists have used **yield**, to measure the success of a chemical synthesis. Yield is defined as a percentage of the degree to which a chemical reaction or chemical synthesis goes to completion. For example, if a chemical reaction shows that 100 grams of product should be produced, but only 85 grams is produced, the yield is 85%. A synthesis with a high yield is not necessarily green because it may still generate significant quantities of useless byproducts if the reaction does so as part of the synthesis process. Instead of yield, green chemistry emphasizes **atom economy**, the fraction of reactant material that actually ends up in final product. With 100 percent atom economy, all of the material that goes into the synthesis process is incorporated into the product. For efficient utilization of raw materials, a 100% atom economy process is most desirable. Figure 1.4 illustrates the concepts of yield and atom economy.

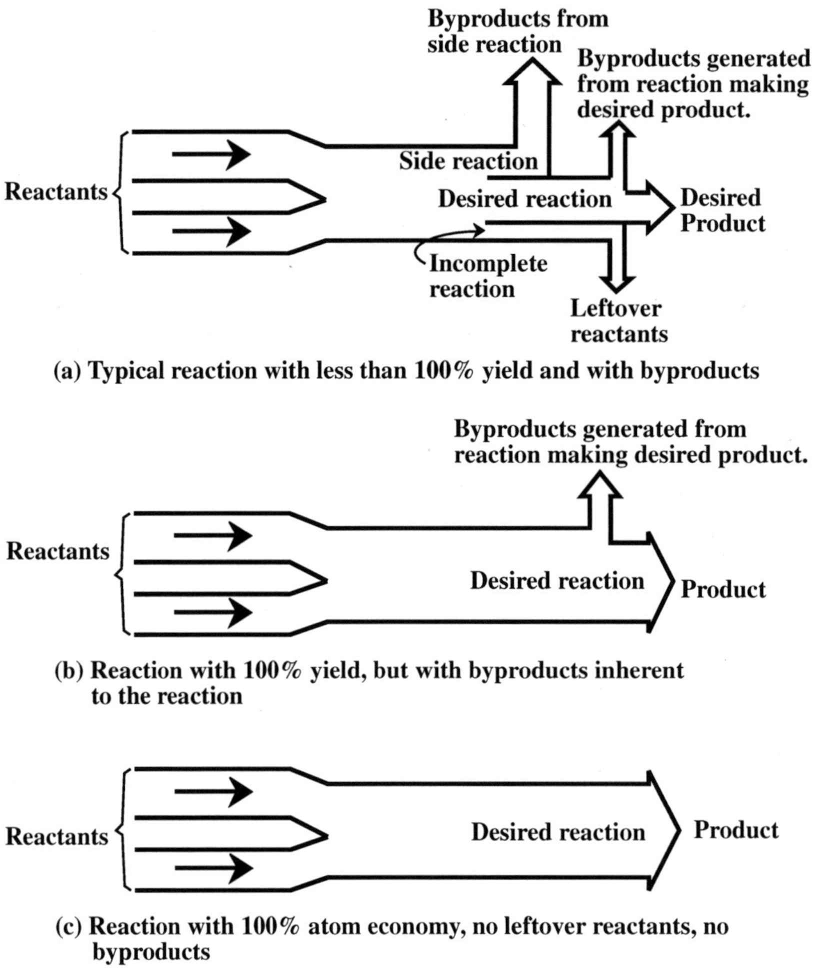

Figure 1.4. Illustration of percent yield and atom economy.

1.7. REDUCTION OF RISK: HAZARD AND EXPOSURE

A major goal in the manufacture and use of commercial products, and, indeed, in practically all areas of human endeavor, is the reduction of risk. There are two major aspects of risk — the hazard presented by a product or process and exposure of humans or other potential targets to those hazards.

$$\text{Risk} = F\{\text{hazard} \times \text{exposure}\} \quad (1.7.1)$$

This relationship simply states that risk is a function of hazard times exposure. It shows that risk can be reduced by a reduction of hazard, a reduction of exposure, and various combinations of both.

The command and control approach to reducing risk has concentrated upon reduction of exposure. Such efforts have used various kinds of controls and protective measures to limit exposure. The most common example of such a measure in the academic chemistry laboratory familiar to all chemistry students is the wearing of goggles to protect the eyes. Goggles will not by themselves prevent acid from splashing into the face of a student, but they do prevent the acid from contacting fragile eye tissue. Explosion shields will not prevent explosions, but they do retain glass fragments that might harm the chemist or others in the vicinity.

Reduction of exposure is unquestionably effective in preventing injury and harm. However, it does require constant vigilance and even nagging of personnel, as any laboratory instructor charged with making laboratory students wear their safety goggles at all times will attest. It does not protect the unprotected, such as a visitor who may walk bare-faced into a chemical laboratory ignoring the warnings for required eye protection. On a larger scale, protective measures may be very effective for workers in a chemical manufacturing operation but useless to those outside the area or the environment beyond the plant walls who do not have protection. Protective measures are most effective against acute effects, but less so against long-term chronic exposures that may cause toxic responses over many years period of time. Finally, protective equipment can fail and there is always the possibility that humans will not use it properly.

Where feasible, hazard reduction is a much more certain way of reducing risk than is exposure reduction. The human factors that play so prominently in successfully limiting exposure and that require a conscious, constant effort are much less crucial when hazards have been reduced. Compare, for example, the use of a volatile, flammable, somewhat toxic organic solvent used for cleaning and degreasing of machined metal parts with that of a water solution of a nontoxic cleaning agent used for the same purpose. To safely work around the solvent requires an unceasing effort and constant vigilance to avoid such hazards as formation of explosive mixtures with air, presence of ignition sources that could result in a fire, and excessive exposure by inhalation or absorption through skin that might cause peripheral neuropathy (a nerve disorder) in workers. Failure of protective measures can result in a bad accident or serious harm to worker health. The water-based cleaning solution, however, would not present any of these hazards so that failure of protective measures would not create a problem.

Normally, measures taken to reduce risk by reducing exposure have an economic cost that cannot be reclaimed in lower production costs or enhanced product value. Of course, failure to reduce exposure can have direct, high economic costs such as in higher claims for worker compensation. In contrast, hazard reduction often has the potential to substantially reduce operating costs. Safer feedstocks are often less costly as raw materials. The elimination of costly control measures can lower costs overall. Comparing an organic solvent to a water-based cleaning solution, the organic solvent is almost certain to cost more than the aqueous solution containing relatively low concentrations of detergents and other additives. Whereas the organic solvent will at least require purification for recycle and perhaps even expensive disposal as a hazardous waste, the water solution may be purified by relatively simple processes, and perhaps even biological treatment, then safely discharged as wastewater to a municipal wastewater treatment facility. It should be kept in mind, however, that not all low-hazard materials are cheap, and may be significantly more expensive than their more hazardous alternatives. And, in some cases, nonhazardous alternatives simply do not exist.

1.8. THE RISKS OF NO RISKS

There are limits to the reduction in risk beyond which efforts to do so become counterproductive. As in other areas of endeavor, there are circumstances in which there is no choice but to work with hazardous substances. Some things that are inherently dangerous are rendered safe by rigorous training, constant attention to potential hazards, and understanding of hazards and the best way to deal with them. Consider the analogy of commercial flight. When a large passenger aircraft lands, 100 tons of aluminum, steel, flammable fuel, and fragile human flesh traveling at a speed of twice the legal interstate speed limits for automobiles come into sudden contact with an unforgiving concrete runway. That procedure is inherently dangerous! But it is carried out hundreds of thousands of times per year throughout the world with but few injuries and fatalities. Such safe landings stand as tributes to the generally superb design, construction, and maintenance of aircraft and the excellent skills and training of aircrew. The same principles that make commercial air flight generally safe also apply to the handling of hazardous chemicals by properly trained personnel using the safest possible practices under carefully controlled conditions.

So, although much of this book is about risk reduction as it relates to chemistry, we must also be aware of the risks of not taking risks. If we become so timid in all of our enterprises that we refuse to take risks, scientific and economic progress will stagnate. The U.S. space program is an example of an area in which progress has been made only by a willingness to take risks, and progress has probably been slowed because of risk aversion resulting from previous accidents, especially the 1987 Challenger space shuttle tragedy. If a point is reached that no chemical can be made if its synthesis involves the use of a potentially toxic or otherwise hazardous substance, the progress of chemical science and the development of such beneficial products as new life-saving drugs or innovative chemicals for treating water pollutants may be held back. It may be argued that thermonuclear fusion entails significant risks as an energy source and that research on controlled thermonuclear fusion must therefore be stopped. But, balanced against the virtually certain risk of continued fossil fuel use that produces climate-warming greenhouse gases, it seems sensible to at least continue research on controlled thermonuclear fusion energy sources. Another example is the use of thermal processes for treating hazardous wastes, somewhat risky because of the potential generation of toxic substances or air pollutants, but still the best way to convert many kinds of hazardous wastes to innocuous materials.

1.9. WASTE PREVENTION

Waste prevention is better than having to treat or clean up wastes. In the earlier years of chemical manufacture the direct costs associated with producing large quantities of wastes were very low because such wastes were simply discarded into waterways, onto the ground, or in the air as stack emissions. With the passage and enforcement of environmental laws after about 1970, costs for waste treatment increased steadily. By the year 2000 in the United States, costs of complying with environmental and occupational health regulations had grown to a magnitude similar to that of research and development for industry as a whole. Although the costs of such things as engineering controls, regulatory compliance, personnel protection, wastewater treatment, and safe disposal of hazardous solid wastes have certainly been worthwhile for society and the environment, they have become a large fraction of the overall cost of doing business. From a purely economic standpoint, therefore, a green chemistry approach that avoids these costs is very attractive, in addition to its large environmental benefits.

1.10. BASIC PRINCIPLES OF GREEN CHEMISTRY

From the preceding discussion, it should be obvious that there are certain basic principles of green chemistry. Some publications recognize “the twelve principles of green chemistry.”[2] This section addresses the main ones of these.

As anyone who has ever spilled the contents of a food container from a refrigerator onto the floor well knows, it is better to not make a mess than to clean it up once made. As applied to green chemistry, this basic rule means that *waste prevention is much better than waste cleanup*. Failure to follow this simple rule has resulted in most of the troublesome hazardous waste sites that are causing problems throughout the world today.

One of the most effective ways to prevent generation of wastes is to make sure that insofar as possible *all materials involved in making a product should be incorporated into the final product.* Therefore, the practice of green chemistry is largely about incorporation of all raw materials into the product, if at all possible. We would not likely favor a food recipe that generated a lot of inedible byproduct. The same idea applies to chemical processes. In that respect, the concept of atom economy discussed in Section 1.6 is a key component of green chemistry.

The use or generation of substances that pose hazards to humans and the environment should be avoided. Such substances include toxic chemicals that pose health hazards to workers. They include substances that are likely to become air or water pollutants and harm the environment or organisms in the environment. Here the connection between green chemistry and environmental chemistry is especially strong.

Chemical products should be as effective as possible for their designated purpose, but with minimum toxicity. The practice of green chemistry is making substantial progress in designing chemicals and new approaches to the use of chemicals such that effectiveness is retained and even enhanced while toxicity is reduced.

Chemical synthesis as well as many manufacturing operations make use of auxiliary substances that are not part of the final product. In chemical synthesis, such a substance consists of solvents in which chemical reactions are carried out. Another example consists of separating agents that enable separation of product from other materials. Since these kinds of materials may end up as wastes or (in the case of some toxic solvents) pose health hazards, *the use of auxiliary substances should be minimized and preferably totally avoided.*

Energy consumption poses economic and environmental costs in virtually all synthesis and manufacturing processes. In a broader sense, the extraction of energy, such as fossil fuels pumped from or dug out of the ground, has significant potential to damage the environment. Therefore, *energy requirements should be minimized.* One way in which this can be done is through the use of processes that occur near ambient conditions, rather than at elevated temperature or pressure. One successful approach to this has been the use of biological processes, which, because of the conditions under which organisms grow, must occur at moderate temperatures and in the absence of toxic substances. Such processes are discussed further in Chapter 12.

Raw materials extracted from earth are depleting in that there is a finite supply that cannot be replenished after they are used. So, wherever possible, *renewable raw materials should be used instead of depletable feedstocks.* As discussed further in Chapter 12, biomass feedstocks are highly favored in those applications for which they work. For depleting feedstocks, recycling should be practiced to the maximum extent possible.

In the synthesis of an organic compound (see Chapter 5), it is often necessary to modify or protect groups on the organic molecule during the course of the synthesis. This often results in the generation of byproducts not incorporated into the final product, such as occurs when a protecting group is bonded to a specific location on a molecule, then removed when protection of the group is no longer needed. Since these processes generate byproducts that may require disposal, *the use of protecting groups in synthesizing chemicals should be avoided insofar as possible.*

Reagents should be as selective as possible for their specific function. (in chemical language, a preference for selective catalytic reagents over nonselective stoichiometric reagents).

Products that must be dispersed into the environment should be designed to break down rapidly into innocuous products. One of the oldest, but still one of the best, examples of this is the modification of the surfactant in household detergents 15 or 20 years after they were introduced for widespread consumption to yield a product that is biodegradable. The poorly biodegradable surfactant initially used caused severe problems of foaming in wastewater treatment plants and contamination of water supplies. Chemical modification to yield a biodegradable substitute solved the problem.

Careful "real-time" control of chemical processes is essential for efficient, safe operation of such processes with minimum production of wastes. To an ever-increasing degree this goal has been made much more attainable by modern computerized controls. However, it requires accurate knowledge of the concentrations of materials in the system measured on a continuous basis. Therefore, *the successful practice of green chemistry requires real-time, in-process monitoring techniques coupled with process control.*

Accidents, such as spills, explosions, and fires, are a major hazard in the chemical industry. Not only are these incidents potentially dangerous in their own right, they tend to spread toxic substances into the environment and increase exposure of humans and other organisms to these substances. For this reason, it is best to *avoid the use or generation of substances that are likely to react violently, burn, build up excessive pressures, or otherwise cause unforeseen incidents in the manufacturing process.*

The principles outlined above are developed to a greater degree in the remainder of the book. They should be kept in mind in covering later sections.

1.11. SOME THINGS TO KNOW ABOUT CHEMISTRY BEFORE YOU EVEN START

Chapters 2-5 explain the basic principles of chemistry as they relate to green chemistry. However, at this point, it is useful to have a brief overview of chemistry, in a sense a minicourse in chemistry that provides the basic definitions and concepts of aspects of chemistry such as chemical compounds, chemical formulas, and chemical reactions before they are covered in detail in the later chapters.

All chemicals are composed of fewer than 100 naturally-occurring fundamental kinds of matter called **elements**. Humans have succeeded in making about 30 artificial elements since the late 1930s, but the amounts of these are insignificant compared to the total of known chemicals. (However, one of these "artificial" elements, plutonium, is very significant because of its use in nuclear weapons.) Elements, in turn, are composed of very small entities called **atoms**. Atoms of the same element may differ a bit in their masses, but all atoms of the same element behave the same chemically. So we can logically begin the study of chemistry with the atoms that make up the elements of which all matter is composed.

Each atom of a particular element is chemically identical to every other atom. Each element is given an **atomic number** specific to the element, ranging from 1 to more than 100. The atomic number of an element is equal to the number of extremely small, positively charged **protons** contained in the **nucleus** located in the center of each atom of the element. Each electrically neutral atom has the same number of **electrons** as it has protons. The electrons are negatively charged and are in rapid motion around the nucleus, constituting a cloud of negative charge that makes up most of the volume of the atom. In addition to its atomic number, each element has a name and a **chemical symbol**, such as carbon, C; potassium, K (for its Latin name kalium); or cadmium, Cd. In addition to atomic number, name, and chemical symbol, each element has an **atomic mass** (atomic weight). The atomic mass of each element is the average mass of all atoms of the element, including the various isotopes of which it consists.

1.12. COMBINING ATOMS TO MAKE MOLECULES AND COMPOUNDS

About the only atoms that exist alone are those of the noble gases, a group of elements including helium, neon, argon, and radon located in the group on the far right of the periodic table. Even the simple hydrogen atom in the elemental state is joined together with another hydrogen atom. Two or more uncharged atoms bonded together are called a **molecule**. The molecule of hydrogen is illustrated in Figure 1.5. It consists of 2 atoms of hydrogen. This fact is denoted by the **chemical formula** of elemental hydrogen, H_2. This formula states that a *molecule* of elemental hydrogen consists of 2 *atoms* of hydrogen, shown by the subscript of 2. The atoms are joined together by a **chemical bond**. Recall from Figure 1.1 that the hydrogen atom has 1 electron. But, hydrogen atoms are more "content" with 2 electrons. So two hydrogen atoms share their two electrons constituting the chemical bond in the hydrogen molecule. A bond composed of shared electrons is called a **covalent bond**.

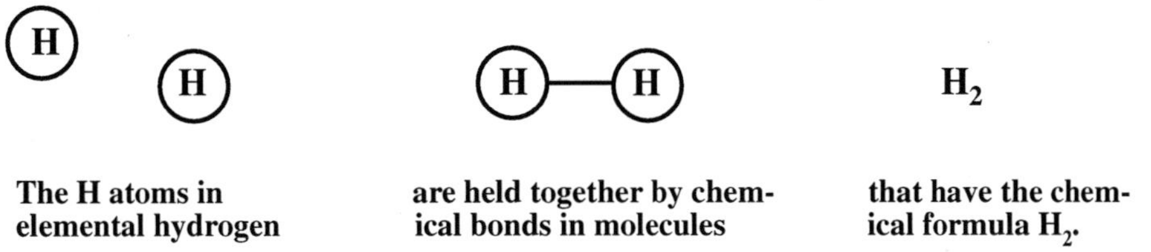

Figure 1.5. Molecule of H_2.

Chemical Compounds

The example just discussed was one in which atoms of the same element, hydrogen, join together to form a molecule. Most molecules consist of atoms of different elements joined together. An example of such a molecule is that of **water**, chemical formula H_2O. This formula states that the water molecule consists of *two* hydrogen atoms bonded to *one* oxygen atom, O, where the absence of a subscript number after the O indicates that there is 1 oxygen atom. The water molecule is shown in Figure 1.6. Each of the hydrogen atoms is held to the oxygen atom in the water molecule by two shared electrons in a covalent bond. A material such as water in which two or more elements are bonded together is called a **chemical compound**. It is because of the enormous number of combinations of two or more atoms of different elements that it is possible to make 20 million or more chemical compounds from fewer than 100 elements.

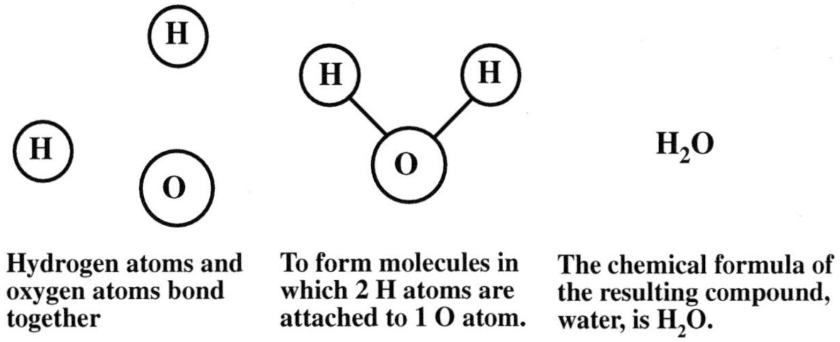

Figure 1.6. A molecule of water, H_2O, formed from 2 H atoms and 1 O atom held together by chemical bonds.

Ionic Bonds

Two different molecules have just been discussed in which atoms are joined together by *covalent bonds* consisting of *shared* electrons. Another way in which atoms can be joined together is by *transfer* of electrons from one atom to another. Recall that a single *neutral* atom has an equal number of electrons and protons. But, if the atom loses one or more negatively charged electrons, it ends up with a net *positive* electrical charge and becomes a positively charged **cation**. An atom that has gained one or more negatively charged electrons ends up with a net *negative* charge and is called an **anion**. Cations and anions are attracted together in an **ionic compound** because of their opposite electrical charges. The oppositely charged ions are joined by **ionic bonds** in a **crystalline lattice**.

Figure 1.7 shows the best known ionic compound, sodium chloride, NaCl. The chemical formula implies that there is 1 Na for each Cl. In this case these consist of Na^+ cations and Cl^- anions. For ionic compounds such as NaCl, the first part of the name is simply that of the metal forming the cation, in this case sodium. The second part of the name is based upon the anion, but has the ending *ide*. So the ionic compound of sodium and chlorine is sodium chlor*ide*. As shown by the preceding example, ionic compounds may consist of ions composed of atoms that have lost electrons (producing positively

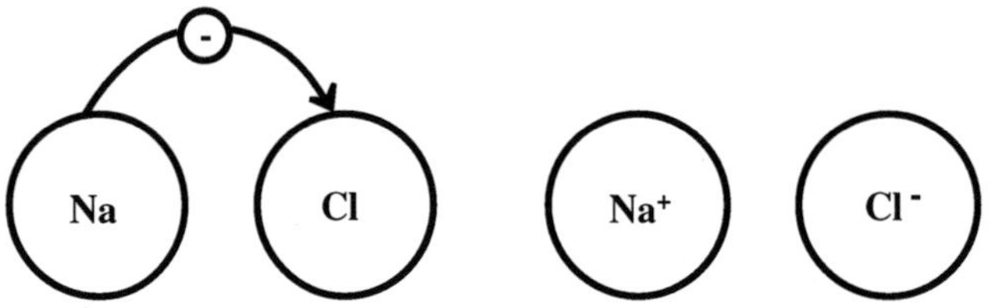

The transfer of a single negatively charged electron from a neutral sodium atom (Na) to a neutral chlorine atom (Cl) **produces a positively charged sodium cation and a negatively charged chloride anion.**

Sodium cations held together with equal numbers of chloride anions by the attraction of their opposite charges (ionic bonds) constitute the ionic compound NaCl.

Figure 1.7. Ionic bonds are formed by the transfer of electrons and the mutual attraction of oppositely charged ions in a crystalline lattice.

charged cations) and other atoms that have gained electrons (producing negatively charged anions). However, ions may also consist of groups of several atoms with a net charge. Ammonium ion, NH_4^+, is such an ion. As shown below, the NH_4^+ cation consists of 4 H atoms covalently bonded (by 2 shared electrons) to a central N atom, with the group of 5 total atoms having a net electrical charge of +1.

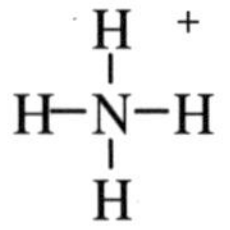

Ammonium ion, NH_4^+

1.13. THE PROCESS OF MAKING AND BREAKING CHEMICAL BONDS: CHEMICAL REACTIONS

The preceding section has discussed chemical compounds and the two major kinds of bonds — covalent bonds and ionic bonds — that hold them together. Next is discussed the process of making and taking apart chemical compounds, **chemical reactions**. A chemical reaction occurs when chemical bonds are broken and formed and atoms are exchanged to produce chemically different species.

First consider two very simple chemical reactions involving only one element, oxygen. In the very thin air high in the stratosphere more than 10 kilometers above Earth's surface (above the altitudes where jet airliners normally cruise), high-energy ultraviolet radiation from the sun, represented by the symbol $h\nu$, splits apart molecules of elemental oxygen, O_2,

$$O_2 + h\nu \rightarrow 2O \qquad (1.13.1)$$

to produce oxygen atoms. As with most single atoms, the O atoms are reactive and combine with oxygen molecules to produce ozone, O_3:

$$O + O_2 \rightarrow O_3 \qquad (1.13.2)$$

Both of these processes are chemical reactions. In a chemical reaction, the substances on the left of the arrow (read as "yields") are the **reactants** and those on the right of the arrow are **products**. The first of these reactions states that the chemical bond holding together a molecule of O_2 *reactant* is split apart by the high energy of the ultraviolet radiation to produce two oxygen atom *products*. In the second reaction, an oxygen atom reactant, O, and an oxygen molecule reactant, O_2, form a chemical bond to yield an ozone product, O_3. Are these very simple chemical reactions important to us? Emphatically yes. They produce a shield of ozone molecules in the stratosphere which in turn absorb ultraviolet radiation that otherwise would reach Earth's surface, destroying life, causing skin cancer and other maladies that would make our existence on Earth impossible. As discussed in Chapter 8, the use of chlorofluorocarbon refrigerants (Freons) has seriously threatened the stratospheric ozone layer. It is a triumph of environmental chemistry that this threat was realized in time to do something about it. The development of relatively safe substitutes for ozone-threatening chemicals stands as an excellent example of the practice of green chemistry.

Many chemical reactions are discussed in this book. At this point a very common chemical reaction can be considered, that of elemental hydrogen with elemental oxygen to produce water. A first approach to writing this reaction is

$$H_2 + O_2 \rightarrow H_2O \qquad (1.13.3)$$

stating that elemental hydrogen and elemental oxygen react together to produce water. This is not yet a proper *chemical equation* because it is not balanced. A **balanced chemical equation** has the same number of each kind of atom on both sides of the equation. As shown above, there are 2 H atoms in the single H_2 molecule on the left and 2 H atoms in the single molecule H_2O product. That balances hydrogen, but leaves 2 O atoms in the O_2 molecule on the left with only 1 O atom in the single H_2O molecule product. But, writing the reaction as

$$2H_2 + O_2 \rightarrow 2H_2O \qquad (1.13.4)$$

gives a balanced chemical equation with a total of 4 H atoms in 2 H_2 molecules on the left, 4 H atoms in 2 H_2O molecules on the right, and a total of 2 O atoms in the 2 H_2O molecules on the right. So the equation as now written is *balanced.* A balanced chemical equation always has the same number of each kind of atom on both sides of the equation.

1.14. THE NATURE OF MATTER AND STATES OF MATTER

We are familiar with matter in different forms. We live in an atmosphere of gas that is mostly N_2 with about 1/4 as much oxygen, O_2, by volume. We only become aware of this gas when something is wrong with it, such as contamination by irritating air pollutants. A person stepping into an atmosphere of pure N_2 would not notice anything wrong immediately, but would die within a few minutes, not because N_2 is toxic, but because the atmosphere lacks life-giving oxygen. The same atmosphere that we breathe contains water in the gas form as water vapor. And we are also familiar, of course, with liquid water and with solid ice.

Like most substances, air is a **mixture** consisting of two or more substances. It is a **homogeneous mixture** meaning that the molecules of air are mixed together at a molecular level. There is no way that we can take air apart by simple mechanical means, such as looking at it under a magnifying glass and picking out its individual constituents. Another common substance that is a homogeneous mixture is drinking water, which is mostly H_2O molecules, but which also contains dissolved O_2 and N_2 from air, dissolved calcium ions (Ca^{2+}), chlorine added for disinfection, and other materials.

A **heterogeneous mixture** is one that contains discernable and distinct particles that, in principle at least, can be taken apart mechanically. Concrete is a heterogeneous mixture. Careful examination of a piece of broken concrete shows that it contains particles of sand and rock embedded in solidified Portland cement.

A material that consists of only one kind of substance is known as a **pure substance**. Absolutely pure substances are almost impossible to attain. Hyperpure water involved in semiconductor manufacturing operations approaches absolute purity. Another example is 99.9995% pure helium gas used in a combination gas chromatograph/mass spectrometer instrument employed for the chemical analysis of air and water pollutants.

Mixtures are very important in the practice of green chemistry. Among other reasons why this is so is that separation of impurities from mixtures in the processing of raw materials and in recycling materials is often one of the most troublesome and expensive aspects of materials utilization. Impurities may make mixtures toxic. For example, toxic arsenic, which is directly below phosphorus in the periodic table and has chemical properties similar to phosphorus, occurs as an impurity in the phosphate ores from which elemental phosphorus is extracted. This is not a problem for phosphorus used as fertilizer because the small amount of arsenic added to the soil is negligible compared to the arsenic naturally present in the soil. But, if the phosphorus is to be made into phosphoric acid and phosphate salts to be added to soft drinks or to food, impurity arsenic cannot be tolerated and must be removed at considerable expense.

Many byproducts of manufacturing operations are mixtures. For example, organochlorine solvents used to clean and degrease machined parts are mixtures that contain grease and other impurities. As part of the process for recycling these solvents, the impurities must be removed by expensive processes such as distillation.

States of Matter

The three common states of matter are gases, liquids, and solids. These are readily illustrated by water, the most familiar form of which is liquid water. Ice is a solid and water vapor in the atmosphere or in a steam line is a gas.

Gases, such as those composing the air around us, are composed mostly of empty space through which molecules of the matter composing the gas move constantly, bouncing off each other or the container walls millions of times per second. A quantity of gas expands to fill the container in which it is placed. Because they are mostly empty space, gases can be significantly **compressed**; squeeze a gas and it responds with a decreased volume. Gas **temperature** is basically an expression of the tendency of the gas molecules to move more rapidly; higher temperatures mean faster molecular movement and more molecules bounding off each other or container walls per second. The constant impact of gas molecules on container walls is the cause of gas **pressure**. Because of the free movement of molecules relative to each other and the presence of mostly empty space, a quantity of gas takes on the volume and shape of the container in which it is placed. The physical behavior of gases is described by several gas laws relating volumes of gas to quantities of the gas, pressure, and temperature. Calculations involving these laws are covered at the beginning of Chapter 8.

Molecules of **liquids** can move relative to each other, but cannot be squeezed together to a significant extent, so liquids are not compressible. Liquids do take on the shape of the part of a container that they occupy. Molecules of **solids** occupy fixed positions relative to each other. Therefore, solids cannot be significantly compressed and retain their shapes regardless of the container in which they are placed.

LITERATURE CITED

1. Manahan, Stanley E., *Environmental Chemistry*, 7th ed., CRC Press/Lewis Publishers, Boca Raton, FL, 2000.

2. Anastas, Paul T., and John C. Warner, *Green Chemistry Theory and Practice*, Oxford University Press, 1998.

QUESTIONS AND PROBLEMS

1. What is chemistry? Why is it impossible to avoid chemistry?
2. What is green chemistry?
3. Match the following pertaining to major areas of chemistry:

A. Analytical chemistry	1. Occurs in living organisms
B. Organic chemistry	2. Underlying theory and physical phenomena
C. Biochemistry	3. Chemistry of most elements other than carbon
D. Physical chemistry	4. Chemistry of most carbon-containing compounds
E. Inorganic Chemistry	5. Measurement of kinds and quantities of chemicals

4. What are the five environmental spheres? Which of these did not exist before humans evolved on Earth?
5. Discuss why you think the very thin "skin" of Earth ranging from perhaps two or three kilometers in depth to several kilometers (several miles) in altitude has particular environmental importance.
6. What is environmental chemistry?
7. Which event may be regarded as the beginning of the modern environmental movement?
8. What is the command and control approach to pollution control?
9. What is the Toxics Release Inventory, TRI. How does it reduce pollution?
10. Why are incremental command and control measures taken for pollution control relatively much less effective now than they were when pollution control laws were first enacted and enforced?
11. What is the special relationship of green chemistry to synthetic chemistry?
12. What does Figure 1.1 show?
13. In which important respects is green chemistry sustainable chemistry?
14. With respect to raw materials, what are two general and often complementary approaches to the practice of green chemistry?
15. What is the distinction between yield and atom economy?
16. What is shown by Figure 1.4?
17. What are two factors that go into assessing risk?
18. What are the risks of no risks?
19. What are the major basic principles of green chemistry?
20. What is shown by the formula O_3? What about H_2O_2?
21. How does a covalent bond differ from an ionic bond?
22. What is the name given to a material in which two or more different elements are bonded together?

23. Considering the compound shown in Figure 1.7, what is the name of Na_2O?

24. Summarize the information given by $3H_2 + O_3 \rightarrow 3H_2O$.

25. In addition to showing the correct reactants and products, a correct chemical equation must be ____________.

26. Name three kinds of matter based upon purity. Which of these is extremely rare?

27. In terms of molecules, how are gases, liquids, and solids distinguished?

28. Describe gas pressure and temperature in terms of molecular motion.

2. THE ELEMENTS: BASIC BUILDING BLOCKS OF GREEN CHEMICALS

2.1. ELEMENTS, ATOMS, AND ATOMIC THEORY

Chemistry is the science of matter. The fundamental building blocks of matter are the atoms of the various elements, which are composed of **subatomic particles**, the positively charged **proton** (+), the negatively charged **electron** (-), and the electrically neutral **neutron** (n). It is the properties of these atoms that determine matter's chemical behavior. More specifically, it is the arrangement and energy levels of electrons in atoms that determine how they interact with each other, thus dictating all chemical behavior. One of the most fundamental aspects of chemistry is that elemental behavior varies periodically with increasing atomic number. This has enabled placement of elements in an orderly arrangement with increasing atomic number known as the **periodic table**. The periodic behavior of elements' chemical properties is due to the fact that, as atomic number increases, electrons are added incrementally to atoms and occupy so-called shells, each filled with a specific number of electrons. As each shell is filled, a new shell is started, thus beginning a new period (row) of the periodic table. This behavior is most easily understood for the first 20 elements using dots to represent electrons, enabling construction of an abbreviated 20-element periodic table. Although simple, this table helps understand and explain most of the chemical phenomena discussed in this book.

The chapter also emphasizes some of the green aspects of the first 20 elements. Included among these elements are the nitrogen, oxygen, carbon (contained in carbon dioxide), and hydrogen and oxygen (in water vapor) that make up most of the air in the "green" atmosphere; the hydrogen and oxygen in water, arguably the greenest compound of all; the sodium and chlorine in common table salt; the silicon, calcium, and oxygen that compose most mineral matter, including the soil that grows plants supplying food to most organisms; and the hydrogen, oxygen, carbon, nitrogen, phosphorus, and sulfur that are the predominant elements in all living material.

Long Before Subatomic Particles Were Known, There Was Dalton's Atomic Theory

Atomic theory describes the atoms in relation to chemical behavior. With the sophisticated tools now available to chemists, the nature of atoms, largely based upon the subatomic particles of which they are composed, is well known. But long before these sophisticated tools were even dreamed about, almost two centuries ago in 1808, an English schoolteacher named John Dalton came up with the atomic theory that bears his name. To a large extent, this theory is the conceptual basis of modern chemistry. Key aspects of Dalton's atomic theory are the following:

- The matter in each element is composed of extremely small particles called atoms. (Dalton regarded atoms as indivisible, unchanging bodies. We now know that they exchange and share electrons, which is the basis of chemical bonding.)

- Atoms of different elements have different chemical properties. (These differences may range from only slightly different, such as those between the noble gases neon and argon, to vastly different, such as those between highly metallic sodium and strongly nonmetallic chlorine.)
- Atoms cannot be created, destroyed, or changed to atoms of other elements. (In modern times, the provision is added that these things do not happen in ordinary chemical processes, since atoms can be changed to atoms of other elements by nuclear reactions, such as those that occur in nuclear reactors.)
- Chemical compounds are formed by the combination of atoms of different elements in definite, constant ratios that usually can be expressed as integers or simple fractions.
- Chemical reactions involve the separation and combination of atoms. (This phenomenon was surmised before anything was known about the nature of chemical bonds.)

Three Important Laws

Dalton's atomic theory explains the three important laws listed below. Evidence for these laws had been found prior to the publishing of Dalton's atomic theory, and the atomic theory is largely based upon them.

1. **Law of Conservation of Mass**: *There is no detectable change in mass in an ordinary chemical reaction.* (This law, which was first stated in 1798 by "the father of chemistry," the Frenchman Antoine Lavoisier, follows from the fact that in ordinary chemical reactions no atoms are lost, gained, or changed; in chemical reactions, mass is conserved.)
2. **Law of Constant** Composition: *A specific chemical compound always contains the same elements in the same proportions by mass.*
3. **Law of Multiple Proportions**: *When two elements combine to form two or more compounds, the masses of one combining with a fixed mass of the other are in ratios of small whole numbers.* A common illustration of this law is provided by the simple hydrocarbon compounds of carbon and hydrogen, which include CH_4, C_2H_2, C_2H_4, and C_2H_6. In these compounds the relative masses of C and H are in ratios of small whole numbers.

The Nature of Atoms

At this point it is useful to note several characteristics of atoms, which were introduced in Section 1.11. Atoms are extremely small and extremely light. Their individual masses are expressed by the miniscule *atomic mass unit, u.* The sizes of atoms are commonly expressed in picometers, where a picometer is 0.000 000 001 millimeters (a millimeter is the smallest division on the metric side of a ruler). Atoms may be regarded as spheres with diameters between 100 and 300 picometers.

As noted at the beginning of this chapter, atoms are composed of three basic *subatomic particles*, the positively charged *proton*, the electrically neutral *neutron*, and the much lighter negatively charged *electron*. Each proton and neutron has a mass of essentially 1 atomic mass unit, whereas the mass of the electron is only about 1/2000 as much The protons and neutrons are located in the *nucleus* at the center of the atom and the electrons compose a "fuzzy cloud" of negative charge around the nucleus. Essentially all the mass of an atom is in the nucleus and essentially all the volume is in the cloud of electrons. Each atom of a specific element has the same number of protons in its nucleus. This is the *atomic number* of the element. Each element has a name and is represented by a **chemical symbol** consisting of one or two letters. Atoms of the same element that have different numbers of neutrons and, therefore, different masses, are called **isotopes**. Isotopes may be represented by symbols such as ${}^{12}_{6}C$ where the subscript is the atomic number and the superscript is the **mass number**, which is the sum of the numbers of protons and neutrons in an atom.

The average mass of all the atoms of an element is the **atomic mass**. Atomic masses are expressed relative to the ${}^{12}_{6}C$ isotope of carbon taken as a mass of exactly 12 u. Atomic masses normally are not integers, in part because atoms of most elements consist of two or more isotopes with different masses.

Electrons in Atoms

The behavior of electrons in the cloud of negative charge making up most of the volume of atoms, particularly their energy levels and orientations in space, are what determine chemical behavior. Arrangements of electrons are described by *electron configuration.* A detailed description of electron configuration is highly mathematical and sophisticated, but is represented in a very simplified fashion in this chapter. Because of their opposite charges, electrons are strongly attracted to positively charged nuclei, but they do not come to rest on it.

The placement of electrons in atoms determines the configuration of the periodic table, a complete version of which showing all known elements is illustrated on the inside back cover of this book. Elements are listed across this table in *periods* such that elements located in the same vertical *groups* have generally similar chemical behavior. The derivation of the complete periodic table showing more than 100 elements is too complicated for this book. So, in the remainder of this chapter, the first 20 elements will be discussed in order and the placement of electrons in the atoms of these elements will illustrate how these elements can be placed in the periodic table. From this information a brief 20-element periodic table will be constructed that should be very useful in explaining chemical behavior.

2.2. HYDROGEN, THE SIMPLEST ATOM

Hydrogen, H, is the element with atomic number 1. Most hydrogen atoms consist of a single proton forming the nucleus with 1 electron per hydrogen atom. Recall from Section 1.12 and Figure 1.5 that elemental hydrogen exists as molecules with 2 H atoms, chemical formula H_2, in which the 2 H atoms are joined together by a *covalent bond* consisting of 2 shared electrons. Molecules consisting of 2 atoms so joined are called **diatomic** molecules. As will be seen, several important elements among the first 20 elements consist of diatomic molecules in their elemental forms.

Showing Electrons in Atomic Symbols and Molecular Formulas

In discussing chemical behavior related to atomic structure, it is particularly useful to have a means of showing the electrons in the atoms (more specifically, the most recently added so-called outer shell electrons). This is done with **Lewis symbols** (named after G. N. Lewis) also called **electron-dot** symbols. The Lewis symbol of the hydrogen atom is

$$\mathbf{H\cdot}$$

Although hydrogen atoms do exist very high in the rarefied upper atmosphere and as transient species in chemical reactions, under normal conditions elemental hydrogen exists as H_2 molecules. It is convenient to have a means of showing the electrons in molecules. This is done with **Lewis formulas** (**electron-dot formulas**). The Lewis formula of H_2 is shown in Figure 2.1.

$H\cdot \rightarrow \leftarrow \cdot H$	H_2	H:H
Elemental hydrogen does not exist as individual H atoms.	**Instead, it exists as molecules made up of 2H atoms with the chemical formula H_2.**	**The covalent bond holding the two H atoms together consists of 2 shared electrons shown in the Lewis formula of H_2 above.**

Figure 2.1. Lewis formula for hydrogen molecules.

Properties and Uses of Elemental Hydrogen

Pure elemental H_2 under normal conditions is a colorless, odorless gas that has the lowest density of any pure substance. Liquified H_2 boils at a very cold -253°C and solidifies at -259°C. Hydrogen gas is widely used in the chemical industry to react chemically with a large number of substances. It burns readily with a large release of energy and mixtures of hydrogen with oxygen or air are extremely explosive. The chemical reaction for elemental hydrogen burning with oxygen (O_2) in air is

$$2H_2 + O_2 \rightarrow 2H_2O + \text{energy} \qquad (2.2.1)$$

The product of this reaction is water. Used as a fuel, elemental hydrogen is a very green element because when it is burned or otherwise reacted to provide energy, the reaction product is simply water, H_2O. Furthermore, given a source of electrical energy, elemental hydrogen and elemental oxygen can be produced at two separate electrodes by passing a direct current through water in which an appropriate salt has been dissolved to make the water electrically conducting:

$$2H_2O \xrightarrow[\text{current}]{\text{Electrical}} 2H_2 + O_2 \qquad (2.2.2)$$

So elemental hydrogen generated by the application of electrical energy to water provides a source of energy that can be moved from one place to another and utilized to produce electricity in fuel cells (see below) or for other beneficial purposes.

Elemental hydrogen is widely used for chemical synthesis and other industrial applications. Its preparation by electrolysis of water was mentioned above. It is now most commonly prepared from methane, CH_4, that is the main ingredient of natural gas by **steam reforming** at high temperatures and pressures:

$$CH_4 + H_2O \xrightarrow[\text{30 atm P}]{\text{800°C T}} CO + 3H_2 \qquad (2.2.3)$$

Hydrogen is used to manufacture a number of chemicals. Two of the most abundantly produced chemicals that require hydrogen for synthesis are ammonia, NH_3, and methanol (methyl alcohol, CH_3OH). The latter is generated by the reaction between carbon monoxide and hydrogen:

$$CO + 2H_2 \rightarrow CH_3OH \qquad (2.2.4)$$

Methanol used to be made by heating wood in the absence of air and condensing methanol from the vapor given off, a process known as destructive distillation. Generation of so-called wood alcohol made by this relatively green process from biomass has the potential to supply at least a fraction of the methanol now needed, thus reducing the consumption of scarce natural gas.

Methanol has some important fuel uses. During the 1930s it was used instead of gasoline to run internal combustion engines to power a significant fraction of automobiles in France. At present it is blended with gasoline as an oxygenated additive; engines using this blended fuel produce less pollutant carbon monoxide. Now the most common use of methanol as a fuel is to break it down to elemental hydrogen and carbon dioxide to produce hydrogen used in fuel cells.

In addition to its uses in making ammonia and methanol, hydrogen is added chemically to some fractions of gasoline to upgrade the fuel value of gasoline. Hydrogen can be added directly to coal or reacted with carbon monoxide to produce synthetic petroleum. It is also combined with unsaturated vegetable oils to make margarine and other hydrogenated fats and oils. This application is controversial because of suspected adverse long-term health effects of these products.

Hydrogen in Fuel Cells

Fuel cells, discussed further in Chapter 6, are devices that enable hydrogen to "burn" at around room temperature and to produce electricity directly without going through some sort of internal combustion engine and electricity generator. A fuel cell (Figure 2.2) consists of two electrically conducting electrodes, an anode and a cathode that are contacted with elemental H_2 and O_2, respectively. As shown in the diagram, at the anode H_2 loses electrons (it is said to be oxidized) to produce H^+ ion. At the cathode O_2 gains electrons (it is said to be reduced) and reacts with H^+ ions to produce water, H_2O. The H^+ ions required for the reaction at the cathode are those generated at the anode and they migrate to the cathode through a solid membrane permeable to protons (the H^+ ion is a proton). The net reaction is

$$2H_2 + O_2 \rightarrow 2H_2O + \text{electrical energy} \qquad (2.2.5)$$

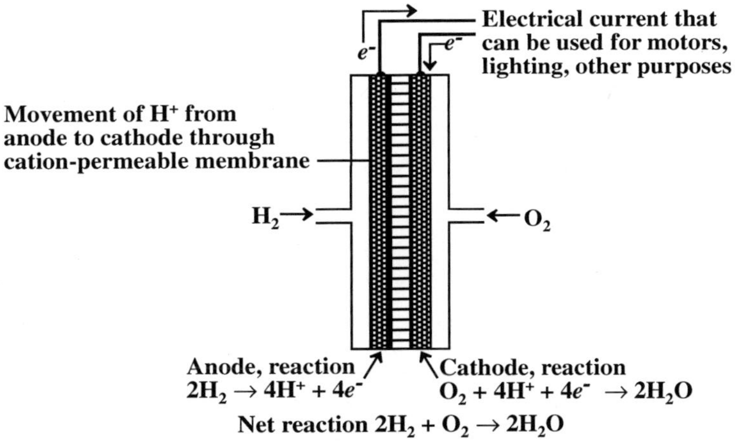

Figure 2.2. Cross-sectional diagram of a fuel cell in which elemental hydrogen can be reacted with elemental oxygen to produce electricity directly with water as the only chemical product.

2.3. HELIUM, THE FIRST NOBLE GAS

The second element in the periodic table is helium, He, atomic number 2. All helium atoms have 2 protons in their nuclei and 2 electrons. There are two isotopes of helium of which ${}^4_2He$ that contains 2 neutrons in the nucleus is by far the predominant one, with much smaller numbers of the lighter isotope, ${}^3_2He$, which has 2 protons and 1 neutron in its nucleus and a mass number of 3.

Helium is a *noble gas* meaning that it exists only as atoms of the elements that are never bonded to other atoms. Figure 2.3 is a representation of the helium atom showing its 2 electrons. The Lewis symbol of helium is simply He with 2 dots. This shows a very important characteristic of atoms. As electrons are added to atoms with increasing atomic number, they are added at various levels known as **electron shells**. The one electron in hydrogen, H, goes into the first electron shell, the one with the lowest possible energy. The second electron added to make the helium atom also goes into the first electron shell. This lowest electron shell can contain a maximum of only 2 electrons, so helium has a **filled electron shell**. Atoms with filled electron shells have no tendency to lose, gain, or share electrons and, therefore, do not become involved with other atoms through chemical bonding. Such atoms exist alone in the gas phase and the elements of which they consist are called noble gases. Helium is the first of the noble gases.

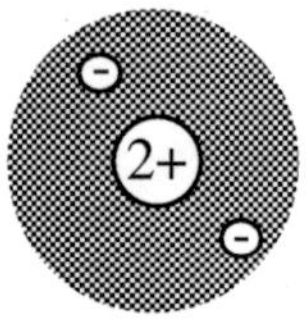

He:

A helium atom has a filled electron shell containing 2 electrons.

It can be represented by the Lewis symbol above.

Figure 2.3. Two representations of the helium atom having a filled electron shell

Helium gas has a very low density of only 0.164 g/L at 25°C and 1 atm pressure. Elemental helium is the second least dense substance next to hydrogen gas. It is this low density characteristic that makes helium so useful in balloons, including weather balloons, which can stay aloft for days, reaching very high altitudes.

Helium is pumped from the ground with some sources of natural gas, some of which contain up to 10% helium by volume. Helium was first observed in the sun by the specific wavelengths of light emitted by hot helium atoms. Underground sources of helium were discovered by drillers searching for natural gas in southwestern Kansas who tried to ignite gas from a new well and were disappointed to find that it would not burn, since it was virtually pure helium!

Chemically unreactive, helium has no chemical uses, except to provide a chemically inert atmosphere. A nontoxic, odorless, tasteless, colorless gas, helium is used because of its unique physical properties. Applications in weather balloons and airships were mentioned previously. Because of its low solubility in blood, helium is mixed with oxygen for breathing by deep-sea divers and persons with some respiratory ailments. Use of helium by divers avoids the very painful condition called "the bends" caused by bubbles of nitrogen forming from nitrogen gas dissolved in blood.

The greatest use of helium is as the super-cold liquid, which boils at a temperature of only 4.2 K above absolute zero (-269°C), especially in the growing science of **cryogenics**, which deals with very low temperatures. Some metals are superconductors at such temperatures so that helium is used to cool electromagnets enabling relatively small magnets to develop very powerful magnetic fields. Such magnets are components of the very useful chemical tool known as nuclear magnetic resonance (NMR). The same kind of instrument modified for clinical applications and called MRI is used as a medical diagnostic tool for scanning sections of the body for evidence of tumors and other maladies.

Hydrogen Wants to be Like Helium

Examination of the Lewis symbol of helium (left, below) and the Lewis formula of elemental hydrogen, H_2, shows that each of the two hydrogen atoms in the H_2 molecule can lay claim to 2 electrons and thereby come to resemble the helium atom. Recall that helium is a noble gas that is very content with its 2 electrons. Each of the H atoms in H_2 is satisfied with 2 electrons, though they are shared. This indicates a basic rule of chemical bonding that atoms of an element tend to acquire the same electron configuration as that of the nearest noble gas. In this case, hydrogen, which comes just before helium in the periodic table, gains the noble gas configuration of helium by sharing electrons.

2.4. LITHIUM, THE FIRST METAL

The element with atomic number 3 is lithium (Li), atomic mass 6.941. The most abundant lithium isotope is ${}^{7}_{3}\mathrm{Li}$ having 4 neutrons in its nucleus. A few percent of helium atoms are the ${}^{6}_{3}\mathrm{Li}$ isotope, which has only 3 neutrons. The third electron in lithium cannot fit in the lowest energy shell, which, as noted above, is full with only 2 electrons. Therefore, the third electron in lithium goes into a second shell, that is, an **outer shell**.

As a consequence of its electronic structure, lithium is the lowest atomic number element that is a **metal**. In a general sense, metals are elements that normally have only 1–3 electrons in their outer shells. These electrons can be lost from metals to produce positively charged *cations* with charges of +1, +2, or +3. In the pure elemental state metals often have a characteristic **luster** (shine), they are **malleable** (can be flattened or pushed into various shapes without breaking) and they **conduct electricity**. Although some metals, notably lead and mercury, are very dense, lithium is the least dense metal at only 0.531 g/cm^3.

Two of lithium's 3 electrons are **inner electrons** contained in an **inner shell** as in the immediately preceding noble gas helium. Inner shell electrons such as these stay on average relatively close to the nucleus, are very tightly held, and are not exchanged or shared in chemical bonds. As mentioned above, the third electron in lithium is an **outer electron** farther from, and less strongly attracted to, the nucleus. The outer electron is said to be in the atom's *outer shell.* These concepts are illustrated in Figure 2.4.

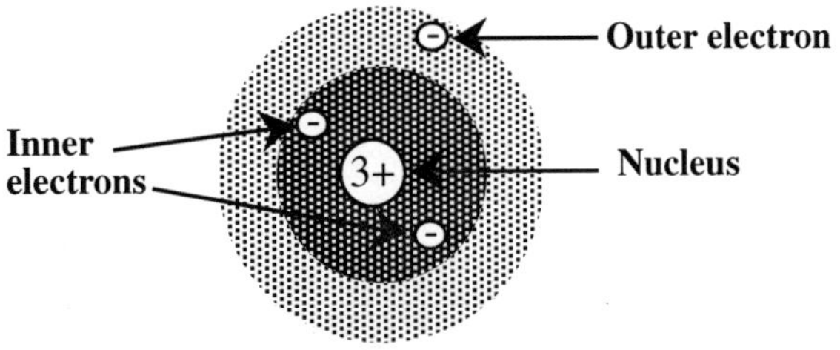

Figure 2.4. An atom of lithium, Li, has 2 inner electrons and 1 outer electron. The latter can be lost to another atom to produce the Li$^+$ ion, which is present in ionic compounds (see Section 1.12).

The Lewis symbol for atoms such as lithium that have both inner shell and outer shell electrons normally shows just the latter. (Inner shell electrons can be shown on symbols to illustrate a point, but normally this takes too much space and can be confusing.) Since lithium has only one outer shell electron, its Lewis symbol is

$$\mathrm{Li}\cdot$$

Consider that the lithium atom has an inner shell of 2 electrons, just like helium. Being only 1 electron away from the helium noble gas structure, lithium has a tendency to lose its extra electron so it can be like helium as shown by the following:

$$\mathrm{Li}\!:\!\cdot \longrightarrow \mathrm{Li}\!:^{+} + e^{-} \qquad (2.3.1)$$

Note that the product of this reaction is no longer a neutral atom, but is a positively charged Li^+ *cation*. In losing an electron to become a cation, the lithium atom is said to be **oxidized**. When lithium forms chemical compounds with other elements, it does so by losing an electron from each lithium atom to become Li^+ cations. These, then, are attracted to negatively charged *anions* in ionic compounds.

Lithium compounds have some important uses. One of the most important lithium compounds is lithium carbonate, Li_2CO_3. It is widely prescribed as a pharmaceutical to alleviate the symptoms of mania in manic-depressive and schizo-affective mental disorders. Lithium carbonate is the most common starting material for the preparation of other lithium compounds and is an ingredient of specialty glasses and enamels. Lithium carbonate is also used in making special kinds of ceramic ware that expand only minimally when heated. Lithium hydroxide, LiOH, is used to formulate some kinds of lubricant greases and it is also used in some long-life alkaline storage batteries.

A particularly interesting application of lithium is its use in combination with iodine to make cells used as sources of electricity for cardiac pacemakers. Implanted in the patient's chest, some of these pacemakers have lasted for 10 years before having to be replaced.

2.5. THE SECOND PERIOD OF THE PERIODIC TABLE

The first period of the periodic table is a short one consisting of only two elements, hydrogen and helium. Lithium, atomic number 3 begins the second period, which has 8 elements. Elements with atomic numbers 4-10, which complete this period, are discussed in this section.

Beryllium, Atomic Number 4

Like atoms of all the elements in the second period of the periodic table, **beryllium**, atomic number 4, atomic mass 9.012, has 2 inner shell electrons. Beryllium also has 2 outer shell electrons, so its Lewis symbol is

$$\text{Be:}$$

In addition to 4 protons in their nuclei, beryllium atoms also have 5 neutrons. When the beryllium atom is oxidized to form a beryllium cation, the reaction is

$$\text{Be:} \rightarrow Be^{2+} + 2e^- \text{ (lost to another atom)} \qquad (2.5.1)$$

Since the beryllium atom needs to lose 2 electrons to reach the two-electron helium electron configuration, it produces doubly charged Be^{2+} cations.

Beryllium has some important uses in metallurgy. Melted together with other metals, a process that produces **alloys**, beryllium yields metal products that are hard and corrosion-resistant. Beryllium alloys can be blended that are good electrical conductors and that are nonsparking when struck, an important characteristic in applications around flammable vapors. Among the devices for which beryllium alloys are especially useful are various specialty springs, switches, and small electrical contacts. Beryllium has found widespread application in aircraft brake components where its very high melting temperature (about 1290˚C) and good heat absorption and conduction properties are very advantageous.

In a sense, beryllium is somewhat the opposite of a green element. This is because of its adverse health effects, including **berylliosis**, a disease marked by lung deterioration. Because of the extreme inhalation hazard of Be, allowable atmospheric levels are very low. Many workers were occupationally exposed to beryllium as part of the nuclear reactor and weapons industry in the U.S. in the decades following World War II. In recognition of the adverse health effects of occupational exposure to beryllium, in the late 1990s the U.S. Government agreed to compensate workers suffering occupational exposure to this metal.

Boron, a Metalloid

Boron, B, atomic number of 5, atomic mass 10.81, consists primarily of the isotope with 6 neutrons in addition to 5 protons in its nucleus; a less common isotope has 5 neutrons. Two of boron's 5 electrons are in a helium core and 3 are outer electrons as denoted by

$$\dot{\mathrm{B}}:$$

Boron is the first example of an element with properties intermediate between those of metals and nonmetals called **metalloids**. In addition to boron, the metalloids include silicon, germanium, arsenic, antimony, and tellurium, of which the most notable, silicon, is among the first 20 elements. In the elemental state, metalloids have a luster like metals, but they do not readily form simple cations. Unlike metals, which generally conduct electricity well, metalloids usually conduct electricity poorly, if at all, but can become conductors under certain conditions. Such materials are called **semiconductors** and are of crucial importance because they form the basis of the world's vast semiconductor industry, which has given us small, powerful computers and a huge array of other electronic products.

Boron is a high-melting substance (2190°C) that is alloyed with copper, aluminum, and steel metals to improve their properties. As a good absorber of neutrons, boron is used in nuclear reactor control rods and in the water that circulates through a reactor core as a heat transfer medium. Boron nitride, BN, is extraordinarily hard, as are some other compounds of boron. Boron oxide, B_2O_3, is an ingredient of heat-insulating fiber glass and boric acid, H_3BO_3, is used as a flame retardant in cellulose insulation in houses.

The Element of Life, Carbon

Carbon, C, atomic number 6, brings us to the middle of the second period of the periodic table. In addition to its 2 inner electrons, the neutral carbon atom has 4 outer electrons as shown by the Lewis symbol

$$\cdot\dot{\mathrm{C}}:$$

The predominant carbon isotope making up 98.9% of all naturally occurring carbon is ${}^{12}_{6}C$ with 6 neutrons in their nuclei. Essentially all the remaining naturally occurring carbon, 1.1% of all carbon atoms, is the ${}^{13}_{6}C$ isotope. Miniscule, but detectable amounts of the radioactive ${}^{14}_{6}C$ are produced by nuclear processes high in the atmosphere.

Carbon is definitely the "element of life." This is because its unique bonding abilities and versatility enable it to provide the framework for the huge number of unique molecules that exist in living systems. Carbon atoms can bond to each other. In so doing, they can produce straight chains, branched chains and rings in a limitless variety of three-dimensional structures. In addition to all biological molecules, carbon is involved in **organic compounds**, thus forming the basis of **organic chemistry**, discussed in detail in Chapter 5. Figure 2.5 shows three organic compounds composed only of carbon and hydrogen (hydrocarbons), each containing 8 carbon atoms. These structures illustrate the bonding diversity of carbon.

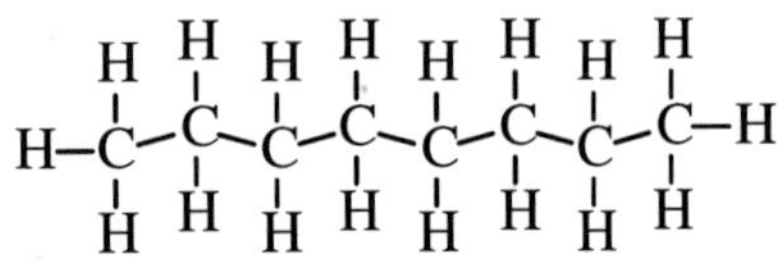

An 8-carbon straight-chain hydrocarbon, formula C_8H_{18}

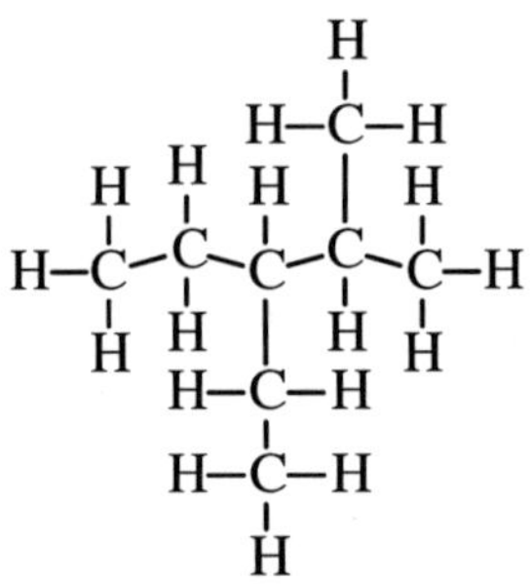

An 8-carbon branched-chain hydrocarbon, formula C_8H_{18}

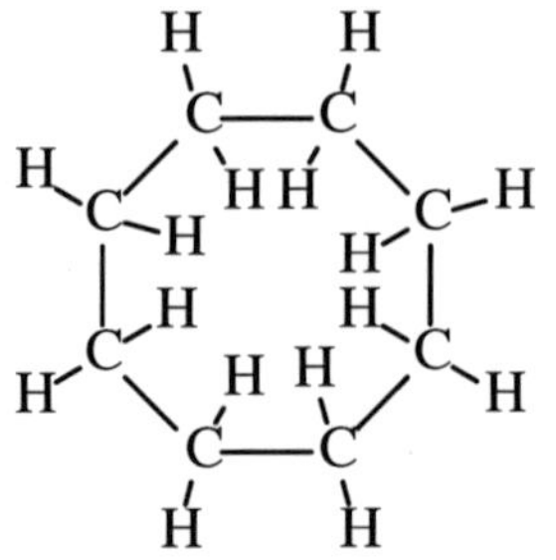

An 8-carbon cyclic hydrocarbon, formula C_8H_{16}

Figure 2.5. Illustration of the bonding versatility of carbon with three carbon-containing hydrocarbon compounds arranged as straight (continuous) chains, branched chains, and in a cyclic compound.

The ability of carbon atoms to bond with each other determines the properties of the several important and useful forms of elemental carbon. Very fine carbon powder composes **carbon black**, which is used in tires, inks, and printer toner. Carbon atoms bonded in large flat molecules compose **graphite**, so soft and slick that it is used as a lubricant. Carbon treated with steam or carbon dioxide at elevated temperatures develops pores that give the carbon an enormous surface area. This product is **activated carbon** that is very useful in purifying foods, removing organic pollutants from water and removing pollutant vapors from air. Elemental carbon fibers are bonded together with epoxy resins to produce light composites so strong that they are used for aircraft construction. Bonded together in a different way that gives a very hard and rigid structure, carbon atoms produce diamond.

A particularly interesting form of elemental carbon discovered only in 1985 consists of aggregates of 60 carbon atoms bonded together in 5- and 6-membered rings that compose the surface of a sphere. This structure resembles the geodesic domes designed as building structures by Buckminster Fuller, a visionary designer. Therefore, the discoverers of this form of carbon named it buckminsterfullerene and the C_{60} balls, which resemble soccer balls in their structure, are commonly called "buckyballs."

Green Carbon from the Air

Carbon is present in the air as gaseous carbon dioxide, CO_2. Although air is only about 0.037% CO_2 by volume, it serves as the source of carbon for the growth of green plants. In so doing, the chlorophyll in plants captures solar energy in the form of visible light, represented $h\nu$, and use it to convert atmospheric carbon dioxide to high-energy glucose sugar, $C_6H_{12}O_6$, as shown by the following reaction:

$$6CO_2 + 6H_2O \xrightarrow{h\nu} C_6H_{12}O_6 + 6O_2 \qquad (2.5.2)$$

The carbon fixed in the form of $C_6H_{12}O_6$ and related compounds provides the basis of the food chains that sustain all organisms. Organic carbon produced by photosynthesis in eons past also provided the raw material for the formation of petroleum, coal, and other fossil fuels. Now, as supplies of these scarce resources dwindle and as the environmental costs of their extraction, transport, and utilization mount, there is much renewed interest in photosynthetically produced carbon compounds as raw materials and even fuels. Despite the low levels of carbon dioxide in the atmosphere and the relatively low efficiency of photosynthesis, rapidly growing plants, such as some varieties of hybrid poplar trees, can produce enormous quantities of carbon compounds very rapidly and in a sustainable manner.

Nitrogen from the Air

Nitrogen, N, atomic number 7, atomic mass 14.01, composes 78% by volume of air in the form of diatomic N_2 molecules. The nitrogen atom has 7 electrons, 2 contained in its inner shell and 5 in its outer shell. So its Lewis symbol is the following:

Like carbon, nitrogen is a nonmetal and, like several elements among the first 20, it exists as diatomic N_2 molecules. Comprising 78% by volume of air, elemental nitrogen can be isolated from air by distilling cold liquid air. The molecules of elemental nitrogen are extremely stable because the two N atoms are bonded together by a triple bond consisting of six shared electrons as shown in Figure 2.6.

Nitrogen gas does not burn and does not support chemical reactions; it is generally chemically unreactive. This has led to uses of nitrogen gas in applications where a nonreactive gas is needed to prevent fires and explosions. People have died of asphyxiation by entering areas filled with nitrogen gas in which oxygen is absent. The absence of odor means that nitrogen gas does not warn of its presence.

Huge quantities of liquid nitrogen, which boils at a very cold -190°C, are used in areas where cold temperatures are needed. This frigid liquid is employed to quick-freeze foods and for drying materials in freeze-drying processes, Biological materials, such as semen used in artificial breeding of animals, can be preserved in liquid nitrogen.

There is an inexhaustible source of nitrogen in the atmosphere, but it is hard to get into a chemically combined form. This is because of the extreme stability of the N_2 molecule, mentioned above. The large-scale chemical fixation of atmospheric nitrogen over a catalytic surface at high temperatures and pressure as represented by the reaction

$$N_2 + 3H_2 \rightarrow 2NH_3 \qquad (2.5.3)$$

was a major accomplishment of the chemical industry about a century ago. It enabled the large-scale production of relatively cheap nitrogen fertilizers, as well as the manufacture of enormous quantities of nitrogen-based explosives that made possible the unprecedented carnage of World War I. Despite the extreme conditions required for the preparation of nitrogen compounds by humans in the anthrosphere, humble bacteria accomplish the same thing under ambient conditions of temperature and pressure, converting N_2 from the air into organically bound nitrogen in biomass. Prominent among the bacteria that do this are *Rhizobium* bacteria that grow symbiotically on the roots of legume plants, fixing atmospheric nitrogen that the plants need and drawing nutrients from the plants. Because of this ability, legumes, such as soybeans and clover grow well with less artificial nitrogen fertilizer than that required by other plants. One of the exciting possibilities with genetically modified plants is the potential to develop nitrogen-fixing varieties of corn, wheat, rice, and other crops that now lack the capability to fix nitrogen.

Nitrogen is an essential life element that is present in all proteins, hemoglobin, chlorophyll, enzymes, and other life molecules. It circulates through nature in the **nitrogen cycle** by which elemental nitrogen is incorporated from the atmosphere into biological material. Nitrogen-containing biomass is converted during biodegradation by bacteria to inorganic forms, which may be utilized as nutrient nitrogen by plants. Eventually, bacterial processes convert the nitrogen back to elemental N_2, which is returned to the atmosphere to complete the cycle.

Oxygen, the Breath of Life

Oxygen, atomic number 8, atomic mass 16.00 is required by humans and many other living organisms. A diatomic nonmetal, elemental oxygen consists of O_2 molecules and makes up 21% of the volume of air. Of its 8 electrons, the oxygen atom has 6 in the outer shell as represented by the Lewis formula

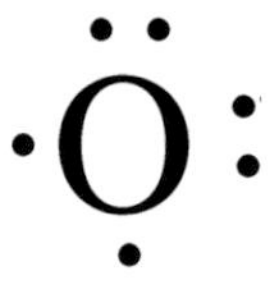

Oxygen can certainly be classified as a green element for a number of reasons, not the least of which is that O_2 in the atmosphere is there for the taking. Elemental oxygen is transferred from the atmosphere to the anthrosphere by liquifying air and distilling the liquid air, the same process that enables isolation of pure nitrogen. Regeneratable absorbent materials that selectively absorb oxygen or nitrogen from air can also be used to produce relatively pure oxygen, such as for people with breathing difficulties. Pure oxygen has a number of applications including use as a gas to breathe by people with lung deficiencies, in chemical synthesis, and in oxyacetylene torches employed for welding and cutting metals.

Although the elemental oxygen molecule is rather stable, at altitudes of many kilometers in the stratosphere, it is broken down to oxygen atoms by the absorption of ultraviolet radiation from the sun as shown in Chapter 1, Reaction 1.13.1 As illustrated by Reaction 1.13.2, the oxygen atoms formed by the photochemical dissociation of O_2 combine with O_2 molecules to produce molecules of ozone, O_3. The result is a layer of highly rarefied air containing some ozone over an altitude range of many kilometers located high in the stratosphere. There is not really much ozone in this layer. If it were pure ozone under the conditions of pressure and temperature that occur at ground level, the ozone layer would be only about 3 millimeters thick! This stratospheric ozone, sparse though it is, serves an essential function in protecting organisms on Earth's surface from the devastating effects of ultraviolet radiation from the sun. Were it not for stratospheric ozone, life as it is now known could not exist on Earth.

As a green form of oxygen, ozone definitely has a split personality. As discussed above, ozone in the stratosphere is clearly beneficial and essential for life. But it is toxic to inhale at levels less than even one part per million by volume. Ozone is probably the most harmful constituent of air polluted by the formation of photochemical smog in the atmosphere at ground levels.

The most notable chemical characteristic of oxygen is its ability to combine with other materials in energy-yielding reactions. One such reaction with which everyone is familiar is the burning of hydrocarbon gasoline in an automobile,

$$2C_8H_{18} + 25O_2 \rightarrow 16CO_2 + 18H_2O + \text{energy} \qquad (2.5.4)$$

performed so efficiently that the combustion of only 1 gallon of gasoline can propel a full-size automobile more than 25 miles at highway speeds. Along with many other organisms, we use oxygen in our bodies to react with nutrient carbohydrates,

$$C_6H_{12}O_6 \text{ (glucose)} + 6O_2 \rightarrow 6CO_2 + 6H_2O + \text{energy} \qquad (2.5.5)$$

to provide energy that we use. Whereas the combustion of a fuel such as gasoline occurs at red-hot temperatures, the "burning" of carbohydrates in our body occurs through the action of enzymes in the body at body temperature of only 37°C.

The Most Nonmetallic Element, Fluorine

Fluorine, F, atomic number 9, atomic mass 19.00 has 7 outer electrons as shown by its Lewis symbol

Elemental fluorine consists of diatomic F_2 molecules constituting a greenish-yellow gas. Fluorine is the most nonmetallic of all the elements. It reacts violently with metals, organic matter, and even glass! Elemental fluorine is a very corrosive poison that attacks flesh and forms wounds that heal very poorly. Because of its hazards, the practice of green chemistry seeks to minimize the generation or use of F_2.

Fluorine is used in chemical synthesis. It was once widely employed to make Freons, chlorofluorocarbons such as dichlorodifluoromethane, Cl_2CF_2, that were used as refrigerant fluids, spray can propellants, and plastic foam blowing agents. As discussed in Chapter 8, these compounds were found to be a threat to the vital stratospheric ozone layer mentioned in the discussion of oxygen above. They have now been replaced with fluorine-containing substitutes such as HFC-134a, CH_2FCF_3, which either do not contain the chlorine (Cl) that destroys stratospheric ozone or are destroyed by atmospheric chemical processes near Earth's surface, and thus never reach the stratosphere.

2.6. THE MAGIC OCTET OF 8 OUTER SHELL ELECTRONS

Only one element remains to complete the second period of the abbreviated periodic table. This element is **neon**, atomic number 10, atomic mass 20.18. Although most neon atoms have 10 neutrons in addition to the 10 protons in their nuclei, some have 12 neutrons, and very few have 11. Neon is a gas consisting of individual Ne atoms that constitutes about 2 parts per thousand of the volume of air. Neon is recovered by distillation of liquid air. Its most common use is as a gas in tubes through which an electrical discharge is carried in glowing neon signs.

The total of 10 electrons in the neon atom are contained in two shells with 8 in the outer shell. So the Lewis symbol of neon is

With 8 electrons, the second shell of electrons is a *filled electron shell.* (Recall that helium has a filled electron shell with only 2 electrons.) Because it has a filled outer electron shell, neon is a noble gas that exists as single gas-phase atoms and forms no chemical compounds.

The Special Significance of the Octet

In addition to helium and neon, there are four other noble gas elements. These are argon (atomic number 18), krypton (atomic number 36), xenon (atomic number 54), and radon (atomic number 86). Other than helium, these all share a common characteristic of 8 outer-shell electrons. Such an electron configuration can be shown by the general Lewis symbol,

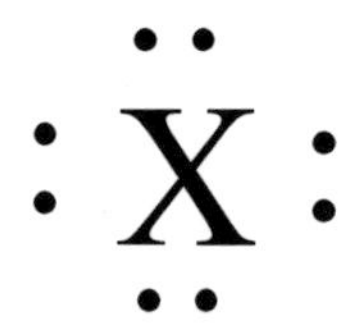

and is known as an **octet** of electrons, where X is the chemical symbol of the noble gas.

Although only atoms of noble gases have octets as single atoms, many other atoms acquire octets by forming chemical bonds. This tendency of atoms to acquire stable octets through chemical bonding is the basis of the **octet rule**, which is used in this book to predict and explain the chemical bonding in a number of compounds, such as those discussed in Chapter 3. To see a simple application of the octet rule in chemical bonding, consider the bonds involved in molecular elemental nitrogen, N_2. Recall the Lewis symbol of the N atoms showing 5 dots to represent the 5 outer-shell electrons in each N atom. Figure 2.6 shows bond formation between 2 N atoms.

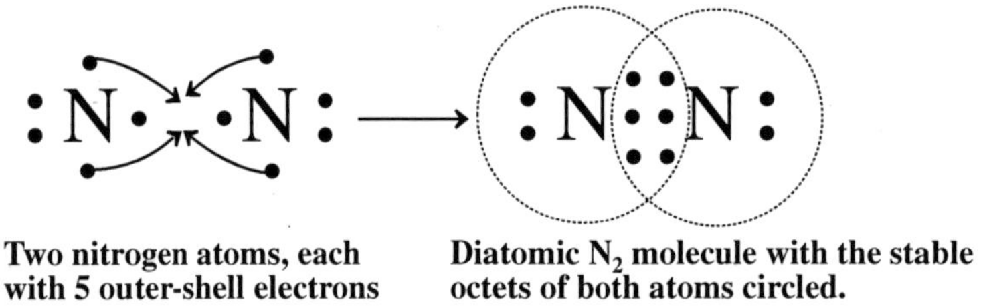

Figure 2.6. Illustration of the octet rule in the covalent bonding together of the 2 N atoms in N_2.

Each of the two N atoms needs 8 outer-shell electrons, but only 10 electrons are available to provide these electrons (the two inner-shell electrons in each N atom are not available to form bonds). This means that a lot of the electrons will have to be shared, and, in fact, 6 of the 10 available electrons are shared between the N atoms to give a *triple bond* consisting of 3 pairs of shared electrons as shown by the 3 pairs of dots between the 2 N atoms in the molecule of N_2 in Figure 2.6.

2.7. COMPLETING THE 20-ELEMENT PERIODIC TABLE

So far in this chapter 10 elements have been defined and discussed. A total of 10 more are required to complete the 20-element abbreviated periodic table. Their names and properties are summarized briefly here. The table is given in Figure 2.7.

Sodium, **Na**, atomic number 11, atomic mass 22.99, comes directly below lithium in the periodic table and is very similar to lithium in being a soft, chemically very reactive metal. There is one major isotope of sodium containing 12 neutrons in the atom's nucleus. Sodium has 10 inner-shell electrons contained in its first inner shell of 2 electrons and its second one of 8 electrons. The 11th electron in the sodium atom is in a third shell, which is an outer shell. This is shown as a single dot in the Lewis symbol of Na in Figure 2.7. The electrons in sodium can be represented as shown in Figure 2.7.

Magnesium, **Mg**, atomic number 12, atomic mass 24.31, has 12 electrons per atom so it has 2 outer shell electrons. There are three isotopes of magnesium containing 12, 13, and 14 neutrons. Magnesium is a relatively strong, very lightweight metal that is used in aircraft, extension ladders, portable tools, and other applications where light weight is particularly important.

Aluminum, **Al**, atomic number 13, atomic mass 26.98, has 3 outer-shell electrons in addition to its 10 inner electrons. Aluminum is a lightweight metal used in aircraft, automobiles, electrical transmission lines, building construction and many other applications. Although it is chemically reactive, the oxide coating formed when aluminum on the surface of the metal reacts with oxygen in air is self-protecting and prevents more corrosion.

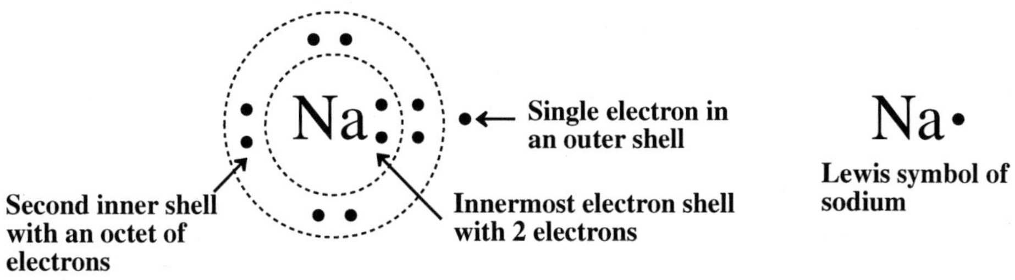

Figure 2.7. Representation of the electrons in 2 inner shells and 1 outer shell of sodium and the Lewis symbol of sodium.

In some important respects aluminum can be regarded as a green metal. This is because aluminum enables construction of strong lightweight components which, when used in aircraft and automobiles, require relatively less energy to move. So aluminum is important in energy conservation. Aluminum cables also provide an efficient way to transmit electricity. Although the ores from which aluminum is made are an extractive resource dug from the earth, aluminum is an abundant element. And there are alternative resources that can be developed, including aluminum in the fly ash left over from coal combustion. Furthermore, aluminum is one of the most recyclable metals, and scrap aluminum is readily melted down and cast into new aluminum goods.

If an "element of the century" were to be named for the 1900s, humble **silicon, Si**, atomic number 14, atomic mass 28.09, would be a likely candidate. This is because silicon is the most commonly used of the semiconductor elements and during the latter 1900s provided the basis for the explosion in electronics and computers based upon silicon-based semiconductor devices. Despite the value of these silicon-based products, silicon is abundant in soil and rocks, ranking second behind oxygen as a constituent of Earth's crust. The silicon atom has 4 outer-shell electrons, half an octet, and it is a metalloid, intermediate in behavior between the metals on the left of the periodic table and the nonmetals on the right.

By vastly reducing the bulk of electronic components relative to performance, silicon has contributed to a huge saving of materials used in radios, televisions, communications equipment, and other electronic devices. Furthermore, the silicon-based semiconductor devices used in solid-state electronics consume only a fraction of the electricity once used by vacuum tube based devices. The bulky wires made of relatively scarce copper formerly employed for transmitting communications signals electrically have been largely replaced by fiber optic devices consisting of transparent silica, SiO_2, which transfer information as pulses of light. A hair-like optical fiber can transmit many times the amount of information per unit time as the thick copper wire that it replaces. And the energy required for transmission of a unit of information by a fiber optic cable is miniscule compared to that required to send the same information by electrical impulse over copper wire. So silicon is truly a green element that, although cheap and abundant, performs electronic and communications functions much faster and better than the copper and other metals that it has replaced.

Phosphorus, **P**, atomic number 15, atomic mass 30.97, has 5 outer-shell electrons. So it is directly below nitrogen in the periodic table and resembles nitrogen in its chemical behavior. Pure elemental phosphorus occurs in several forms, the most abundant of which is white phosphorus. White phosphorus is a chemically very reactive nonmetal that may catch fire spontaneously in the atmosphere. It is toxic and causes deterioration of bone. The jawbone is especially susceptible to the effects of phosphorus and develops a condition known as "phossy jaw" in which the bone becomes porous and weak and may break from the strain of chewing. Chemically combined phosphorus is an essential life element, however, and is one of the components of DNA, the basic molecule that directs molecular life processes. Phosphorus is also an essential plant fertilizer and is an ingredient of many industrial chemicals including some pesticides.

Arsenic is in the same group of the periodic table as phosphorus and occurs as an impurity with phosphorus processed from ore. If this phosphorus is to be used for food, the arsenic has to be removed.

Sulfur, **S**, atomic number 16, atomic mass 32.06, has 6 outer-shell electrons. It is a brittle, generally yellow nonmetal. It is an essential nutrient for plants and animals occurring in the amino acids that compose proteins. Sulfur is a common air pollutant emitted as sulfur dioxide, SO_2, in the combustion of fossil fuels that contain sulfur. Much of the large quantities of sulfur required for industrial production of sulfuric acid and other sulfur-containing chemicals is reclaimed from the hydrogen sulfide, H_2S, that contaminates much of the natural gas (methane, CH_4) that is such an important fuel and raw material in the world today. In keeping with the best practice of green chemistry, the hydrogen sulfide is separated from the raw natural gas and about 1/3 of it is burned,

$$2H_2S + 3O_2 \rightarrow 2SO_2 + 2H_2O \qquad (2.7.1)$$

producing sulfur dioxide, SO_2. The sulfur dioxide produced is then reacted with the remaining hydrogen sulfide through the Claus reaction

$$2H_2S + SO_2 \rightarrow 3S + 2H_2O \qquad (2.7.1)$$

yielding an elemental sulfur product that is used to synthesize sulfuric acid and other sulfur chemicals.

Chlorine, Cl, atomic number 17, atomic mass 35.453, has 7 outer-shell electrons, just 1 electron short of a full octet. Elemental chlorine is a greenish-yellow diatomic gas consisting of Cl_2 molecules. In these molecules the Cl atoms attain stable octets of outer-shell electrons by sharing two electrons in a covalent bond as illustrated in Figure 2.8. The chlorine atom can also accept an electron to attain a stable octet in the Cl^- anion as shown in the ionic compound sodium chloride, NaCl, in Figure 2.8.

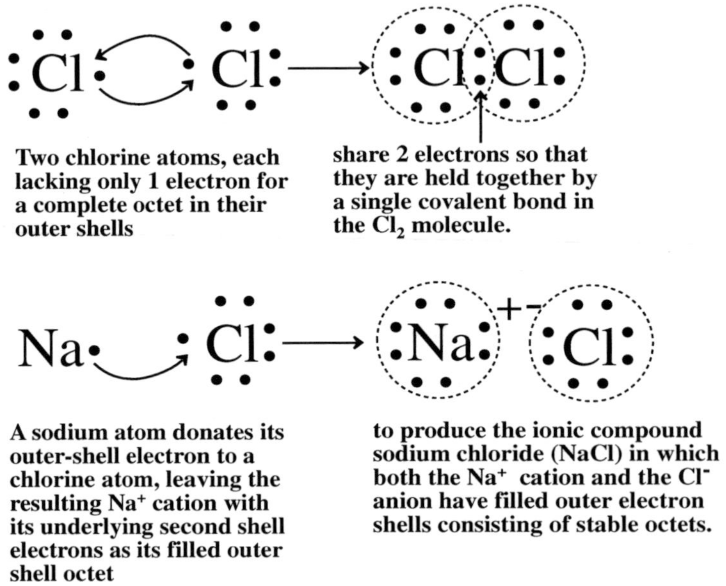

Figure 2.8. Chlorine atoms can attain a stable octet of outer-shell electrons by sharing electrons in covalent bonds, as shown for molecular diatomic elemental Cl_2, or by accepting electrons to produce Cl^- anion as shown for ionic sodium chloride, NaCl.

Elemental chlorine can be deadly when inhaled and was the first military poison gas used in World War I. Despite its toxic nature, chlorine gas has saved many lives because of its use for approximately the last 100 years as a drinking water disinfectant that has eradicated deadly water-borne diseases, such as cholera and typhoid. Chlorine is an important industrial chemical that is used to make plastics and solvents. There is no possibility of a shortage of chlorine and it can even be made by passing an electrical current through seawater, which contains chlorine as dissolved sodium chloride.

The green aspects of chlorine depend upon its application. Elemental chlorine is certainly a dangerous material whose manufacture and use are generally to be avoided where possible in the practice of green chemistry. But, as noted above, elemental chlorine has saved many lives because of its uses to disinfect water. A number of persistent pesticides including DDT are organic compounds composed of chlorine along with carbon and hydrogen. In addition to the ecological damage done by these pesticides, the waste byproducts of their manufacture and of the production of other organochlorine compounds are among the most abundant contaminants of troublesome hazardous waste chemical dumps. A common plastic, polyvinyl chloride (PVC), contains chlorine. This plastic is widely used in water pipe and drain pipe, in the former application replacing relatively scarce and expensive copper metal and toxic lead. But the material used to make PVC is volatile vinyl chloride. It is one of the few known human carcinogens, having caused documented cases of a rare form of liver cancer in workers formerly exposed to very high levels of vinyl chloride vapor in the workplace. Because of the dangers of elemental chlorine and the problems caused by organochlorine compounds, the practice of green chemistry certainly tries to minimize the production and use of elemental chlorine and generally attempts to minimize production of organochlorine compounds and their dispersion in the environment.

Element number 18, **argon, Ar**, atomic mass 39.95, brings us to the end of the third period of the abbreviated periodic table. It has a complete octet of outer-shell electrons and is a noble gas. No true chemical compounds of argon have been isolated and no chemical bonds involving this element were known until formation of a very unstable transient bond involving Ar atoms was reported in September, 2000. Argon composes about 1% by volume in the atmosphere. Largely because of its chemically inert nature, argon has some uses. It is employed as a gas to fill incandescent light bulbs. In this respect it helps prevent evaporation of white-hot tungsten atoms from the glowing lamp filament, thus significantly extending bulb life. It is also used as a plasma medium in instruments employed for inductively coupled plasma atomic emission analysis of elemental pollutants. In this application a radiofrequency signal is used to convert the argon to a gaseous plasma that contains positively charged Ar^+ ions and negatively charged electrons and is heated to an incredibly hot 10,000°C.

Completing the Periodic Table

The next element to be added to the abbreviated periodic table is element number 19. This begins a fourth period of the periodic table. This period actually contains 18 elements, but we will take it only as far as the first two. That is because element number 21 is the first of the transition metals and to explain their placement in the periodic table on the basis of the electrons in them gets a little more complicated and involved than is appropriate for this book. The reader needing more details is referred to other standard books on beginning chemistry or to the book *Fundamentals of Environmental Chemistry*.[1]

The element with atomic number 19 is **potassium, K**, having an atomic mass of 39.10. Most potassium consists of the isotope with 20 neutrons, $^{39}_{19}K$. However, a small fraction of naturally occurring K is in the form of $^{40}_{19}K$. This is a radioactive isotope of potassium and since we all have potassium (an essential element for life) in our bodies, we all are naturally radioactive! Muscle mass contains more potassium than adipose (fat) tissue, so more muscular people are more radioactive. But not to worry, the levels of radioactivity from potassium in the body are too low to cause concern and, under any circumstances, cannot be avoided. (One proponent of nuclear energy has pointed out that sleeping with a muscular person exposes one to more radioactivity than does living close to a nuclear power reactor.)

The same things that can be said of sodium, element number 11, are generally true of potassium. In the pure elemental state, potassium is a very reactive *alkali metal*. As an essential element for life, it is a common fertilizer added to soil to make crops grow well. Chemically, potassium loses its single outer-shell electron to produce K^+ ion.

Calcium, Ca, atomic number 20, atomic mass 40.08, has 2 outer-shell electrons, two beyond a full octet. The calcium atom readily loses its 2 "extra" electrons to produce Ca^{2+} cation. Like other elements in its group in the periodic table, calcium is an alkaline earth metal. Elemental calcium metal is chemically reactive, though not so much so as potassium. Calcium has chemical properties very similar to those of magnesium, the alkaline earth metal directly above calcium in the periodic table.

Calcium is essential for life, although most soils contain sufficient calcium to support optimum crop growth. Calcium is very important in our own bodies because as hard mineral hydroxyapatite, $Ca_5OH(PO_4)_3$, it is the hard material in teeth and bones. Calcium deficiency can cause formation of poor teeth and the development of disabling osteoporosis a condition characterized by weak bones that is especially likely to afflict older women.

2.8. THE BRIEF PERIODIC TABLE IS COMPLETE

With element number 20, all of the elements required for the abbreviated periodic table have been described. As noted above, the placement of electrons in elements with atomic number 21 and higher is a little too complicated to explain here. However, these elements are important and they are all shown in the complete periodic table on the inside front cover. Among the heavier elements in the complete periodic table are the transition metals, including the important metals chromium, manganese, iron, cobalt, nickel, and copper. Also included are the lanthanides and the actinides. Among these elements are thorium, uranium, and plutonium, which are important in nuclear energy and nuclear weaponry.

The abbreviated periodic table with the first 20 elements is illustrated in Figure 2.9. In addition to atomic number and atomic mass, this table shows the Lewis symbol of each element. It is seen that the symbols of the elements in the same vertical columns have the same number of dots showing identical configurations for their outer-shell electrons. This very simple, brief table contains much useful information, and it is recommended that the reader become familiar with it and be able to reproduce the Lewis symbols for each of the 20 elements. As will be seen in later chapters, the chemistry of the first 20 elements tends to be straightforward and easily related to the atomic structures of these elements.

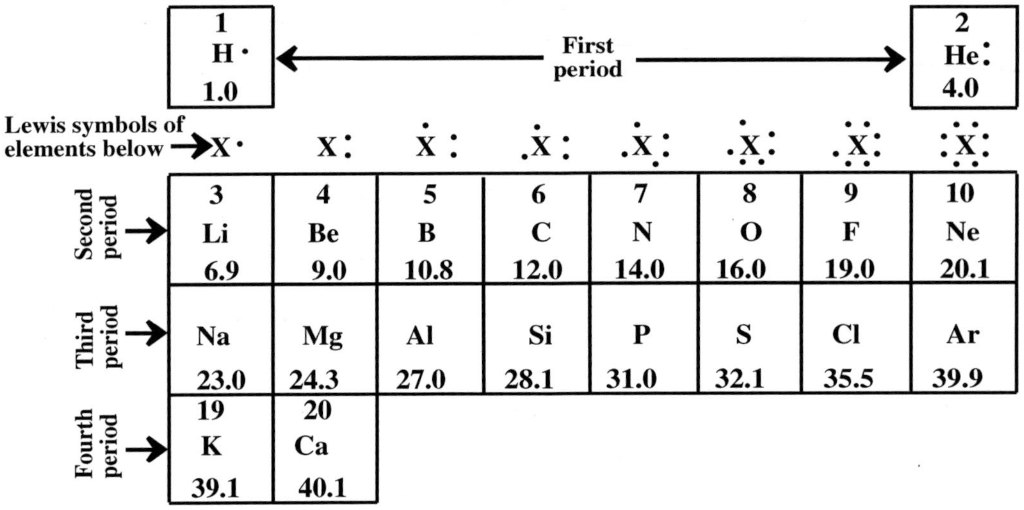

Figure 2.9 Abbreviated 20-element version of the periodic table showing Lewis symbols of the elements.

In examining the periodic table, hydrogen should be regarded as having unique properties and not belonging to a specific group. Otherwise, the elements in vertical columns belong to *groups* with similar chemical properties. Excluding hydrogen, the elements in the first group on the left of the table — lithium, sodium, and potassium — are *alkali metals*. In the elemental state alkali metals have a very low density and are so soft that they can be cut with a knife. Freshly cut, an alkali metal surface has a silvery-white color which almost instantaneously turns to a coating of gray metal oxide with exposure to air. The alkali metals (represented by M, below) react violently with water,

$$2M + 2H_2O \rightarrow 2MOH + H_2 \qquad (2.5.4)$$

to produce the metal hydroxides, strongly basic substances that can be very destructive to flesh that they contact. The alkali metals react with elemental chlorine to produce the ionic chloride salts including, in addition to NaCl shown in Figure 2.8, LiCl and KCl. The second group of the abbreviated periodic table contains beryllium, magnesium, and calcium, all known as *alkaline earth* metals. Freshly exposed surfaces of these metals have a grayish-white luster. These metals are highly reactive to form doubly charged cations (Be^{2+}, Mg^{2+}, Ca^{2+}) by the loss of 2 electrons per atom. The second group from the right, which in the abbreviated periodic table consists of fluorine and chlorine, is known as the **halogens**. These elemental halogens are diatomic gases in which the two atoms of F_2 or Cl_2 are held together by a single covalent bond consisting of two shared electrons. These elements are the most nonmetallic of the elements. Rather than losing electrons to produce positively charged cations, as is common with metals, the halogens readily gain electrons to complete their outer shell electron octets, producing F^- and Cl^- anions. The far right group of the abbreviated periodic table is composed of the *noble gases*, helium, neon, and argon. These elements have complete outer shells, exhibit no tendency to enter into chemical bonds, and consist of individual gas-phase atoms.

LITERATURE CITED

1. Manahan, Stanley E., *Fundamentals of Environmental Chemistry*, 2nd ed., CRC Press/Lewis Publishers, Boca Raton, FL, 2000.

QUESTIONS AND PROBLEMS

1. Match the law or observation denoted by letters below with the portion of Dalton's atomic theory that explains it denoted by numbers:

A. Law of Conservation of Mass	B. Law of Constant Composition
C. Law of Multiple Proportions	D. The reaction of C with O_2 does not produce SO_2.

   1. Chemical compounds are formed by the combination of atoms of different elements in definite constant ratios that usually can be expressed as integers or simple fractions.
   2. During the course of ordinary chemical reactions, atoms are not created or destroyed
   3. During the course of ordinary chemical reactions, atoms are not changed to atoms of other elements.
   4. Illustrated by groups of compounds such as $CHCl_3$, CH_2Cl_2, or CH_3Cl.

2. Explain why it is incorrect to say that atomic mass is the mass of any atom of an element. How is atomic mass defined?

3. Define what is meant by the notation ${}^{y}_{x}A$. What do *y*, *x*, and A mean?

4. What is the Lewis symbol of hydrogen and what does it show? What is the Lewis formula of H_2 and what does it show?
5. Why should hydrogen be considered in a separate category of the periodic table?
6. Consider the Lewis symbol of helium and explain how the helium atom illustrates the concepts of electron shell, filled electron shell, and noble gases.
7. What does helium have to do with cyrogenics?
8. Use three dots to show all the electrons in the lithium atom, Li. What does this show about inner and outer electrons and why Li produces Li^+ cation?
9. In what respect may it be argued that beryllium is definitely not a green element?
10. What are two elemental oxygen species other than molecular O_2 found at very high altitudes in the stratosphere? How do they get there?
11. In what respects may carbon be classified as the "element of life"?
12. How are a specific kind of fluorine compounds related to stratospheric ozone? What does this have to do with green chemistry?
13. Neon illustrates some important points about the octet and the octet rule. Explain.
14. Of the following, the **untrue** statement pertaining to matter, atoms, and elements is

    A. All matter is composed of only about a hundred fundamental kinds of matter called elements.
    B. Each element is made up of very small entities called atoms
    C. All atoms of the same element have the same number of protons and neutrons and the same mass.
    D. All atoms of the same element behave identically chemically.
    E. All atoms of the same element have the same number of protons in their nuclei.

15. Given that in the periodic table elements with atomic numbers 2, 10, and 18 are gases that do not undergo chemical reactions and consist of individual molecules, of the following, the statement most likely to be true is

    A. Elements with atomic numbers 3, 11, and 19 are also likely to be gases.
    B. Elements with atomic numbers 3, 11, and 19 would not undergo chemical reactions.
    C. Elements with atomic numbers 10 and 18 would be at opposite ends of the table.
    D. The element with atomic number of 11 may well be an unstable, highly reactive metal.
    E. The properties of elements with atomic number 3, 11, and 19 would be expected to have chemical properties that are much different from each other.

16. Of the following, the statement that is **untrue** regarding chemical bonding and compounds is

    A. Chemical bonds occur only in compounds, not in pure elements.
    B. Molecules of H_2 are held together by bonds consisting of shared electrons.
    C. Ionic compounds consist of charged atoms or groups of atoms.
    D. Both pure elemental hydrogen, H_2, and the compound water, H_2O, have covalent bonds.
    E. An atom that has more electrons than protons is an anion.

17. The two atoms represented below

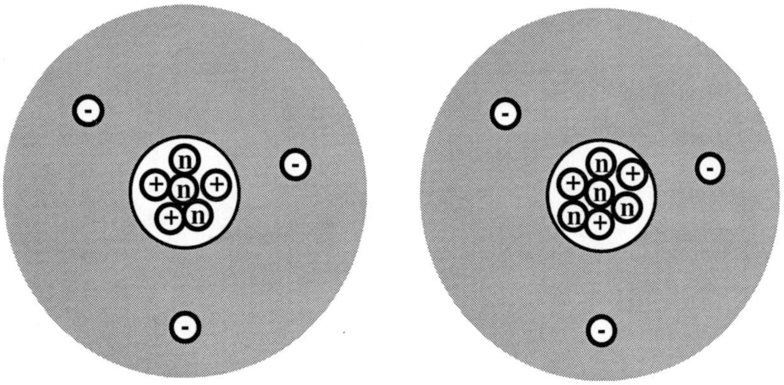

A. Are of different elements.
B. Are atoms of deuterium, a form of hydrogen, H.
C. Are of the same element.
D. Are **not** isotopes of the same element.
E. Are of two elements with atomic numbers 6 and 7.

18. Suggest a material that is a source of electrons in a fuel cell used as a source of electricity. What may accept electrons?

19. What are semiconductors? What is the most important semiconductor discussed in this chapter.

20. What is the most notable chemical characteristic of elemental oxygen?

21. What are some reasons that aluminum can be regarded as a green metal?

22. What are some of the toxic elements or elemental forms among the first 20 elements?

23. What is a common air pollutant gas that contains sulfur?

24. Why does the abbreviated periodic table stop at atomic number 20?

25. Suggest why calcium might be particularly important in the diet of (A) children and (B) older people?

26. Which elements among the first 20 are commonly present in fertilizers used to enhance the growth of food crops?

27. What is the special significance of the carbon isotope with 6 neutrons in its nucleus.

28. What is the single exception to the rule that all atoms contain at least 1 neutron?

29. What is the single exception to the rule that noble gases contain stable octets of electrons?

30. What is the outer-shell electron configuration of metals? What does this have to do with their chemical behavior?

31. What is it about the carbon atom that enables millions of organic compounds to exist?

32. What are some of the forms of elemental carbon and their uses? Which of these was discovered only relatively recently?

33. What is the major chemical characteristic of elemental nitrogen? What is a major advantage afforded by this characteristic? In what respect is this a problem?

34. What are two applications that elemental magnesium and aluminum have in common?

35. How do copper and silica differ in the way that they transfer communications signals?

36. Using the octet rule, propose a Lewis formula for O_2.

3. COMPOUNDS: SAFER MATERIALS FOR A SAFER WORLD

3.1. CHEMICAL BONDS AND COMPOUND FORMATION

Chemical compounds consist of molecules or aggregates of ions composed of two or more elements held together by chemical bonds. Several examples of chemical compounds including water (H_2O), ammonia (NH_3), and sodium chloride (NaCl) were given in earlier chapters. This chapter addresses chemical compounds in more detail, including aspects of their green chemistry.

A crucial aspect of chemical compounds consists of the kinds of bonds that hold them together. As noted earlier, these may be **covalent bonds** composed of shared electrons or **ionic bonds** consisting of positively charged **cations** and negatively charged **anions**. The strengths of these bonds vary and are important in determining compound behavior. For example, *chlorofluorocarbons*, such as dichlorodifluoromethane, Cl_2CF_2, are so stable that they persist in the atmosphere and do not break down until reaching very high altitudes in the stratosphere, where the release of chlorine atoms destroys stratospheric ozone. The extreme stabilities of the chlorofluorocarbons are due to the very high strengths of the C-Cl and C-F bonds by which chlorine and fluorine are bonded to a central carbon atom. The proper practice of green chemistry requires that substances that get released to the environment break down readily. Since Cl_2CF_2 is so stable when released to the atmosphere, it cannot be regarded as being a very good green chemical.

Another important aspect of chemical compounds is **molecular structure**, which refers to the shape of molecules. Consider the Cl_2CF_2 compound just mentioned in which the Cl and F atoms are bonded to a single carbon atom. To represent this molecule as the flat structure,

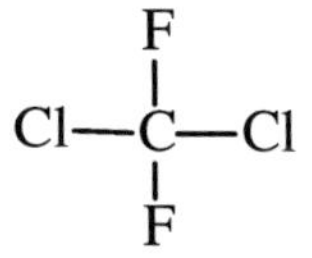

is not totally correct because not all of the 5 atoms in the compound lie in the same plane. Instead, the F and Cl atoms can be visualized as being distributed as far apart as possible in three dimensions around a sphere, at the center of which is the C atom. This can be represented as

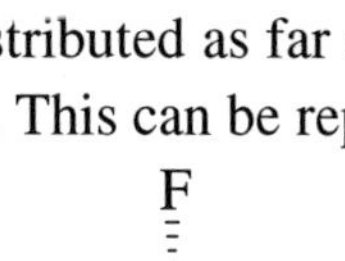

where the two Cl atoms are visualized as being above the plane of the book page toward the reader and the two F atoms are visualized as being below the plane of the page away from the reader. The shape of molecules is very important in determining the ways in which they interact with other molecules. For example, the molecules of enzymes that enable metabolism to occur in living organisms recognize the substrate molecules upon which they act by their complementary shapes.

What Are Green Chemical Compounds?

Chemical compounds vary markedly in the degree to which they are "green." Dichlorodifluoromethane, Cl_2CF_2, the chlorofluorocarbon discussed above, is definitely not green. That is not because it is toxic — it is one of the least toxic synthetic compounds known — but because it is so extremely stable and persistent in the atmosphere and can cause stratospheric ozone destruction. The compounds that have replaced it, the hydrofluorocarbons and hydrochlorofluorocarbons, are much more green because they do not last long when released in the atmosphere or do not contain ozone-damaging chlorine.

There are several characteristics of compounds that meet the criteria of being green. These are the following:

- Preparation from renewable or readily available resources by environmentally friendly processes
- Low tendency to undergo sudden, violent, unpredictable reactions such as explosions that may cause damage, injure personnel, or cause release of chemicals and byproducts to the environment.
- Nonflammable or poorly flammable
- Low toxicity
- Absence of toxic or environmentally dangerous constituents, particularly heavy metals
- Facile degradability, especially biodegradability, in the environment.
- Low tendency to undergo bioaccumulation in food chains in the environment

An example of a green compound is sodium stearate, common hand soap. This common substance is prepared by reacting byproduct animal fat with sodium hydroxide, which is prepared by passing an electrical current through saltwater. Flushed down the drain, sodium stearate reacts with calcium in water to form a solid, calcium stearate, the white solid that composes "bathtub ring," and is removed from the water. The nontoxic calcium stearate readily undergoes biodegradation so that it does not persist in the environment.

3.2. ELECTRONS INVOLVED IN CHEMICAL BONDS AND OCTETS OF ELECTRONS

The electrons in the *outermost shell* of atoms are those that become involved in chemical bonds. These are called **valence electrons**. Refer back to the *Lewis symbols* of the elements shown in Figure 2.9. Note that the three elements on the right of the table are *noble gases* that are chemically content with their filled outer electron shells containing 2 electrons in the case of helium and 8 each for neon and argon. As a basis for the understanding of chemical bonds, consider that the other elements try to attain the filled electron shells of their nearest-neighbor noble gases by sharing, losing, or gaining electrons. Of these elements, the only one that we will consider in detail that attains a helium-like electron configuration is hydrogen, H, each atom of which almost always has access to 2 electrons shared in covalent bonds. The other elements that we will consider, carbon and higher, attain 8 electrons in their outer shells by chemical bonding. This is the basis of the **octet rule**, the tendency of atoms to attain stable outer shells of 8 electrons by forming chemical bonds. The octet rule is immensely useful in explaining and predicting chemical bonds and the formulas and structures of chemical compounds and will be emphasized in this chapter.

Some examples of the kinds of bonding arrangements discussed above have already been illustrated in Chapter 2. Figure 2.1 illustrates that, even in the elemental form, H_2, hydrogen atoms have 2 valence electrons in the diatomic molecule. Examples of elements that have 8 valence electrons as the result of chemical bonding were also shown. Figure 2.6 illustrates the two N atoms in the N_2 molecule sharing 6 electrons in a covalent bond so that each of the atoms may have an octet. Figure 2.8 shows that 2 Cl atoms, each with 7 valence electrons, share 2 electrons in the covalent bond of the Cl_2 molecule to attain octets. The same figure shows that Na loses its single valence electron in forming ionic NaCl to produce the Na^+ ion, which has an octet of electrons in its outer shell. In forming the same ionic compound, Cl gains an electron to become the Cl^- anion, which also has a stable octet of outer-shell electrons.

In the remainder of this chapter, the octet rule will be used in explaining the formation of chemical compounds consisting of two or more different elements bonded together. It was already used to show the bonding in ionic sodium chloride in Figure 2.8. One of the best compounds for showing the octet rule in covalent compounds is methane, CH_4, shown in Figure 3.1. The molecule of CH_4 is produced when an atom of carbon with 4 outer electrons (see Figure 2.9) attains an octet of 8 electrons by sharing with H atoms.

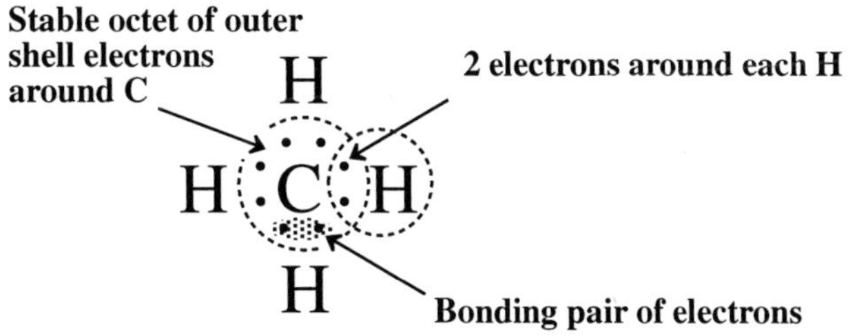

Figure 3.1. Stable outer electron shells from covalent bonding in compounds.

Each H atom has 1 electron to donate to the sharing arrangement, so by each of 4 H atoms contributing an electron the carbon atom can gain an octet. Each of the H atoms has access to 2 electrons in the single covalent bond that connects it to the C atom. Examination of Figure 3.1 implies that the 4 H atoms and the C atom all lie in the same plane in a flat structure. But that is not the case because of the tendency for the 4 electron pairs composing the 4 covalent bonds to be oriented as far as possible from each other around the sphere of the carbon atom. The resulting structure is a little hard to illustrate on paper, but one way to approximate it is with a ball and stick model that represents the atoms as balls and the chemical bonds as sticks connecting the atoms. Figure 3.2 is an illustration of the ball and stick model of CH_4.

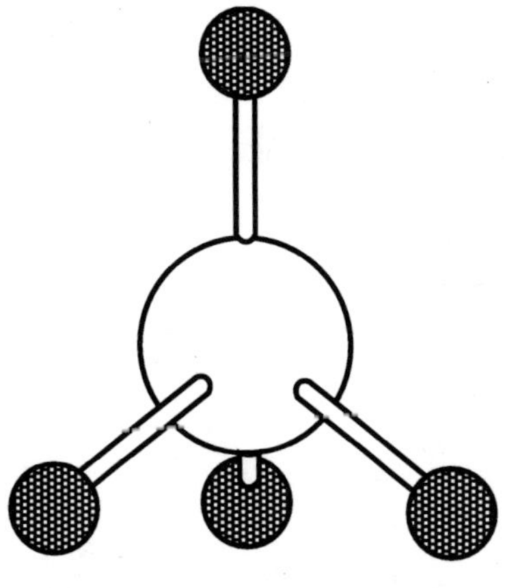

Figure 3.2. Ball and stick model of CH_4 in which 4 H atoms (shaded balls) are arranged in three dimensions around an oxygen atom.

3.3. SODIUM CHLORIDE AND IONIC BONDS

Many atoms and groups of atoms in chemical compounds are **ions** that have an electrical charge because of their unequal numbers of protons and electrons. **Cations** are positively charged ions and **anions** are negatively charged ions. Compounds consisting of ions are **ionic compounds** and the bonds holding them together are **ionic bonds**. Ionic bonds depend upon the mutual attraction between positive cations and negative anions for their bond strength (oppositely charged bodies attract each other, whereas negatively charged bodies repel each other).

The formation of ions based upon the octet rule is readily seen for the well-known ionic compound, sodium chloride, NaCl, as illustrated in Figure 3.3. By losing an electron to become the Na^+ cation, sodium's underlying shell of 8 electrons becomes the ion's outer shell with a stable octet. Chlorine attains a stable octet of 8 outer-shell electrons by gaining 1 electron per atom to produce Cl^- ion.

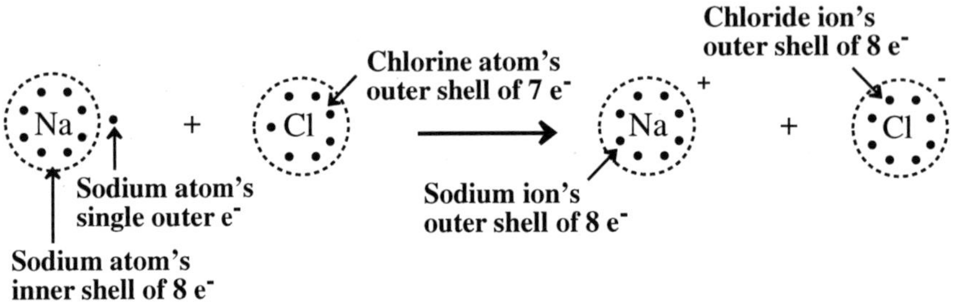

Figure 3.3. Formation of ionic sodium chloride from neutral Na and Cl atoms

Sodium chloride is a very stable compound because of the mutual attraction of oppositely charged ions. But the ions have to be arranged in an optimum manner for this attraction to be effective. Since oppositely charged ions attract each other, but ions with the same charge are mutually repulsive, the ions in an ionic compound such as sodium chloride have to be packed to maximize attraction and minimize repulsion. The arrangement that does this for NaCl is shown by a ball and stick model in Figure 3.4. Although it may be a little hard to imagine for a model represented on paper, the <u>six</u> nearest neighbors of each negatively charged Cl^- anion are Na^+ cations. And the six nearest neighbors of each positively charged Na^+ cation are negatively charged Cl^- anions.

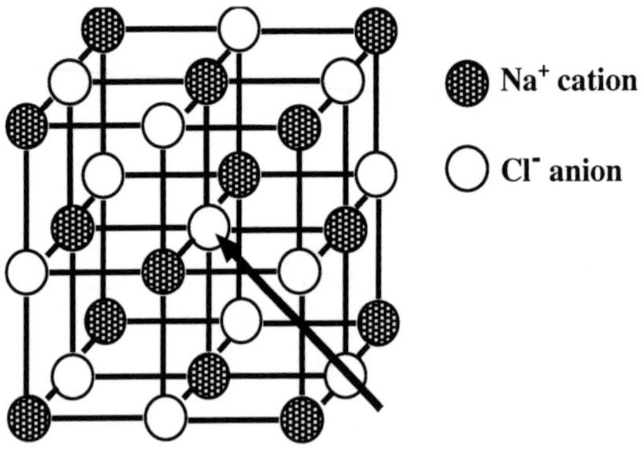

Figure 3.4. Representation of the structure of NaCl. This structure can be visualized in three dimensions as "balls" with interconnecting "sticks." The balls are Na^+ and Cl^{--} ions. The closest neighbors to the Cl^- ion marked with an arrow are 6 Na^+ ions.

In reality, ions are more accurately represented in an ionic structure as spheres that touch. The Na^+ cation is significantly smaller than the Cl^- anion, so a more accurate representation of NaCl would show rather large Cl^- spheres between which are nestled barely visible Na^+ spheres. But the imperfect ball and stick model shown in Figure 3.4 shows several important points regarding ionic NaCl. It illustrates the relative positions of the ions. These positions, combined with ionic charge and size, determine the **crystal structure** of the solid crystal of which the ionic compound is composed. Furthermore, examination of the figure shows that no single Cl^- anion belongs to a specific Na^+ cation, and no single Na^+ cation belongs to a specific Cl^- anion. So, although the chemical formula NaCl accurately represents the relative numbers of Na and Cl atoms in sodium chloride it *does not* imply that there are discrete molecules consisting of 1 Na and 1 Cl. So it is not correct to refer to a *molecule* of sodium chloride because molecules of ionic compounds do not exist as such. Instead, reference is made to **formula units** of ionic compounds, where a formula unit of NaCl consists of 1 Na^+ ion and 1 Cl^- ion, the smallest quantity of a substance that can exist and still be sodium chloride.

The stabilities of chemical compounds are all about energy. In general, the more energy released when a compound forms, the more stable the compound is. Sodium chloride could be formed by reacting elemental solid sodium with elemental Cl_2 gas,

$$2\text{Na}(solid) + \text{Cl}_2(gas) \rightarrow 2\text{NaCl}\ (solid) \qquad (3.3.1)$$

to produce solid sodium chloride. This reaction releases a large amount of energy and elemental sodium burns explosively in chlorine gas. The reaction can be viewed in terms of the following steps.

1. The atoms in solid Na are taken apart, which requires energy.
2. Each molecule of Cl_2 is taken apart, which requires energy.
3. An electron is taken from each Na atom to produce Na^+ ion, which requires energy.
4. An electron is added to each Cl atom to produce a Cl^- ion, which releases energy.
5. All the Na^+ cations and 1 Cl^- anion are assembled in a 1/1 ratio in a crystal lattice to produce NaCl, which releases a very large quantity of energy.

The very large amount of energy involved in Step 5 is called the **lattice energy** and is primarily responsible for the high stability of ionic compounds. A general picture of the energy involved is shown in Figure 3.5.

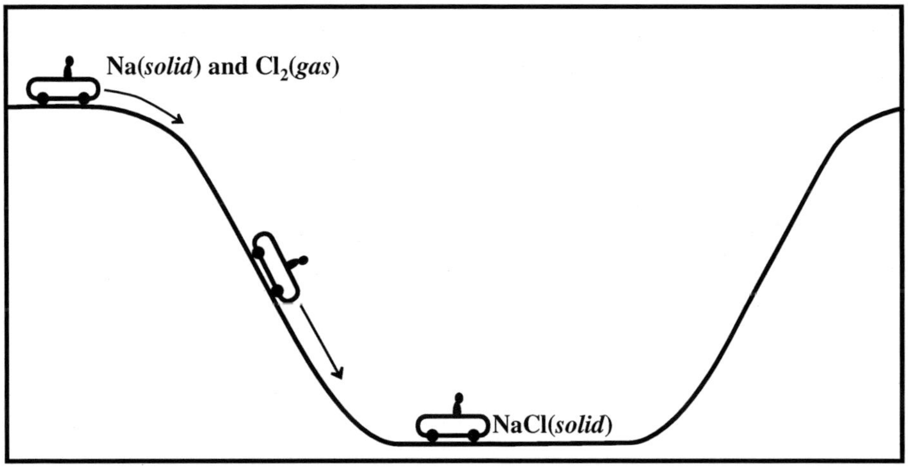

Figure 3.5. The energy involved in forming a compound is much like that involved in rolling a cart down a hill. In the case of sodium chloride, a lot of energy is released in forming solid crystalline ionic NaCl from solid Na metal and gaseous Cl_2.

The differences in ionic size noted above are represented in Figure 3.6 for several monatomic (1-atom) ions from elements close to each other in the periodic table. The figure shows that negative ions are generally larger than positive ions formed from elements that are nearby in the periodic table. Thus, the negative F^- ion is larger than the positive Na^+ ion, although both ions have the same number of electrons (10) and the atomic number of Na is higher than that of F. It is seen that for ions in the same group of elements that have the same charge, the ion from the element with higher atomic number is larger. Figure 3.6 shows the Cl^- ion larger than the F^- ion and the K^+ ion is larger than the Na^+ ion. As electrons are removed from elements in the same period of the periodic table to produce progressively more highly charged cations, ion size shrinks notably as shown by the order of ion sizes $Na^+ > Mg^{2+} > Al^{3+}$, each of which has 10 e^-. This occurs because as the charge of the nucleus becomes larger relative to the charge of the negative electron cloud around it, the cloud is drawn closer to the nucleus, shrinking ion size. As electrons are added to atoms to produce more highly charged anions, the anion size <u>increases</u> because more electrons occupy more space. So S^{2-} ion is larger than Cl^- ion.

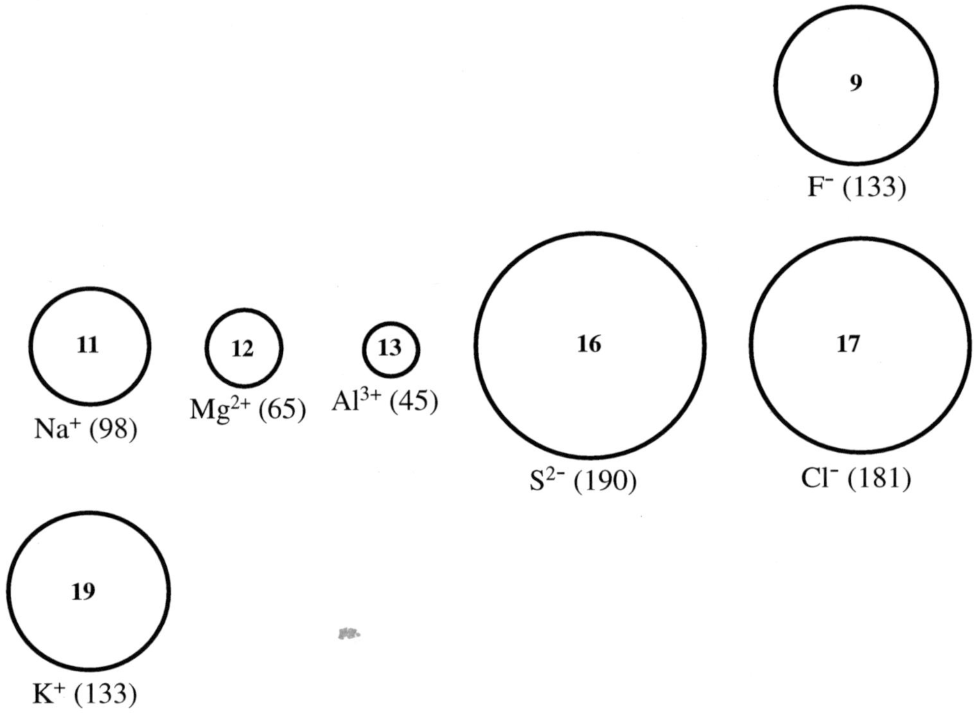

Figure 3.6. Relative sizes of several ions formed by removing electrons from or adding electrons to several elements shown in the abbreviated periodic table. The ion diameters are given in picometers (pm) in parentheses after the symbol for each ion. The atomic numbers of the elements from which each ion has been formed are shown inside the circles representing the ions.

In order to further understand ion formation, several more examples can be considered. Calcium and chlorine react,

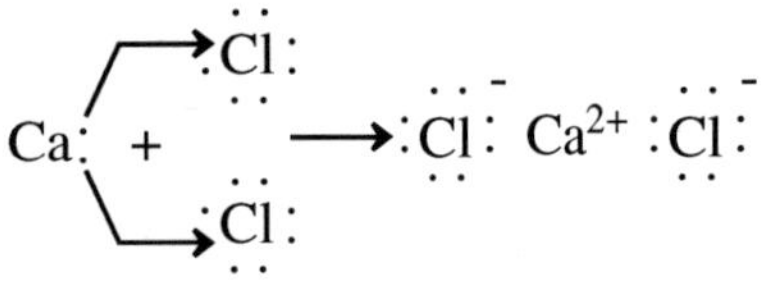

(3.3.2)

to form calcium chloride, $CaCl_2$. This compound is a byproduct of some industrial processes, from which its disposal can be a problem. It is commonly used as road salt to melt ice and snow on streets and highways. Although calcium chloride is effective in this respect, it is corrosive to automobiles and calcium chloride is a pollutant salt that can contribute to excess salt levels in bodies of water. A "greener," though more costly substitute is calcium acetate, $Ca(C_2H_3O_2)_2$. This compound is composed of Ca^{2+} ions and acetate ($C_2H_3O_2^-$) anions. Its advantage is that bacteria on soil and in water readily cause biodegradation of the acetate anion as shown by the reaction,

$$Ca(C_2H_3O_2)_2 + 4O_2 \xrightarrow{\text{Bacteria}} CaCO_3 + 3CO_2 + 3H_2O \qquad (3.3.3)$$

from which the calcium ends up as calcium carbonate, a common component of rock and soil.

Another example of the formation of an ionic compound is,

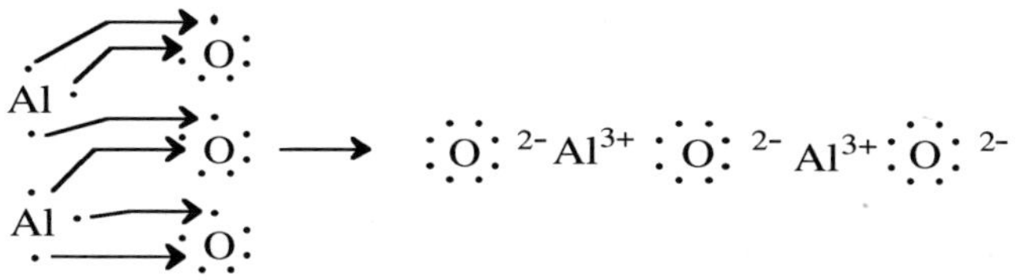

(3.3.4)

to produce aluminum oxide, Al_2O_3. This compound is the source of aluminum in bauxite, the ore from which aluminum is produced. Called alumina, aluminum oxide is used to make abrasives, ceramics, antacids, and antiperspirants.

Exercise: Fill in the table below for the ionic products of the reaction of the metal and nonmetal indicated.

Metal	Nonmetal	Cation formed	Anion formed	Ionic compound
Na	**Cl**	**(a)**________	**(b)**________	**(c)**________
K	**O**	**(d)**________	**(e)**________	**(f)**________
Ca	**Cl**	**(g)**________	**(h)**________	**(i)**________

Answers: (a) Na^+, (b) Cl^-, (c) NaCl, (d) K^+, (e) O^{2-}, (f)K_2O, (g) Ca^{2+}, (h) Cl^-, (i) $CaCl_2$

In addition to ions formed from single atoms losing or gaining electrons, many ions consist of groups of atoms covalently bound together, but having a net electrical charge because of an excess or a deficiency of electrons. An example of such an ion is the acetate ion shown above in the formula of calcium acetate, $Ca(C_2H_3O_2)_2$. The structural formula of the acetate anion, $C_2H_3O_2^-$, is,

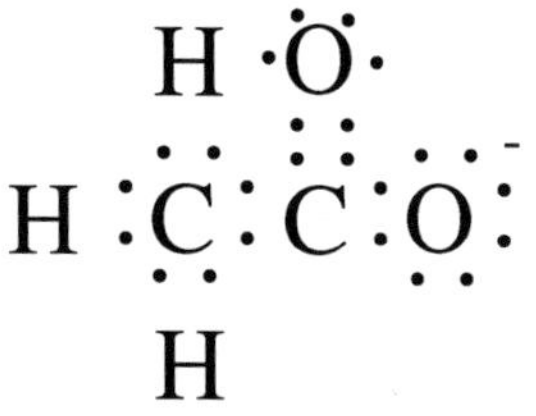

in which the two carbon atoms are joined with a single covalent bond consisting of two shared electrons, each of the three H atoms are joined to one of the carbon atoms by a single covalent bond and the other carbon atom is joined to one oxygen with a single covalent bond and to the other by a double covalent bond consisting of 4 shared electrons. The net charge on the ion is -1.

Ionic Liquids and Green Chemistry

Most common ionic compounds such as sodium chloride are hard solids because the ions of which they are composed are relatively small and packed tightly together in the crystalline lattice. These ionic compounds must be heated to very high temperatures before they melt, 801°C for NaCl, for example. In recent years, ionic compounds have been developed that are liquids under ordinary conditions. The ions in these **ionic liquids** are composed of large organic molecules composed of skeletons of numerous carbon atoms bonded to other atoms and having a net negative charge. The charges on the ions in such compounds is much less concentrated than in simple inorganic compounds like NaCl, the large ions move readily relative to each other in the ionic crystal, and the compound is liquid at low temperatures. A common example of an ionic liquid compound is decylmethylimidazolium hexafluorophosphate, which is a liquid at temperatures above 40°C, the temperature of a very hot summer's day.

There has been a lot of interest in the application of ionic liquids to green chemistry. This is because many chemical reactions including those for preparing polymers used in synthetic fabrics or pharmaceutical compounds are carried out in liquid solvents, which tend to evaporate and contaminate air and to pose disposal problems. Furthermore, although the solvent properties in chemical synthesis often play a strong role in determining the course of the synthesis, the number of available solvents is very limited. In the case of ionic liquids, however, there is a vast variety of both cations and liquids which, combined together, could give as many as a trillion different ionic liquids. This versatility enables fine-tuning the properties of the ionic liquids for specialized uses in synthesis and other applications. The ionic liquids are rather easy to recycle, adding to their green character. In addition to their applications as solvents for chemical synthesis media, ionic liquids may be useful for isolating pollutants. For example, an appropriate ionic liquid may be shaken with water contaminated with toxic heavy metals, such as lead or cadmium, which bind with the ionic liquid. Since such a liquid is normally not soluble in water, it can be physically separated from the water, carrying the heavy metals with it and leaving purified water.

3.4. COVALENT BONDS IN H_2 AND OTHER MOLECULES

From their Lewis symbols it is easy to see how some atoms of elements on the left side of the table with only one or two electrons would have a strong tendency to lose those electrons to form cations such as Na^+ or Ca^{2+}. It is also easily seen that atoms from groups near the right side of the periodic table can accept one or two electrons to gain stable octets and become anions such as Cl^- or O^{2-}. But, it is difficult to impossible to take more than two electrons away from an atom to form cations with charges greater than +2 or to add 3 or more electrons to form anions with charges of -3 or even more negative, although ions such as Al^{3+} and N^{3-} do exist. So atoms of elements in the middle of the periodic table and the nonmetals on the right have a tendency to share electrons in covalent bonds, rather than becoming ions.

It is easy to see how mutually attracting oppositely charged ions are held together in a crystalline lattice. In covalent bonds, shared electrons between the nuclei of atoms joined by the bond act to reduce the forces of repulsion between the positively charged nuclei of the atoms that they join together. That is most easily seen for the case of the hydrogen molecule, H_2. The nuclei of H atoms consist of single protons, and these two H atom nuclei in the H_2 molecule repel each other. However, their 2 shared electrons compose a cloud of negative charge between the two H nuclei, shielding the nuclei from each other's repelling positive charge and enabling the molecule to exist as a covalently bound molecule (Figure 3.7).

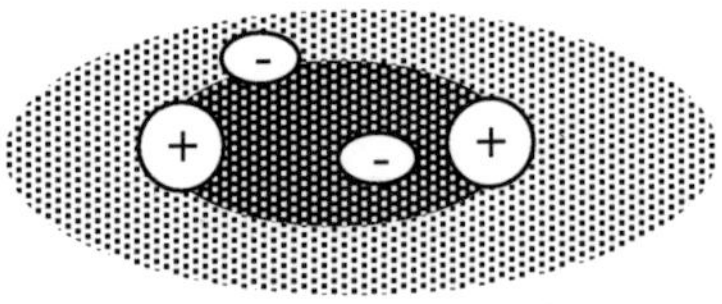

Figure 3.7. Electrons in motion between and around mutually repelling H nuclei shield the nuclei from each other constituting a single covalent bond that holds the H_2 molecule together.

3.5. COVALENT BONDS IN COMPOUNDS

Consider next some example covalent bonds between atoms of some of the lighter elements. These are best understood in reference to Figure 2.9, the abbreviated version of the periodic table showing the Lewis symbols (outer shell valence electrons) of the first 20 elements. As is the case with ions, atoms that are covalently bonded in molecules often have an arrangement of outer shell electrons like that of the noble gas with an atomic number closest to the element in question. It was just seen that covalently bonded H atoms in molecules of H_2 have 2 outer shell electrons like the nearby noble gas helium. For atoms of many other elements, the tendency is to acquire 8 outer shell electrons— an octet — in sharing electrons through covalent bonds. This tendency forms the basis of the *octet rule* discussed in Section 3.2. In illustrating the application of the octet rule to covalent bonding, Section 3.7 considers first the bonding of atoms of hydrogen to atoms of elements with atomic numbers 6 through 9 in the second period of the periodic table. These elements are close to the noble gas neon and tend to attain a "neon-like" octet of outer shell electrons when they form covalently bonded molecules.

Covalent bonds are characterized according to several criteria. The first of these is the number of electrons involved. The most common type of covalent bond consists of 2 shared electrons and is a *single bond.* Four shared electrons as shown for the bond joining an O atom to one of the C atoms in the structure of the acetate anion above constitute a *double covalent bond.* And 6 shared electrons as shown for the very stable covalent bond joining the two N atoms in the N_2 molecule illustrated in Chapter 2, Figure 2.6 make up a *triple covalent bond.* These bonds are conventionally shown as lines in the structural formulas of molecules (large numbers of dots in a formula can get a little confusing). So the single covalent bond in H_2 is shown as

$$H-H$$

The double bond consisting of 4 shared electrons holding the two carbon atoms together in C_2H_4 (ethylene, a hydrocarbon used to make polyethylene plastic) are shown by the following:

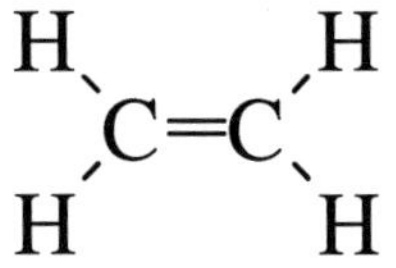

And the very strong triple bond joining the two N atoms in the N_2 molecule are shown by three lines:

$$N \equiv N$$

Recall from Figure 2.6 that each N atom has a pair of electrons that are not part of any chemical bonds. These are omitted from the structure above, and are not ordinarily shown when bonds are represented by lines.

Covalent bonds have a characteristic **bond length**. Bond lengths are of the general magnitude of the size of atoms, so they are measured in units of picometers (pm). The H-H bond in H_2 is 75 pm long.

A third important characteristic of bonds is **bond energy**. Bond energy is normally expressed in kilojoules (kJ) required to break a mole (6.02×10^{23}) of bonds. (See Section 3.8 for a detailed definition of the mole.) The bond energy of the H-H bond in H_2 is equal to 435 kJ/mole. This means that an amount of energy required to break all the H-H bonds in a mole of H_2 (2.0 grams of H_2, 6.02 $\times 10^{23}$ molecules) is a very substantial 435 kJ.

3.6. COVALENT BONDS AND GREEN CHEMISTRY

The nature of covalent bonds is strongly related to green chemistry. Some reasons why this is so include the following:

- High-energy bonds in raw materials require a lot of energy and severe conditions, such as those of temperature and pressure, to take apart in synthesizing chemicals. The practice of green chemistry tries to avoid such conditions.
- Especially stable bonds may make substances unduly persistent in the environment.

- Relatively weak bonds may allow molecules to come apart too readily, contributing to reactive species in the atmosphere or in biological systems.
- Unstable bonds or arrangements of bonds may lead to excessive reactivity in chemicals making them prone to explosions and other hazards.
- Some arrangements of bonds contribute to chemical toxicity.

An example of a substance that has a very high bond stability making it an energy-intensive source of raw material is N_2. As mentioned earlier, large amounts of energy and severe conditions are required to take this molecule apart in the synthesis of ammonia, NH_3, the compound that is the basis for most synthetic nitrogen compounds. As discussed with nitrogen in Section 2.5, *Rhizobium* bacteria growing on the roots of leguminous plants such as soybeans convert N_2 to chemically fixed nitrogen. The amount of nitrogen fixed by this totally green route is certainly a welcome addition to the pool of fertilizer nitrogen.

An example of a compound in which especially stable bonds contribute to persistence and ultimate environmental harm is provided by dichlorodifluoromethane, Cl_2CF_2, a chlorofluorocarbon implicated in the destruction of stratospheric ozone (see Chapter 2, Section 2.5, and Chapter 8). The chemical bonds in this compound are so strong that nothing attacks them until the molecule has penetrated many kilometers high into the stratosphere where extremely energetic ultraviolet radiation breaks the C-Cl bond in the molecule. This produces Cl atoms that bring about the destruction of protective stratospheric ozone.

A somewhat opposite condition occurs in the case of atmospheric nitrogen dioxide, NO_2, near ground level in the atmosphere. Here the NO bond is relatively weak so that the relatively low-energy ultraviolet radiation ($h\nu$) that is close to the wavelength of visible light and penetrates to the atmosphere at or near ground level can break apart NO_2, molecules:

$$NO_2 + h\nu \rightarrow NO + O \qquad (3.6.1)$$

The O atoms released are very reactive and interact with pollutant hydrocarbons, such as those from gasoline, in the atmosphere resulting in the disagreeable condition of photochemical smog.

Some bonding arrangements are notable for instability. These include bonds in which two N atoms are adjacent or very close in a molecule and are bonded with double bonds. Also included are arrangements in which N and O atoms are adjacent and double bonds are also present.

The presence of some kinds of bonds in molecules can contribute to their biochemical reactivity and, therefore, to their toxicity. An organic compound with one or more C=C double bonds in the molecule is often more toxic than a similar molecule without such bonds.

By avoiding generation, use, or release to the environment of compounds with the kinds of bonds described above as part of a program of green chemistry, the practice of chemistry and the chemical industry can be made much safer. Green chemistry also considers bonds that may have to be protected in chemical synthesis. Often steps must be added to attach protecting groups to bonding groups to prevent their reacting during a step of a synthesis. After the desired step is performed, the protecting groups must be removed to give the final product. Materials are consumed and byproducts are generated by these steps, so the practice of green chemistry attempts to avoid them whenever possible.

3.7. PREDICTING COVALENTLY BOUND COMPOUNDS

It is often possible to predict the formulas of molecules using the Lewis symbols of the elements in the compound with the octet rule for chemical bonding. This is shown in Figure 3.8 for the hydrogen compounds of several elements in the second period of the periodic table. The prediction of chemical bonds in compounds in which H is bonded to another atom is very simple because each H atom has to be involved in sharing two electrons and the other kind of atom has to have a total of 8 electrons in its valence shell octet; these may be shared in bonds or present as unshared pairs.

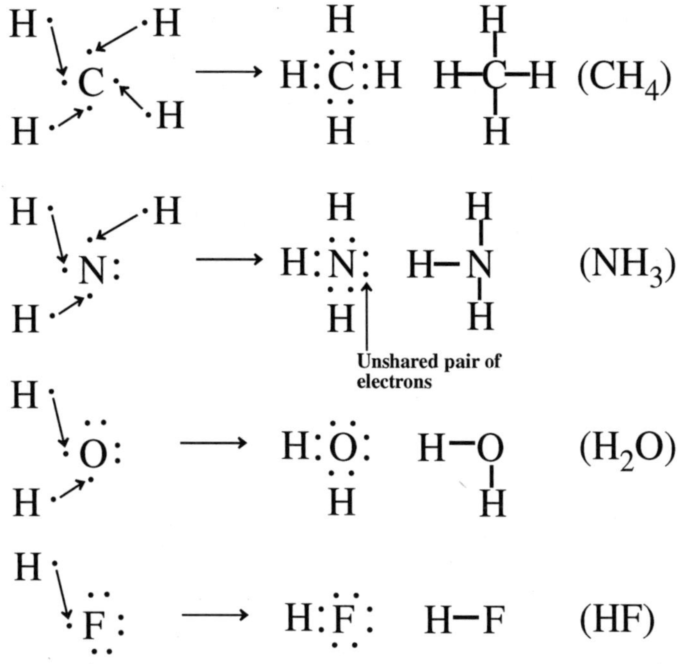

Figure 3.8. Formation of covalent hydrogen compounds with elements in the second row of the periodic table. All of carbon's valence electrons are involved in bonds in CH_4, NH_3 has 1 unshared pair of valence electrons, H_2O has 2 unshared pairs of valence electrons, and HF has 3 unshared pairs of valence electrons. The unshared pairs of electrons are not shown in the structures in which the bonds are represented by straight lines.

The octet rule can be used to predict bonding between atoms neither of which is H. The atoms may be bonded by single bonds of 2 electrons, double bonds of 4 electrons, and even triple bonds of 6 electrons such that each atom in the molecule has 8 outer-shell electrons. As an example, consider a well-known compound of carbon, carbon dioxide, chemical formula CO_2. The Lewis symbol of C and those of the two O atoms can be used to deduce the Lewis formula of CO_2 as shown in Figure 3.9.

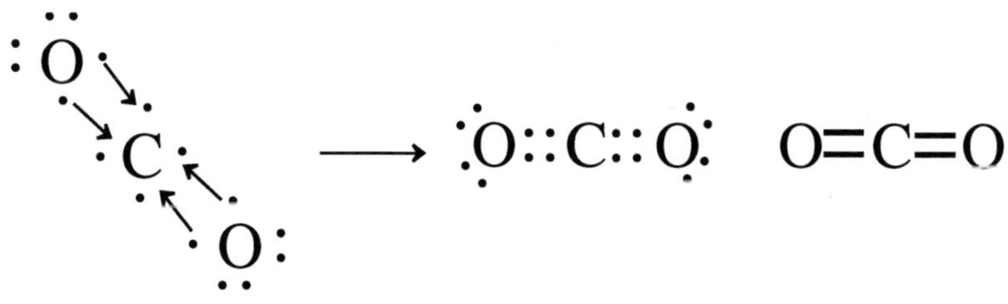

Figure 3.9. Bonding of two oxygen atoms to a central carbon atom to produce carbon dioxide, CO_2. In order to provide an octet of electrons around each of the three atoms with the total of only 16 valence electrons available, it is necessary to have double bonds between C and each of the O atoms.

As another example of the application of the octet rule, consider hydrogen peroxide, H_2O_2. This compound's formula looks a lot like that of water, but it is a lot different from water. Hydrogen peroxide decomposes to release oxygen:

$$2H_2O_2(\textit{liquid}) \rightarrow O_2(\textit{gas}) + 2H_2O(\textit{liquid}) \quad (3.7.1)$$

As a liquid in the form of a concentrated aqueous solution, hydrogen peroxide provides a source of oxygen so potent that it has been used in rockets. It was the treacherous oxidant used along with hydrazine (N_2H_4) fuel in the German Luftwaffe's Messerschmidt 163 rocket plane at the end of World War II. Trailing an exhaust of lethal NO_2 gas, this miniscule manned missile (on the rare occasions when it worked according to plan) was propelled rapidly into the lower stratosphere, then glided down through waves of Allied bombers, attempting to nick them with machine gun fire as it plummeted back to Earth. Few bombers were damaged but many Me-163 pilots died in the attempt, some as the result of explosions, fires, and spills from the hydrogen peroxide oxidant.

In assembling the structure of the hydrogen peroxide molecule, one has simply to work with two O atoms each contributing 6 valence electrons and two H atoms each with 1 valence electron. The Lewis formula of the H_2O_2 molecule is

showing that all of the 14 total valence electrons are involved in chemical bonds and both oxygens have octets of outer-shell electrons.

Despite the evil nature of concentrated solutions of hydrogen peroxide, it can be regarded as a green compound in more dilute solutions, such as the 3% hydrogen peroxide commonly used to kill bacteria in treating wounds. Among its green applications, dilute hydrogen peroxide makes an effective and safe bleaching agent that is much safer to handle than elemental chlorine commonly used for bleaching and that does not produce the potentially toxic byproducts that chlorine generates. And even though it kills bacteria, hydrogen peroxide can be pumped underground to serve as an oxidant for acclimated bacteria that attack wastes that have been placed in or seeped into underground locations.

Molecules That Do Not Obey the Octet Rule

In some cases the octet rule is not obeyed. This occurs when a molecule has an uneven number of electrons so that it is impossible for each atom to have an octet (an even number) of electrons. A simple example of this is nitric oxide, NO, made from an atom of N with 5 valence electrons and one of O with 6 valence electrons. The resulting molecule is shown in Figure 3.10. Since the uneven number of 11 valence electrons cannot provide complete octets of electrons around both the N and O atoms simultaneously, the NO molecule is shown in two forms in which one atom has 8 valence electrons and the other has 7. These are known as **resonance structures**.

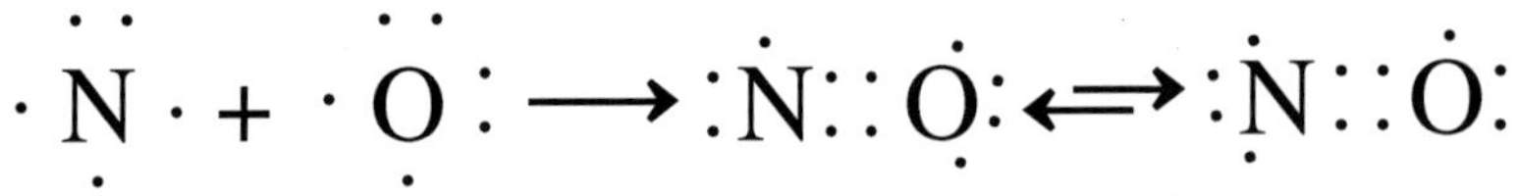

Figure 3.10. Formation of NO. The uneven number of electrons cannot accommodate complete octets around both atoms simultaneously, so the structure is a hybrid resonance structure of the two forms shown.

Unequal Sharing of Electrons

The Lewis formula for water,

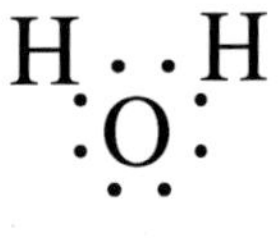

indicates that the molecule is not symmetrical and the two H atoms are located on one side of the molecule and the O atom on the other side. One might think that the electrons shared between H and O are shared equally. But such is not the case because the relatively larger O atom nucleus with its 8 protons has a stronger attraction for the electrons than do the two H atom nuclei, each with only 1 proton. So the shared electrons spend relatively more time around the O atom and less around the H atoms. This gives each H atom a partial positive charge and the O atom a partial negative charge. An unequal distribution of charge such as that makes a body **polar** and the O-H bonds are **polar covalent bonds**. In fact, the whole water molecule is polar and can be represented as the following, where the small spheres stand for H atoms and the large one for the O atom:

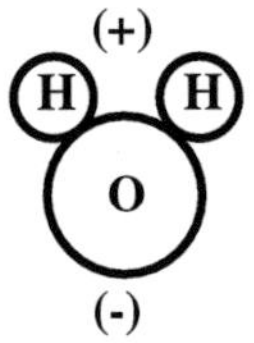

The polar water molecule

The polar nature of the water molecule has a lot to do with water as a solvent and how it behaves in the environment and in living systems. These aspects are covered in more detail in the discussion of aquatic chemistry in Chapter 7.

When Only One Atom Contributes to a Covalent Bond

In some cases only one of the two atoms joined by a covalent bond contributes both the electrons in the bond. This occurs with ammonia, NH_3, dissolved in water. Water contains dissolved H^+ cations, and the more acidic water is, the higher the concentration of H^+. The H^+ cation, would be stabilized by two electrons which it can get by binding with dissolved NH_3 as shown in Reaction 3.7.2. Both of the electrons shared between N and the H^+ cation now bound to it as part of a new species, the ammonium ion, NH_4^+, were contributed by the N atom. Such a covalent bond is called a **coordinate covalent bond** or a **dative bond**. In the case of NH_4^+, once the coordinate covalent N-H bond is formed, it is indistinguishable from all the other N-H bonds.

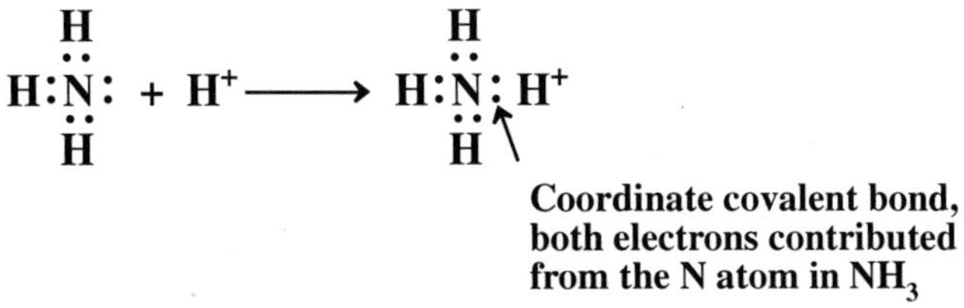

(3.7.2)

The formation of the coordinate covalent bond in NH_4^+ is very useful when soil is fertilized with nitrogen. The most economical way to apply nitrogen fertilizer is by injecting NH_3 into the soil, but NH_3 is a gas that would be expected to rapidly evaporate from soil. Instead, it becomes bound with H^+ ion from the water in the soil and is bound to the soil as the NH_4^+ ion.

Another important example of a coordinate covalent bond occurs in water. As discussed in Section 3.9, acids, which are very important materials commonly dissolved in water, produce the hydrogen ion, H^+, in water. This ion does not exist simply dispersed in water. Instead, it binds strongly to a water molecule to produce the hydronium ion, H_3O^+:

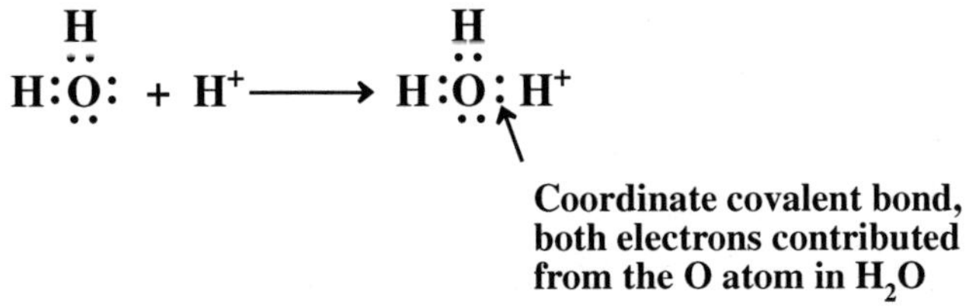

(3.7.3)

CHEMICAL FORMULAS, THE MOLE, AND PERCENTAGE COMPOSITION

Chemical formulas represent the composition of chemical compounds. A number of chemical formulas have been shown so far including H_2O for water and NH_3 for ammonia. A chemical formula of a compound contains a lot of significant information (Figure 3.11) including the following:

- The elements that compose the compound
- The relative numbers of each kind of atom in the compound
- How the atoms are grouped, such as in ions (for example, SO_4^{2-}) present in the compound
- With a knowledge of atomic masses, the molar mass of the compound
- With a knowledge of atomic masses, the percentage composition of the compound

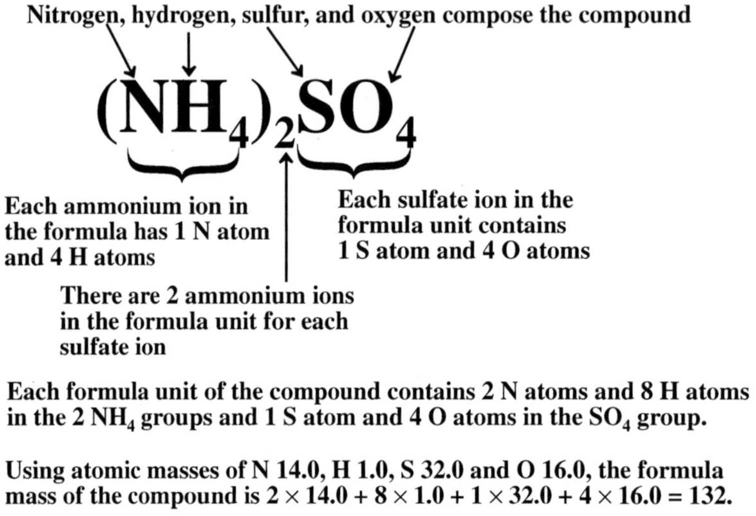

Figure 3.11. Information contained in a chemical formula.

Where the symbols of the elements represent letters in the alphabet of chemical language, the formulas of compounds represent words composed of those letters. As discussed in Chapter 5, formulas are put together in chemical equations that act as sentences in the chemical language to describe what chemical substances do.

The Mole

With a knowledge of atomic masses, the percentage composition of a compound is readily calculated from its formula. Before doing such a calculation for ammonium sulfate, however, it is useful to introduce the concept of the **mole**. Chemists use the mole to express quantities of materials containing a specific number of specified entities, which may be atoms of elements, molecules of elements that exist as diatomic molecules, formula units of ionic compounds, or molecules of covalent compounds. A quantity of a mole of a substance is simply the atomic mass, molecular mass, or formula mass followed by *grams*. This quantity is called the **molar mass**. The masses of a mole of several typical substances are given below:

Atoms of neon, atomic mass 20.1: 20.1 g/mol

Molecules of H_2, atomic mass 1.0, molecular mass 2.0: 2.0 g/mole

Molecules of CH_4, molecular mass 16.0: 16.0 g/mole

Formula units of ionic CaO, formula mass 56.1: 56.1 g/mol

The number of specified entities in a mole of a substance is always the same regardless of the substance. This number is very large, 6.02×10^{23}, and is called **Avogadro's number**. As examples, a mole of neon contains 6.02×10^{23} neon atoms, a mole of elemental hydrogen contains 6.02×10^{23} H_2 molecules (but $2 \times 6.02 \times 10^{23}$ H atoms), and a mole of CaO contains 6.02×10^{23} formula units (pairs of Ca^{2+} and O^{2-} ions) of CaO.

To calculate the percentage composition of $(NH_4)_2SO_4$, it should be noted that its molecular mass is 132, so its molar mass is 132 g/mol. Each mole of this substance contains $2 \times 1 = 2$ mol of N, $2 \times 4 = 8$ mol of H, 1 mol of S, and $4 \times 1 = 4$ mol of O. With this information it is easy to calculate the percentage composition of $(NH_4)_2SO_4$. The mass of each element in a mole of the compound and each element's percentage is given by the following:

$$2 \text{ mol N} \times 14.0 \text{ g N/mol N} = 28.0 \text{ g N, } \%\text{N} = \frac{28.0 \text{ g}}{132 \text{ g}} \times 100 = 21.2\% \text{ N}$$

$$8 \text{ mol H} \times 1.0 \text{ g H/mol H} = 8.0 \text{ g H, } \%\text{H} = \frac{8.0 \text{ g}}{132 \text{ g}} \times 100 = 6.1\% \text{ H}$$

$$1 \text{ mol S} \times 32.0 \text{ g S/mol S} = 32.0 \text{ g S, } \%\text{S} = \frac{32.0 \text{ g}}{132 \text{ g}} \times 100 = 24.2\% \text{ S}$$

$$4 \text{ mol O} \times 16.0 \text{ g O/mol O} = 64.0 \text{ g O, } \%\text{N} = \frac{64.0 \text{ g}}{132 \text{ g}} \times 100 = 48.5\% \text{ O}$$

Example: Given the atomic masses Ca 40.0, C 12.0, and O 16.0, what is the percentage composition of calcium oxalate, CaC_2O_4?

Answer: 31.3% Ca, 18.8% C, 50.0% O

3.9. WHAT ARE CHEMICAL COMPOUNDS CALLED?

The naming of chemical compounds can get a little complicated. This is particularly true of organic compounds, the names of which are discussed in Chapter 9. Some of the simpler aspects of naming inorganic compounds are discussed here.

In naming compounds, prefixes are used to represent the relative numbers of atoms in the formula unit of the compound. These prefixes through number 10 are given below:

1-mono	3-tri	5-penta	7-hepta	9-nona
2-di	4-tetra	6-hexa	8-octa	10-deca

The first class of inorganic compounds to be addressed here are **binary molecular compounds**, those that consist of only 2 kinds of elements and do not contain ions. For these compounds, the first part of the name is simply the name of the first element in the compound formula. The second part of the name is that of the second element in the compound formula modified to have the ending *-ide.* Prefixes are added to indicate how many of each kind of atom are present in the molecule. Consider as an example the name of N_2O_5. The name of the compound is dinitrogen pentoxide where *di* indicates 2 N atoms, *pent* indicates 5 oxygen atoms, and the second element has the *ide* ending. Other examples of this system of naming are $SiCl_4$, silicon tetrachloride; Si_2F_6, disilicon hexafluoride; PCl_5, phosphorus pentachloride; and SCl_2, sulfur dichloride.

A number of compounds, including binary molecular compounds, have **common names** that have been used for so long that they are part of the chemical vocabulary. An especially common example is the name of water for H_2O; its official name is dihydrogen monoxide. Another example is dinitrogen monoxide, N_2O, commonly called nitrous oxide.

Recall that *ionic compounds* are those composed of ions that are held together by ionic bonds, rather than covalent bonds. As noted in the discussion of ionic sodium chloride in Section 3.3, ionic compounds do not consist of discrete molecules, but rather of aggregates of ions whose relative numbers make the compound electrically neutral overall. Therefore, it is not correct to refer to molecules of ionic compounds but instead reference should be made to *formula units* equal to the smallest aggregate of ions that can compose the compound. Consider, for example, the ionic compound composed of Na^+ and SO_4^{2-} ions. Every ionic compound must be electrically neutral with the same number of positive as negative charges. For the compound in question this requires 2 Na^+ ions for each SO_4^{2-} ion. Therefore, the formula of the compound is Na_2SO_4 and a formula unit contains 2 Na^+ ions and 1 SO_4^{2-} ion. Furthermore, a mole of Na_2SO_4 composed of 6.02×10^{23} formula units of Na_2SO_4 contains $2 \times 6.02 \times 10^{23}$ Na^+ ions and 6.02×10^{23} SO_4^{2-} ions. Since the ionic charges determine the relative numbers of ions, prefixes need not be used in naming the compound and it is called simply sodium sulfate.

Exercise: Give the formulas and names of compounds formed from each cation on the top row with each anion on the bottom row, below:

1. NH_4^+	2. Ca^{2+}	3. Al^{3+}
(A) Cl^-	(B) SO_4^{2-}	(C) PO_4^{3-}

Answers: 1(A) NH_4Cl, ammonium chloride; 1(B) $(NH_4)_2SO_4$, ammonium sulfate; 1(C), $(NH_4)_3PO_4$, ammonium phosphate; 2(A) $CaCl_2$, calcium chloride; 2(B) $CaSO_4$, calcium sulfate; 2(C), $Ca_3(PO_4)_2$, calcium phosphate; 3(A) $AlCl_3$, aluminum chloride; 3(B) $Al_2(SO_4)_3$, aluminum sulfate; 3(C), $AlPO_4$, aluminum phosphate;

Prefixes are used in naming ionic compounds where more than 1 cation or more than 1 anion are present in the formula unit. For example, Na_2HPO_4 in which each formula unit is composed of 2 Na^+ ions, 1 H^+ ion, and 1 PO_4^{3-} ion is called disodium monohydrogen phosphate. And KH_2PO_4 is called monopotassium dihydrogen phosphate.

3.10. ACIDS, BASES, AND SALTS

Other than binary molecular compounds, most inorganic compounds can be classified as acids, bases, or salts. These three categories of compounds and their names are addressed briefly here.

Acids

Acids are characterized by the H^+ ion, the presence of which in water, makes the water **acidic**. An acid either contains this ion or produces it when it dissolves in water. Sulfuric acid, H_2SO_4, is an example of a compound that contains H^+ ion. Dissolved in water, a molecule of sulfuric acid exists as 2 H^+ ions and 1 SO_4^{2-} ion. An example of a compound that is classified as acidic because it *produces* H^+ ion when it dissolves in water is carbon dioxide, which undergoes the following reaction in water solution:

$$CO_2 + H_2O \rightarrow H^+ + HCO_3^- \qquad (3.9.1)$$

In this case, only a small fraction of the CO_2 molecules dissolved in water undergo the above reaction to produce H^+ so water solutions of CO_2 are weakly **acidic** and carbon dioxide is classified as a **weak acid**. It is the presence of dissolved CO_2 from the carbon dioxide naturally present in air that makes rainfall coming from even nonpolluted atmospheres slightly acidic and, as discussed in Chapter 7, the weakly acidic properties of CO_2 are very important in natural waters in the environment. Other acids, such as hydrochloric acid, HCl, are completely dissociated to H^+ and an anion (in the case of HCl the Cl^- anion) when they are dissolved in water; such acids are **strong acids**.

The naming of acids follows certain rules. In the case of an acid that contains only H and one other element, the acid is a hydro-ic acid. So HCl is called *hydro*chlor*ic* acid. Somewhat different rules apply when an acid contains oxygen. Some elements form acids in which the anion has different amounts of oxygen; examples are H_2SO_4 and H_2SO_3. The acid with more oxygen is an "-ic" acid, so H_2SO_4 is sulfuric acid. The acid with the lesser amount of oxygen is an "-ous" acid, so H_2SO_3 is sulfurous acid. A greater amount of oxygen than even the "-ic" acid is denoted by the prefix "per-", and a lesser amount of oxygen than even the "-ous" acid is denoted by the prefix "hypo-". These names are shown very well by the names of the oxyacids of chlorine. So the names of $HClO_4$, $HClO_3$, $HClO_2$, and HClO are, respectively, perchloric acid, chloric acid, chlorous acid and hypochlorous acid.

Acids are extremely important as industrial chemicals, in the environment, and in respect to green chemistry. About 40 million metric tons (40 billion kilograms) of sulfuric acid are produced in the United States each year. It is the number 1 synthetic chemical, largely because of its application to treat phosphate minerals to make phosphate crop fertilizers. Sulfuric acid is also used in large quantities to remove corrosion from steel, a process called steel pickling. Other major uses include detergent synthesis, petroleum refining, lead storage battery manufacture, and alcohol synthesis. About 7-8 million tons of nitric acid, HNO_3, are produced in the U.S. each year giving it a rank of 10th, and hydrochloric acid ranks about 25th with annual production around 3 million metric tons.

Acids are important in the environment. Improperly disposed acid has caused major problems around hazardous waste sites. Sulfuric acid along with smaller quantities of hydrochloric and nitric acid are the major constituents of acid rain (see Chapter 8). Acids figure prominently in the practice of green chemistry. Reclamation and recycling of acids are commonly performed in the practice of industrial ecology. As noted earlier, much of the sulfuric acid now manufactured uses a potential waste and pollutant, hydrogen sulfide, H_2S, removed from sour natural gas sources as a source of sulfur.

In cases where a relatively weak acid can be used, acetic acid made by the fermentation of carbohydrates is an excellent green alternative to stronger acids, such as sulfuric acid. Yeasts can convert the carbohydrates to ethanol (ethyl alcohol, which is present in alcoholic beverages) and other microorganisms in the presence of air convert the ethanol to acetic acid by the same process that vinegar, a dilute solution of acetic acid, is made from cider or wine. The structural formula of acetic acid is

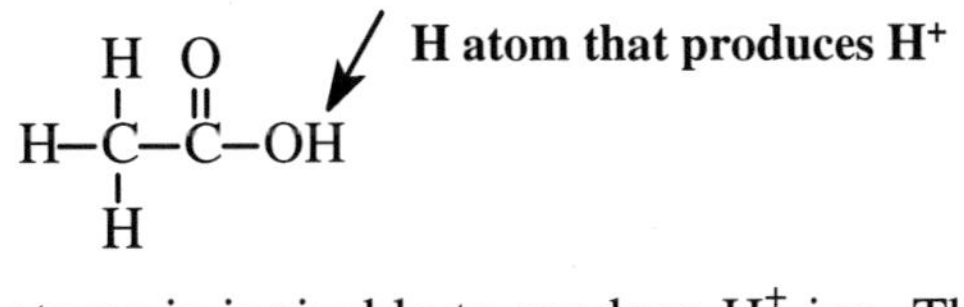

in which only one of the 4 H atoms is ionizable to produce H^+ ion. The production of acetic acid is a green process that uses biological reactions acting upon renewable biomass raw materials. As a weak acid, acetic acid is relatively safe to use, and contact with humans is not usually very dangerous (we ingest dilute acetic acid as vinegar, but pure acetic acid attacks flesh and is used to remove warts from skin). Another advantage of acetic acid is that it is biodegradable, so any of it released to the environment does not persist.

Bases

A **base** either contains hydroxide ion, OH^-, or reacts with water to produce hydroxide. Most bases that contain hydroxide consist of metal cations and hydroxide; examples are sodium hydroxide, NaOH, and calcium hydroxide, $Ca(OH)_2$. The most common basic substance that produces hydroxide in water is ammonia, NH_3, which reacts with water as follows:

$$NH_3 + H_2O \rightarrow NH_4^+ + OH^- \quad (3.9.5)$$

Only a small fraction of the ammonia molecules undergo this reaction in water, so ammonia does not produce much OH^- in water and is known as a **weak base**. The metal hydroxides, such as KOH, that completely dissociate in water are **strong bases**. Metal hydroxides are named by the metal followed by "hydroxide." Therefore, $Mg(OH)_2$ is magnesium hydroxide.

Salts

Acids and bases react to form a **salt**, an ionic compound that has a cation other than H^+ and an anion other than OH^-. This kind of reaction always produces water and is known as a **neutralization reaction**. The most well known salt is sodium chloride, NaCl. Although it is commonly what one means in referring to "salt," there are many other salts as well. These include calcium chloride, $CaCl_2$, used to melt road ice, sodium carbonate, Na_2CO_3, used in cleaning formulations and potassium chloride, KCl, a source of potassium fertilizer for crops. A typical neutralization reaction is the one between NaOH and hydrochloric acid, HCl, to produce sodium chloride:

$$\underset{\text{Base}}{NaOH} + \underset{\text{acid}}{HCl} \rightarrow \underset{\text{a salt, sodium chloride}}{NaCl} + \underset{\text{water}}{H_2O} \quad (3.9.2)$$

Salts are named very simply with just the name of the cation followed by that of the anion. The charges of the ions determine the formulas of the salts, so it is not necessary to add prefixes to denote the relative numbers of each ion. Therefore, $CaCl_2$ is simply calcium hydroxide, not calcium dichloride. As noted earlier in this chapter, prefixes are added in names of salts that contain more than 1 kind of cation or more than 1 kind of anion to show the relative numbers of ions. As an example, KH_2PO_4 is called potassium dihydrogen phosphate.

QUESTIONS AND PROBLEMS

1. What distinguishes the molecules of chemical compounds from those of elements, such as N_2?
2. Several "characteristics of compounds that meet the criteria of being green" were mentioned at the beginning of this chapter. Near the end of the chapter, acetic acid was mentioned as a "green acid." In what respects does it meet the criteria of green compounds?
3. What is sodium stearate? Why is it regarded as being green?
4. Which of the following is **not** usually regarded as a characteristic of green chemical compounds? Why is it not so regarded?

   A. Preparation from renewable resources

   B. Low tendency to undergo sudden, violent, unpredictable reactions

   C. Readily biodegradable

   D. Extremely high stability
5. What are valence electrons? Why are they particularly important?

6. What is the octet rule? Why is it particularly important in chemistry?

7. What does the structure representing CH_4 below say about bonding and octets of electrons around the central C atom?

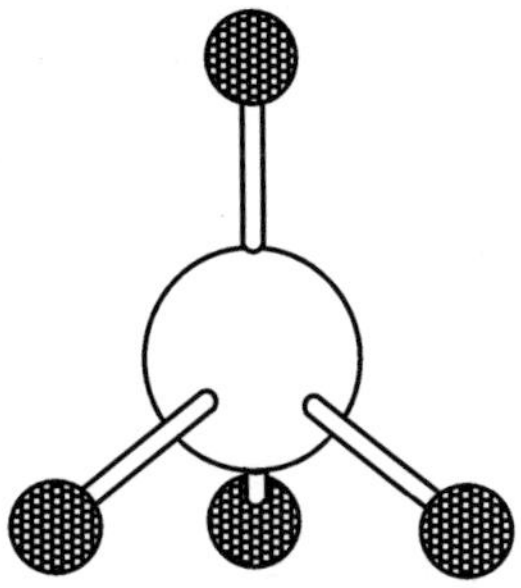

8. Considering that the central nitrogen atom in ammonia, NH_3, has an unshared pair of valence electrons and 3 pairs shared between N and H, propose a structure for the ammonia molecule based upon the structure of the methane molecule in the preceding question. Use a pair of dots to represent the unshared pair of electrons.

9. What is an ionic bond? Why is it not regarded as being between one specific cation and a specific anion in an ionic compound?

10. Do ionic compounds such as NaCl obey the octet rule? Explain.

11. Why is NaCl referred to as a formula unit of the ionic compound rather than a molecule of sodium chloride?

12. Energy is involved in several steps of the process by which an elemental metal and an elemental nonmetal are converted to an ionic compound (salt). Of these, which has the largest energy?

13. Place the following ions in decreasing order of size: Na^+, Cl^-, Al^{3+}, K^+

14. What is a major disadvantage of calcium chloride as a road de-icing agent? Why is calcium acetate a good substitute?

15. List some important characteristics of a covalent bond.

16. What is the major characteristic of ions in ionic liquids that enable these materials to be liquid at around room temperature?

17. Can the atoms in NO_2 obey the octet rule? Suggest the structural formula for this molecule in which the 2 O atoms are bonded to an N atom.

18. Coordinate covalent bonds are normally regarded as those in which each of two atoms contributes electrons to be shared in the bond. Are there any circumstances in which this is not true? If so, give an example.

19. What are three major ways in which covalent bonds are characterized?

20. What are some of the ways in which the characteristics of covalent bonds are related to green chemistry?

21. Why are elements in the middle of periods of the periodic table less likely to form ionic compounds and more likely to form covalent compounds than those near either end of each period?

22. Predict the formula of the compound formed when H reacts with P and explain.

23. Although hydrogen chloride, HCl, exists as a gas, the contest for the two shared electrons in the bond between H and Cl is unequal, with the Cl nucleus having the greater attraction. Suggest the nature of the H-Cl bond and suggest what may happen when HCl gas dissolves in water to produce hydrochloric acid.

24. Discuss the bonding in sulfur dioxide, SO_2, shown below:

25. Summarize the information shown in the formula $Ca_3(PO_4)_2$

26. Name each of the following binary molecular compounds: SO_3, N_2O_5, N_2O_4, NO_2, NO, N_2O, SiO_2, P_2O_4, P_4O_7, P_4O_{10}, Cl_2O_7

27. Give the formulas of each of the following ionic compounds: Sodium fluoride, potassium oxide, sodium nitride, magnesium chloride, aluminum chloride, aluminum oxide.

28. What are acids, bases, and salts? How are these three classes of compounds related to each other?

29. What is the distinction between weak and strong acids? Is NH_3 a weak or a strong base? Explain.

30. Given that the name of HNO_3 is nitric acid, what is the name of HNO_2?

31. The molecular formula of acetic acid is $C_2H_4O_2$. What is wrong with the reaction below in showing acetic acid dissociating in water?

$$C_2H_4O_2 \rightarrow 4H^+ + C_2O_2^-$$

32. Given 40.0, 14.0. and 16.0 for the atomic masses of Ca, N, and O respectively, what is the percent composition of $Ca(NO_3)_2$?

33. Given 39.1, 12.0, and 16.0 for the atomic masses of K, C, and O, respectively, what is the percent composition of K_2CO_3?

34. Using atomic masses given in Figure 2.9, what is the mass of a mole of $Al_2(CO_3)_3$? What is its percentage composition?

35. What is the mass of a mole of $Al(NO_3)_3$? How many moles of Al, N, and O are in exactly 1 mole of this compound?

36. Iron, Fe forms an ionic oxide with oxygen in which each iron atom has lost 3 electrons? What is the chemical formula of this compound?

37. Considering the number of valence electrons in H, C, and N, write the Lewis formulas of (A) hydrazine, N_2H_4, in which 2 N atoms are joined by a covalent bond and each of the N atoms is also bonded to 2 H atoms, and of (B) HCN, in which an H atom and an N atom are each bonded with a central C atom.

38. Complete the following: A mole is the quantity of material that contains ____________________
____.

39. How many Al atoms, N, atoms, and O atoms are contained in 213 g of $Al(NO_3)_3$?

4. CHEMICAL REACTIONS: MAKING MATERIALS SAFELY WITHOUT DAMAGING THE ENVIRONMENT

4.1. DESCRIBING WHAT HAPPENS WITH CHEMICAL EQUATIONS

How far would you have to go to find a diverse chemical factory carrying out hundreds of complex chemical processes? Not far, because your own body is just such a remarkably sophisticated factory that could not be duplicated by the efforts of thousands of chemists and chemical engineers and the expenditure of billions of dollars. As an example of a process that our bodies carry out consider the utilization of glucose sugar, chemical formula $C_6H_{12}O_6$, by metabolic processes in the body:

$$C_6H_{12}O_6 + 6O_2 \rightarrow 6CO_2 + 6H_2O \qquad (4.1.1)$$

This is a **chemical equation** that represents a chemical reaction, something that actually occurs with chemicals. It states that glucose reacts with molecular oxygen to produce carbon dioxide and water. The chemical reaction also produces energy, and the body carries it out to obtain the energy needed to move, work, and grow. The production of energy is sometimes denoted in the equation by adding "+ energy" to the right side.

Just as a chemical formula contains a lot of information about a chemical compound, a chemical equation contains much information about a chemical process. A chemical equation is divided into two parts by the arrow, which is read "yields." On the left of the arrow are the **reactants** and on the right are the **products**. A key aspect of a correctly written chemical equation is that it is **balanced**, with the same number of atoms of each element on the left as on the right. Consider the chemical equation above. The single molecule of $C_6H_{12}O_6$ contains 6 C atoms, 12 H atoms, and 6 O atoms. The 6 O_2 molecules contain 12 O atoms, giving a total of 18 O atoms among the reactants. Adding up all the atoms on the left gives 6 C atoms, 12 H atoms, and 18 O atoms among the reactants. On the right, the products contain 6 C atoms in the 6 CO_2 molecules, 12 H atoms in the 6 H_2O molecules, and 12 O atoms in the 6 CO_2 molecules, as well as 6 O atoms in the 6 H_2O molecules, a total of 18 O atoms. So there are 6 C atoms, 12 H atoms, and 18 O atoms among the products, the same as in the reactants. Therefore, the equation is balanced.

An important exercise is the process of balancing a chemical equation. This consists of putting the correct numbers before each of the reactants and products so that equal numbers of each kind of atom are on both the left and right sides of the equation. The procedure for balancing a chemical equation is addressed in section 4.2.

Learning chemistry is largely an exercise in learning chemical language. In the chemical language the symbols of the elements are the alphabet. The formulas of the compounds are the words. And chemical equations are the sentences that tell what actually happens.

It is often important to know the physical states of reactants and products in chemical reactions. Suppose, for example, that a geologist tested a sample of rock to see if it were limestone by adding some liquid hydrochloric acid to the rock and observing the CO_2 gas coming off. The equation for the chemical reaction that occurred is

$$CaCO_3(s) + 2HCl(aq) \rightarrow CO_2(g) + CaCl_2(aq) + H_2O(l) \quad (4.1.2)$$

Here abbreviations in parentheses are used to represent the physical state of each reaction participant — (*s*) for solid, (*aq*) for a substance in solution, (*g*) for gas, and (*l*) for liquid. The equation above states that solid calcium carbonate reacts with an aqueous solution of hydrochloric acid to produce carbon dioxide gas, a solution of calcium chloride, and liquid water.

Chemical reactions often are **reversible**, that is, they may go either forward or backward. A reversible reaction is shown with a double arrow, $\leftarrow\rightarrow$. As an example, consider the reaction of dissolved ammonia, NH_3, with water to produce ammonium ion, NH_4^+, and hydroxide ion, OH^-:

$$NH_3(aq) + H_2O(l) \leftarrow\rightarrow NH_4^+(aq) + OH^-(aq) \quad (4.1.3)$$

Actually, only a small fraction of NH_3 molecules undergo this reaction at any given time, and those that are converted to NH_4^+ are rapidly converted back to NH_3. The double arrow in the chemical equation shows that both the forward and reverse processes occur.

Another symbol that is sometimes used in chemical equations is Δ. This symbol denotes that heat is applied to make the chemical reaction occur at a more rapid pace. It is normally placed over the arrow in the chemical reaction.

Chemical equations are used to calculate the quantities of chemicals involved in a chemical reaction, either as reactants or as products. This is an important area of chemistry that is addressed by the topic of stoichiometry discussed later in this chapter in Section 4.8.

4.2. BALANCING CHEMICAL EQUATIONS

As noted earlier, a *balanced* chemical equation shows the same number of each kind of atom on both sides of the equation. The process of balancing chemical equations is an important exercise in chemistry and is addressed here.

Consider a simple example of balancing a chemical equation, the reaction of methane, CH_4, with elemental chlorine, Cl_2, to produce dichloromethane, CH_2Cl_2, an important laboratory solvent, and byproduct hydrogen chloride, HCl. The unbalanced chemical equation is

$$CH_4 + Cl_2 \rightarrow CH_2Cl_2 + HCl \quad (4.2.1)$$

Inspection of this equation as it is written shows that it is not balanced because it has 4 H on the left, but just 3 on the right and 2 Cl on the left, but 3 Cl on the right. In order to balance such an equation, consider one element at a time. Carbon is already balanced, so it is best to avoid changing any of the numbers in front of the C-containing compounds. The equation can be balanced for H by putting a 2 in front of HCl:

$$CH_4 + Cl_2 \rightarrow CH_2Cl_2 + 2HCl \quad (4.2.2)$$

Now everything is balanced except for Cl, of which there are 4 on the right, but just 2 on the left. Placing a 2 in front of Cl_2 gives the required 4 Cls on the left:

$$CH_4 + 2Cl_2 \rightarrow CH_2Cl_2 + 2HCl \quad (4.2.3)$$

This equation is now balanced with 1 C, 4 Hs, and 4 Cls on both the left and the right.

A crucial thing to remember in balancing a chemical equation is that *the chemical formulas must not be altered. Only the relative numbers of reactant and product species may be changed.*

Next consider the reaction of methane, CH_4, with iron oxide, Fe_2O_3, to give iron metal, Fe, carbon dioxide, CO_2, and water, H_2O. The unbalanced equation is

$$CH_4 + Fe_2O_3 \rightarrow Fe + CO_2 + H_2O \quad (4.2.4)$$

In this case it is helpful to note that CH_4 is the only source of both C and H and that 4 times as many H atoms as C atoms must appear in the products. That means that for each CO_2 there must be 2 H_2Os. Both C and H are balanced in the following:

$$CH_4 + Fe_2O_3 \rightarrow Fe + CO_2 + 2H_2O \quad (4.2.5)$$

But now O is not balanced. Furthermore, the 3 Os in Fe_2O_3 means that the number of O atoms must be divisible by 3, so try multiplying the three species balanced so far—CH_4, CO_2, and $2H_2O$—by 3:

$$3CH_4 + Fe_2O_3 \rightarrow Fe + 3CO_2 + 6H_2O \quad (4.2.6)$$

That gives a total of 12 O atoms on the right, 6 each in 3 CO_2 and 6 H_2O. Taking 4 times Fe_2O_3 gives 12 Os on the left:

$$3CH_4 + 4Fe_2O_3 \rightarrow Fe + 3CO_2 + 6H_2O \quad (4.2.7)$$

The only species remaining to be balanced is Fe, which can be balanced by putting 8 in front of Fe on the right. The balanced equation is

$$3CH_4 + 4Fe_2O_3 \rightarrow 8Fe + 3CO_2 + 6H_2O \quad (4.2.8)$$

Checking the answer shows on both left and right 3 Cs, 8 Fes, 12 Hs, and 12 Os demonstrating that the equation is in fact balanced.

Exercise: Balance the following:

1. $Fe_2O_3 + CO \rightarrow Fe + CO_2$
2. $FeSO_4 + O_2 + H_2O \rightarrow Fe(OH)_3 + H_2SO_4$
3. $C_2H_2 + O_2 \rightarrow CO_2 + H_2O$
4. $Mg_3N_2 + H_2O \rightarrow Mg(OH)_2 + NH_3$
5. $NaAlH_4 + H_2O \rightarrow H_2 + NaOH + Al(OH)_3$
6. $Zn(C_2H_5)_2 + O_2 \rightarrow ZnO + CO_2 + H_2O$

Answers: (1) $Fe_2O_3 + 3CO \rightarrow 2Fe + 3CO_2$, (2) $4FeSO_4 + O_2 + 10H_2O \rightarrow 4Fe(OH)_3 + 4H_2SO_4$, (3) $2C_2H_2 + 5O_2 \rightarrow 4CO_2 + 2H_2O$, (4) $Mg_3N_2 + 6H_2O \rightarrow 3Mg(OH)_2 + 2NH_3$ (5) $NaAlH_4 + .4H_2O \rightarrow 4H_2 + NaOH + Al(OH)_3$, (6) $Zn(C_2H_5)_2 + 7O_2 \rightarrow ZnO + 4CO_2 + 5H_2O$

4.3. JUST BECAUSE YOU CAN WRITE IT DOES NOT MEAN THAT IT WILL HAPPEN

The fact that a balanced chemical equation can be written does not necessarily mean that the chemical reaction that it represents will occur. As an example, it is known that a number of metals will react with acid to release elemental hydrogen gas and produce a metal salt. For example, if iron wire, Fe, is placed into a solution of sulfuric acid, H_2SO_4, H_2 gas is evolved,

$$Fe(s) + H_2SO_4(aq) \rightarrow H_2(g) + FeSO_4(aq) \qquad (4.3.1)$$

and $FeSO_4$ salt remains in solution. A corresponding copper salt, $CuSO_4$, is also known to exist. So it is logical to believe that it could be prepared by reacting copper metal with sulfuric acid:

$$Cu(s) + H_2SO_4(aq) \rightarrow H_2(g) + CuSO_4(aq) \qquad (4.3.2)$$

This equation is balanced and it looks reasonable. But, placing copper metal into a solution of H_2SO_4 in the laboratory results in — nothing. The reaction simply does not occur. The lesson here is that a balanced chemical equation is not sufficient reason to conclude that a reaction will occur.
Since $CuSO_4$ is known to exist, there has to be a way to prepare it. There are, in fact many ways. One pathway to the preparation of this salt starting with copper metal is to first react the copper with oxygen at a relatively high temperature to produce copper oxide:

$$2Cu(s) + O_2(g) \rightarrow 2CuO(s) \qquad (4.3.4)$$

The CuO product reacts with sulfuric acid to give $CuSO_4$ salt:

$$CuO(s) + H_2SO_4(aq \rightarrow CuSO_4(aq) + H_2O(l) \qquad (4.3.5)$$

Alternate Reaction Pathways in Green Chemistry

Much of the science of green chemistry involves making decisions about alternative chemical reactions to choose a reaction or reaction sequence that provides maximum safety, produces minimum byproduct, and utilizes readily available materials. Consider two ways of preparing iron sulfate, $FeSO_4$. This chemical is commonly used to treat (clarify) water because when it is added to water and air is bubbled through the water, it produces $Fe(OH)_3$, a gelatinous solid that settles in the water and carries suspended mud and other particles with it. Consider two possible ways of making $FeSO_4$. The first of these was shown earlier and consists of the reaction of iron metal with sulfuric acid:

$$Fe(s) + H_2SO_4(aq) \rightarrow H_2(g) + FeSO_4(aq) \qquad (4.3.1)$$

A second pathway would be to react iron oxide, FeO, with sulfuric acid:

$$FeO(s) + H_2SO_4(aq) \rightarrow FeSO_4(aq) + H_2O(aq) \qquad (4.3.6)$$

Which of these reactions would be the better choice? Both would work. The first reaction generates elemental H_2 gas as a byproduct. That has a potential downside because elemental hydrogen is highly explosive and flammable and could cause an explosion or fire hazard. But, in a contained reaction vessel that allowed for capture of H_2, the elemental hydrogen could be put to use as a fuel or reacted directly in a fuel cell to produce electricity (Section 2.2 and Figure 2.2). Furthermore, scrap iron metal and waste sulfuric acid are common materials that should be recycled and the synthesis of $FeSO_4$ by the direct reaction of the two can prepare a useful material from the two recyclable substances.

The second reaction (4.3.6) also gives the desired product. Its only byproduct is innocuous water. And there is no hazard from elemental hydrogen. In principle, the FeO required could be made by reacting scrap iron metal with oxygen from the air,

$$2Fe + O_2 \rightarrow 2FeO \tag{4.3.7}$$

but in practice the reaction tends to produce other oxides of iron, particularly Fe_2O_3 and Fe_3O_4.

4.4. YIELD AND ATOM ECONOMY IN CHEMICAL REACTIONS

A fundamental concept basic to green chemistry that can be illustrated by chemical reactions is the distinction between yield and atom economy. In Chapter 1 *yield* was defined as a percentage of the degree to which a chemical reaction or synthesis goes to completion and *atom economy* was defined as the fraction of reactants that go into final products. Those two ideas are illustrated here for the preparation of HCl gas which, dissolved in water, produces hydrochloric acid. There are several ways in which HCl can be prepared. One of these commonly used in the laboratory is the reaction of concentrated sulfuric acid, H_2SO_4, with common table salt, NaCl, accompanied by heating to drive off the volatile HCl vapor:

$$2NaCl(s) + H_2SO_4(l) \rightarrow 2HCl(g) + Na_2SO_4(s) \tag{4.4.1}$$

This reaction can be performed so that all of the NaCl and H_2SO_4 react, which gives a 100% yield. But it produces Na_2SO_4 byproduct, so the atom economy is less than 100%. The percent atom economy is calculated very simply by the relationship

$$\text{Percent atom economy} = \frac{\text{Mass of desired product}}{\text{Total mass of product}} \times 100 \tag{4.4.2}$$

(We could just as well divide by the total mass of reactants since in a chemical reaction it is equal to the total mass of products.) In this case, the mass of the desired product is that of 2 HCl and the total mass of product is that of 2HCl + Na_2SO_4. Given the atomic masses H 1.0, Cl 35.5, Na 23.0, and O 16.0 gives the following:

$$\text{Mass of desired product} = 2 \times (1.0 + 35.5) = 73.0 \tag{4.4.3}$$

$$\text{Total mass product} = 2 \times (1.0 + 35.5) + (2 \times 23.0 + 32.0 + 4 \times 16.0) = 215 \tag{4.4.4}$$

$$\text{Percent atom economy} = \frac{73.0}{215} \times 100 = 34.0\% \tag{4.4.5}$$

This result shows that even with 100% yield, the reaction is only 34.0% atom economical and if it were used as a means to prepare HCl large quantities of Na_2SO_4, a material with only limited value, would be produced. In contrast, the direct reaction of hydrogen gas with chlorine gas to give HCl gas,

$$H_2(g) + Cl_2(g) \rightarrow 2HCl(g) \tag{4.4.2}$$

can be carried out with 100% atom economy if all of the H_2 reacts with Cl_2. There is no waste byproduct.

4.5. CATALYSTS THAT MAKE REACTIONS GO

Carbon monoxide will certainly burn in the presence of oxygen from air as shown by the reaction

$$2CO + O_2 \rightarrow 2CO_2 \quad (4.5.1)$$

Carbon monoxide is a product of automobile exhausts and an undesirable, toxic air pollutant. One way of ridding automobile exhaust gases of this pollutant is to pump air into the exhaust and convert the carbon monoxide to carbon dioxide as shown by the reaction above. However, even in the presence of oxygen, this reaction does not proceed to completion in an ordinary automobile exhaust system. It is enabled to occur, however, by passing the exhaust mixed with air over a solid honeycomb-like surface of ceramic coated with a metal that enables the reaction to occur, but is not itself consumed in the reaction. Such a substance is called a **catalyst**. Most people who have an automobile are vaguely aware that they have an automotive exhaust catalyst. They become much more acutely aware of this fact if the automobile's exhaust system does not pass a required emissions test and the catalytic converter in it has to be replaced. The cost can easily be several hundred dollars!

We do not have to go any farther than our own bodies to find catalysts. That is because all living organisms have biological catalysts that enable reactions to occur. Such living catalysts consist of specialized proteins called **enzymes**. Enzymes are discussed in Chapter 9.

A common enzyme-catalyzed process is the reaction of glucose (blood sugar, $C_6H_{12}O_6$) with molecular oxygen to produce energy mentioned at the beginning of this chapter:

$$C_6H_{12}O_6 + 6O_2 \rightarrow 6CO_2 + 6H_2O + \text{energy} \quad (4.1.1)$$

This is the important process of **aerobic respiration** carried out by all organisms that live in contact with air and utilize oxygen from air to react with food materials. The actual process requires many steps and several catalytic enzymes are used. Other enzymes are used for various life processes, such as protein synthesis. There are enzymes that detoxify toxic substances, and in some cases they inadvertantly make toxic substances out of nontoxic ones. Some of the more common cancer-causing substances are actually synthesized from other molecules by enzyme action. Obviously enzymes are very important in life processes.

Catalysts speed up reactions. Depending upon the conditions the **rate of reaction** can vary significantly. Rates of chemical reactions are addressed by the area of **chemical kinetics**.

4.6. KINDS OF CHEMICAL REACTIONS

It is useful to place chemical reactions in various categories. The important categories of chemical reactions are addressed here.

The simplest kind of chemical reaction to visualize is a **combination reaction** in which two substances come together to form a new substance. The substances may be two elements, two compounds, or an element and a compound. An example of a combination reaction occurs when elemental carbon burns,

$$C + O_2 \rightarrow CO_2 \quad (4.6.1)$$

to produce carbon dioxide. Since this reaction generates only one product, it occurs with 100% atom economy.

Another combination reaction occurs when calcium oxide, CaO, present in a bed of solid material in a fluidized bed furnace used to burn coal reacts with sulfur dioxide:

$$CaO + SO_2 \rightarrow CaSO_3 \tag{4.6.2}$$

The sulfur dioxide is a potential air pollutant produced from the burning of sulfur present in the coal. By injecting pulverized coal into a bed of CaO and other minerals kept in a fluid-like state by the injection of air, the sulfur dioxide produced has the opportunity to react with CaO and is not emitted as a pollutant with the stack gas.

In addition to being a combination reaction, the reaction above could also be called an **addition reaction** because the SO_2 adds to the CaO. Addition reactions are very desirable in the practice of green chemistry because they are 100% atom economical.

The opposite of a combination reaction is a **decomposition reaction**. An example of such a reaction occurs when a direct electrical current is passed through water to which a salt such as Na_2SO_4 has been added to make the water electrically conducting:

$$2H_2O(l) \xrightarrow{\text{Electrolysis}} 2H_2(g) + O_2(g) \tag{4.6.3}$$

Reactions such as this that occur by the action of electricity passed through a solution are called **electrolysis** reactions. As written, the reaction is 100% atom economical. However, some side reactions may occur that reduce the efficiency. For example, impurity chloride ion, Cl^-, must be avoided in solution because it can produce some Cl_2 gas, a toxic, undesirable byproduct. Another inefficiency occurs because not all of the electricity passed through the solution is utilized to decompose water.

An example of a useful decomposition reaction is the high-temperature decomposition of methane,

$$CH_4(g) \xrightarrow{\Delta} C(s) + 2H_2(g) \tag{4.6.4}$$

to produce elemental C and H_2 gas (where the triangle over the arrow shows that heat is applied — in this case to a temperature of 1260–1425° C — to make the reaction occur). The elemental carbon from this reaction is generated as a fine powder called **carbon black**. Carbon black is an ingredient of the paste in dry cells (such as those used in portable electronic devices); it is used as a filler in tires and to make electrodes for electrolysis processes such as the one by which aluminum metal is prepared.

Decomposition reactions do not always produce elements. For example, sodium bicarbonate mineral, $NaHCO_3$ may be heated,

$$2NaHCO_3(s) \xrightarrow{\Delta} Na_2CO_3(s) + CO_2(g) + H_2O(g) \tag{4.6.5}$$

to produce sodium carbonate, Na_2CO_3, commonly used as an industrial chemical to treat water, in cleaning solutions, and as an ingredient of glass.

Example: Using atomic masses Na 23.0, H 1.0, C 12.0, and O 16.0, calculate the percent atom economy of the above reaction for the production of Na_2CO_3.

Answer: When 2 formula units of $NaHCO_3$ react, 1 formula unit of Na_2CO_3 is produced. The masses involved in atomic mass units, u, are the following:

$$\text{Mass } 2NaHCO_3 = 2 \times (23.0 + 1.0 + 12.0 + 3 \times 16.0) = 168 \text{ u}$$

$$\text{Mass } Na_2CO_3 = 2 \times 23.0 + 12.0 + 3 \times 16.0 = 106 \text{ u}$$

$$\text{Percent atom economy} = \frac{106 \text{ u}}{168 \text{ u}} \times 100 = 63.1\%$$

A **substitution** or **replacement** reaction is one such as the reaction of iron and sulfuric acid,

$$Fe(s) + H_2SO_4(aq) \rightarrow H_2(g) + FeSO_4(aq) \quad (4.3.1)$$

in which Fe replaces H in H_2SO_4, a reaction shown earlier for the preparation of $FeSO_4$. This reaction also falls under the classification of reactions involving **evolution of a gas**, in this case evolution of hydrogen gas. A **double replacement** reaction, also called a **metathesis** reaction, is one in which two compounds trade ions or other groups. When dissolved calcium chloride reacts with dissolved sodium carbonate,

$$CaCl_2(aq) + Na_2CO_3(aq) \rightarrow CaCO_3(s) + 2NaCl(aq) \quad (4.6.6)$$

the Ca^{2+} ion in calcium chloride simply switches places with the Na^+ ions in the sodium carbonate to produce solid calcium carbonate and NaCl in solution. This is also a **precipitation** reaction in which a solid material forms from two substances dissolved in water; the solid formed is a **precipitate**. The removal of calcium from water as shown by this reaction is a common water treatment process called **water softening**. It is done because excessive levels of calcium cause formation of scale that can clog water pipes and damage plumbing apparatus.

Whenever an acid and a base react, as shown here for the reaction of hydrochloric acid with sodium hydroxide,

$$HCl(aq) + NaOH(aq) \rightarrow H_2O(l) + NaCl(aq) \quad (4.6.6)$$

water and a salt are formed. Such a reaction is a **neutralization reaction** or simply an **acid-base** reaction.

Exercise: Classify each of the following reactions as combination, decomposition, substitution, metathesis, neutralization, precipitation, or evolution of a gas. In some cases, a reaction will fit into more than one category.

(a) $2Ca(s) + O_2(g) \rightarrow 2CaO(s)$

(b) $2KClO_3(s) \xrightarrow{\Delta} 2KCl(s) + 3O_2(g)$

(c) $SO_3(g) + H_2O(l) \rightarrow H_2SO_4(aq)$

(d) $MgCO_3(s) + 2HCl(aq) \rightarrow MgCl_2(aq) + H_2O(l) + CO_2(g)$

(e) $Zn(s) + CuCl_2(aq) \rightarrow Cu(s) + ZnCl_2(aq)$

(f) $KOH(aq) + HCl(aq) \rightarrow KCl(aq) + H_2O(l)$

(g) $MgSO_4(aq) + 2KOH(aq) \rightarrow Mg(OH)_2(s) + Na_2SO_4(aq)$

Answers: (a) Combination, (b) decomposition, evolution of a gas, (c) combination, (d) metathesis, evolution of a gas, (e) substitution, (f) neutralization, metathesis, (g) precipitation, metathesis.

4.7. OXIDATION-REDUCTION REACTIONS AND GREEN CHEMISTRY

Many reactions, including some of those given in the preceding section, are **oxidation-reduction** reactions, frequently called **redox** reactions. This name derives from the long standing use of *oxidation* to describe the reaction of a substance with oxygen. Consider the following reaction of elemental calcium with elemental oxygen:

$$\mathrm{Ca}\!: + :\ddot{\underset{\cdot\cdot}{\mathrm{O}}}::\ddot{\underset{\cdot\cdot}{\mathrm{O}}}: + \mathrm{Ca}\!: \rightarrow :\ddot{\underset{\cdot\cdot}{\mathrm{O}}}:^{2-}\ \mathrm{Ca}^{2+} :\ddot{\underset{\cdot\cdot}{\mathrm{O}}}:^{2-}\ \mathrm{Ca}^{2+} \tag{4.7.1}$$

$$2Ca + O_2 \rightarrow 2CaO$$

Because it combines with oxygen, the calcium metal is *oxidized*. Whenever something is oxidized, something else has to be *reduced*. In this case, elemental oxygen is reduced to produce the oxide ion, O^{2-} in CaO. It is seen from this reaction that the calcium atoms lose electrons when they are oxidized and the oxygen atoms gain electrons. This leads to another definition of oxidation-reduction reactions, which is that *when a chemical species loses electrons in a chemical reaction it is oxidized and when a species gains electrons it is reduced.*

Elemental hydrogen is commonly involved in oxidation-reduction. *Whenever a chemical species reacts with elemental hydrogen, it is reduced.* As an example, iron(II) oxide, FeO, can be reacted with elemental hydrogen,

$$FeO + H_2 \rightarrow Fe + H_2O \tag{4.7.2}$$

In this case the Fe in FeO is reduced to iron metal and the hydrogen in elemental H_2 is oxidized to H_2O.

When elemental oxygen reacts to produce chemically combined oxygen, it is acting as an **oxidizing agent** and is reduced. And when elemental hydrogen reacts to produce chemically combined hydrogen, it acts as a **reducing agent** and is oxidized. Consider what happens when the opposite reactions occur. When chemically combined oxygen is released as elemental oxygen from a chemical reaction, the oxygen is *oxidized*. And when elemental hydrogen is released as the result of a chemical reaction, hydrogen is *reduced*. A good illustration of these definitions may be seen when a direct electrical current is passed between two metal electrodes through water made electrically conducting by dissolving in it a salt, such as Na_2SO_4 as shown in Figure 4.1. At the left electrode, electrons are pumped into the system *reducing* the chemically bound H in H_2O to elemental H_2. An electrode at which *reduction* occurs is called the **cathode**. At the other electrode, electrons are removed from the system, elemental O_2 is released, and the oxygen in H_2O is *oxidized*. An electrode at which oxidation occurs is the **anode**.

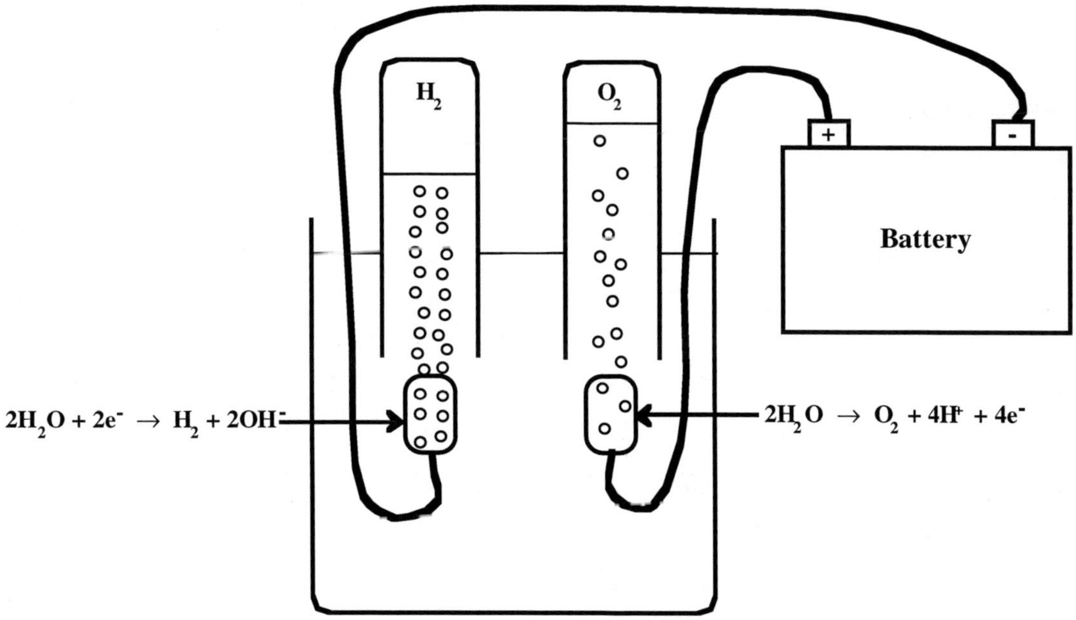

Figure 4.1. Electrolysis of water containing some dissolved salt to make it electrically conducting. At the left electrode (cathode) H in H_2O is reduced by adding electrons releasing H_2 gas. At the right electrode (anode) electrons are removed from chemically bound O in H_2O releasing elemental O_2 and the oxygen is oxidized.

The reaction described above is an **electrolysis** reaction. A significant electrolysis reaction in the practice of green chemistry is the one illustrated in Figure 4.1, which can be used as a means of getting pure hydrogen and pure oxygen from water without the use of any other chemical reagents, except for a small amount of salt to make the water conducting. For example, using a nonpolluting source of energy, such as wind power, elemental hydrogen can be generated for use in nonpolluting fuel cells (see Figure 2.2 and Chapter 6).

Oxidation-reduction reactions are very significant in energy conversion processes. An important example is photosynthesis,

$$6CO_2 + 6H_2O + h\nu \rightarrow C_6H_{12}O_6 + 6O_2 \tag{4.7.3}$$

in which solar energy ($h\nu$) from sunlight is used by plants to produce glucose sugar, $C_6H_{12}O_6$, a high-energy compound that is used by organisms to provide energy for their metabolic needs. Since elemental oxygen is produced, oxygen is oxidized. Although it is not obvious based upon the discussion of oxidation-reduction so far, carbon is reduced; the carbon in the $C_6H_{12}O_6$ product is reduced compared to the carbon in the CO_2 reactant. The reverse of this reaction shown at the beginning of this chapter is

$$C_6H_{12}O_6 + 6O_2 \rightarrow 6CO_2 + 6H_2O + \text{energy} \tag{4.1.1}$$

which occurs when organisms — including humans — utilize glucose sugar to produce energy. In this case, oxygen reacts, an obvious oxidation process. The oxygen is reduced and carbon is oxidized by the action of the elemental oxygen.

A very common oxidation-reduction reaction occurs when fossil fuels are burned to produce energy. One such reaction occurs when natural gas (methane, CH_4) burns,

$$CH_4 + 2O_2 \rightarrow CO_2 + 2H_2O + \text{energy} \tag{4.7.4}$$

to produce carbon dioxide and water, releasing energy. The burning of gasoline, diesel fuel, coal, wood, and even hydrogen gas are oxidation-reduction reactions in which carbon or hydrogen are oxidized by the action of oxygen yielding usable energy.

Oxidation-reduction reactions are the most important kinds of reactions considered in green chemistry. That is true in part because of the central role currently played by the oxidation of fossil fuels and other materials in producing energy needed for chemical processes. Furthermore, the most common raw material currently used for making plastics, synthetic fabrics, and other manufactured materials is petroleum hydrocarbon. There are many hydrocarbon compounds all containing chemically bound carbon and hydrogen. A typical such compound is ethane, C_2H_6. The hydrogen and carbon in a hydrocarbon are in the most chemically reduced form, but required raw materials often are partially oxidized hydrocarbons in which O atoms are bonded to the hydrocarbon (complete oxidation of a hydrocarbon yields CO_2 and H_2O). Ethanol, C_2H_6O, used in chemical synthesis and as an oxygenated additive to make gasoline burn more smoothly with emission of fewer air pollutants is a partially oxidized hydrocarbon.

Large quantities of materials and energy are expended in converting petroleum hydrocarbons to partially oxidized compounds used as raw materials. For example, ethanol can be made from ethane taken from petroleum and natural gas by a series of chemical reactions for which the net process is the following:

$$2C_2H_6 + O_2 \rightarrow 2C_2H_6O \tag{4.7.5}$$

This transformation requires relatively severe conditions and a net loss of energy. A greener alternative is to grow yeasts that produce ethanol on glucose sugar produced by photosynthesis (Reaction 4.7.3),

$$C_6H_{12}O_6 \rightarrow 2C_2H_6O + 2CO_2 \quad (4.7.6)$$

a process that occurs under room temperature conditions. In addition to making ethanol, this fermentation process yields carbon dioxide in a concentrated form that can be used for carbonated beverages, supercritical carbon dioxide solvent, or pumped underground for tertiary petroleum recovery. The protein-rich yeast biomass produced in fermentation makes a good animal feed additive.

4.8. QUANTITATIVE INFORMATION FROM CHEMICAL REACTIONS

Much of green chemistry is involved with calculations of quantities of materials involved in chemical reactions. It is essential to do such calculations in order to deal with the important concepts of percent yield and atom economy. Fortunately, it is easy to calculate quantities of materials if a balanced chemical reaction is known along with the pertinent atomic and formula masses.

To this point, we have been viewing chemical reactions in terms of individual atoms and molecules and have been thinking of masses in atomic mass units, u, used to express the masses of individual atoms and molecules. But that is much too small a scale to use in the laboratory. The chemist conveniently deals with grams and moles where a mole of a substance typically has a mass of several to several hundred grams.

Consider a typical chemical reaction, in this case the combustion of ethane, a hydrocarbon fuel with a chemical formula of C_2H_6, to give heat:

$$2C_2H_6 + 7O_2 \rightarrow 4CO_2 + 6H_2O \quad (4.8.1)$$

Rather than viewing this reaction in terms of individual molecules, it is possible to scale up to moles. Recall that the mole is a fundamental unit for quantity of material and that each mole contains Avogadro's number (6.022×10^{23}) of formula units (molecules of covalently bound compounds). This equation simply says that 2 moles of C_2H_6 react with 7 moles of O_2 to yield 4 moles of CO_2 and 6 moles of H_2O. Now we can examine the equation in more detail to do some quantitative calculations. Before doing that, however, review the following two terms:

Formula mass: The sum of the atomic masses of all the atoms in a formula unit of a compound. Although the average masses of atoms and molecules may be expressed in atomic mass units (amu or u), formula mass is generally viewed as being relative and without units.

Molar mass: Where X is the formula mass, the molar mass is X grams of an element or compound, that is, the mass in grams of 1 mole of the element or compound.

Given the atomic masses H 1.0, C 12.0, and O 16.0 the molar mass of C_2H_6 is $2 \times 12.0 + 6 \times 1.0 = 30.0$ g/mol, that of O_2 is $2 \times 16.0 = 32.0$ g/mol, that of CO_2 is $12.0 + 2 \times 16.0 = 44.0$ g/mol, and that of H_2O is $2 \times 1.0 + 16.0 = 18.0$ g/mol. Now consider the equation

$$2C_2H_6 + 7O_2 \rightarrow 4CO_2 + 6H_2O \quad (4.8.1)$$

in terms of the minimum whole number of moles reacting and produced and the masses in grams of these quantities. The equation states that 2 moles of C_2H_6 with a mass of 2×30.0 g = 60.0 g of C_2H_6 react with 7 moles of O_2 with a mass of 7×32.0 g = 224 g of O_2 to produce 4 moles of CO_2 with a mass of 4×44.0 g = 176 g of CO_2 and 6 moles of H_2O with a mass of 6×18.0 g = 108 g of H_2O. The total mass of reactants is

$$60.0 \text{ g of } C_2H_6 + 224 \text{ g of } O_2 = 284.0 \text{ g of reactants}$$

and the total mass of products is

$$176 \text{ g of } CO_2 + 108 \text{ g of } H_2O = 284 \text{ g of products}$$

As required by the law of conservation of mass, the total mass of products equals the total mass of reactants. In the following section, these mole and mass relationships are used to calculate quantities of chemicals involved in chemical reactions

4.9. STOICHIOMETRY BY THE MOLE RATIO METHOD

The calculation of quantities of materials involved in chemical reactions is addressed by **stoichiometry**. Stoichiometry is based upon the *law of conservation of mass* which states that *the total mass of reactants in a chemical reaction equals the total mass of products*. This law holds true because matter is neither created nor destroyed in chemical reactions.

The **mole ratio method** of stoichiometric calculations is based upon the fact that the relative numbers of moles of reactants and products remain the same regardless of the total quantity of reaction. It is best shown by example. Consider the following reaction:

$$2C_2H_6 + 7O_2 \rightarrow 4CO_2 + 6H_2O \tag{4.9.1}$$

In moles, this equation states that 2 moles C_2H_6 react with 7 moles of O_2 to produce 4 moles of CO_2 and 6 moles of H_2O. The same ratios hold true regardless of how much material reacts. So for 10 times as much material, 20 moles C_2H_6 react with 70 moles of O_2 to produce 40 moles of CO_2 and 60 moles of H_2O.

Suppose that it is given that 18.0 g of C_2H_6 react. What is the mass of O_2 that will react with this amount of C_2H_6? What mass of CO_2 is produced? What mass of H_2O is produced? This problem can be solved by the **mole ratio method**. Mole ratios are, as the name implies, simply the ratios of various moles of reactants and products to each other as shown by a chemical equation. Mole ratios are obtained by simply examining the chemical equation in question; the three that will be used in solving the problem posed are the following:

$$\frac{7 \text{ mol } O_2}{2 \text{ mol } C_2H_6} \qquad \frac{4 \text{ mol } CO_2}{2 \text{ mol } C_2H_6} \qquad \frac{6 \text{ mol } H_2O}{2 \text{ mol } C_2H_6}$$

To solve for the mass of O_2 reacting the following steps are involved:

A. Mass of C_2H_6 reacting   B. Convert to moles of C_2H_6 C. Convert to moles of O_2 D. Convert to mass of O_2

In order to perform the calculation, it will be necessary to have the molar mass of C_2H_6, stated earlier as 30.0 g/mol, the molar mass of O_2 (32.0 g/mol) and the mole ratio relating moles of O_2 reactant to moles of C_2H_6, 7 mol O_2/2 mol C_2H_6. The calculation becomes the following:

$$\text{Mass of } O_2 = 18.0 \text{ g } C_2H_6 \times \frac{1 \text{ mol } C_2H_6}{30.0 \text{ g } C_2H_6} \times \frac{7 \text{ mol } O_2}{2 \text{ mol } C_2H_6} \times \frac{32.0 \text{ g } O_2}{1 \text{ mol } O_2} = 67.2 \text{ g } O_2$$

Note that in this calculation units cancel above and below the line, starting with units of g C_2H_6.

Now that the mass of O_2 reacting has been calculated, it is possible using the appropriate mole ratios and molar masses to calculate the masses of CO_2 and of H_2O produced as follows:

$$\text{Mass of } CO_2 = 18.0 \text{ g } C_2H_6 \times \frac{1 \text{ mol } C_2H_6}{30.0 \text{ g } C_2H_6} \times \frac{4 \text{ mol } CO_2}{2 \text{ mol } C_2H_6} \times \frac{44.0 \text{ g } CO_2}{1 \text{ mol } CO_2} = 52.8 \text{ g } CO_2$$

$$\text{Mass of } H_2O = 18.0 \text{ g } C_2H_6 \times \frac{1 \text{ mol } C_2H_6}{30.0 \text{ g } C_2H_6} \times \frac{6 \text{ mol } H_2O}{2 \text{ mol } C_2H_6} \times \frac{18.0 \text{ g } H_2O}{1 \text{ mol } H_2O} = 32.4 \text{ g } H_2O$$

Are the masses calculated above correct? A good check is to compare the total mass of reactants, 18.0 g C_2H_6 + 67.2 g O_2 = 85.2 g of reactants, with the total mass of products, 52.8 g CO_2 + 32.4 g H_2O = 85.2 g of products. The fact that the total mass of reactants is equal to the total mass of products gives confidence that the calculations are correct.

As one more example consider the reaction of 15.0 g of Al with Cl_2 to give $AlCl_3$:

$$2Al + 3Cl_2 \rightarrow 2AlCl_3$$

What mass of Cl_2 reacts and what is the mass of $AlCl_3$ produced? The atomic mass of Al is 27.0 and that of Cl is 35.5. Therefore, the molar mass of Cl_2 is 71.0 g/mol and the molar mass of $AlCl_3$ is 133.5 g/mole. The mass of Cl_2 reacting is

$$\text{Mass of } Cl_2 = 15.0 \text{ g Al} \times \frac{1 \text{ mol Al}}{27.0 \text{ g Al}} \times \frac{3 \text{ mol } Cl_2}{2 \text{ mol Al}} \times \frac{71.0 \text{ g } Cl_2}{1 \text{ mol } Cl_2} = 59.2 \text{ g } Cl_2$$

$$\text{Mass of } AlCl_3 = 15.0 \text{ g Al} \times \frac{1 \text{ mol Al}}{27.0 \text{ g Al}} \times \frac{2 \text{ mol } AlCl_3}{2 \text{ mol Al}} \times \frac{133.5 \text{ g } AlCl_3}{1 \text{ mol } AlCl_3} = 74.2 \text{ g } AlCl_3$$

As a check, 15.0 g Al + 59.2 g Cl_2 reactant gives a total of 74.2 g of reactants equal to the mass of the $AlCl_3$ product.

Exercise: Calculate the mass of CH_4 that reacts and the masses of the products when 25.0 g of Fe_2O_3 undergo the reaction below. The atomic masses involved are H 1.0, C 12.0, O 16.0, Fe 55.8.

$$4Fe_2O_3 + 3CH_4 \rightarrow 8Fe + 3CO_2 + 6H_2O$$

Answer: 1.88 g CH_4, 17.5 g Fe, 5.2 g CO_2, 4.2 g H_2O

Exercise: Calculate the mass of O_2 that reacts and the masses of the products when 100 g of benzoic acid, $C_7H_6O_2$ undergo the reaction below. The atomic masses involved are H 1.0, C 12.0, and O 16.0.

$$2C_7H_6O_2 + 15O_2 \rightarrow 14CO_2 + 6H_2O$$

Answer: 197 g O_2, 252 g CO_2, 44.3 g H_2O

4.10. LIMITING REACTANT AND PERCENT YIELD

Mixing of exact amounts of reactants such that all are consumed and none left over in a chemical reaction almost never occurs. Instead, one of the reactants is usually a **limiting reactant**. Suppose, for example that 100 g of elemental zinc (atomic mass 65.4) and 80 g of elemental sulfur (atomic mass 32.0) are mixed and heated undergoing the following reaction:

$$Zn + S \rightarrow ZnS \quad (4.10.1)$$

What mass of ZnS, formula mass 97.4 g/mol, is produced? If 100 g of zinc react completely, the mass of S reacting and the mass of ZnS produced would be given by the following calculations:

$$\text{Mass S} = 100.0\text{ g Zn} \times \frac{1\text{ mol Zn}}{65.4\text{ g Zn}} \times \frac{1\text{ mol S}}{1\text{ mol Zn}} \times \frac{32.0\text{ g S}}{1\text{ mol S}} = 48.9\text{ g S}$$

$$\text{Mass ZnS} = 100.0\text{ g Zn} \times \frac{1\text{ mol Zn}}{65.4\text{ g Zn}} \times \frac{1\text{ mol ZnS}}{1\text{ mol Zn}} \times \frac{97.4\text{ g ZnS}}{1\text{ mol ZnS}} = 149\text{ g ZnS}$$

Only 48.9 g of the available S react, so sulfur is in excess and zinc is the limiting reactant. A similar calculation for the amount of Zn required to react with 80 g of sulfur would show that 164 g of Zn would be required, but only 100 g is available.

Exercise: A solution containing 10.0 g of HCl dissolved in water (a solution of hydrochloric acid) was mixed with 8.0 g of Al metal undergoing the reaction

$$2\text{Al} + 6\text{HCl} \rightarrow 2\text{AlCl}_3 + 3\text{H}_2$$

Given atomic masses H 1.0, Al 27.0, and Cl 35.5, which reactant was left over? How much? What mass of $AlCl_3$ was produced?

Answer: HCl was the limiting reactant. Only 2.47 g of Al were consumed leaving 5.53 g of Al unreacted. The mass of $AlCl_3$ *produced was 12.2 g.*

Percent Yield

The mass of product calculated from the mass of limiting reactant in a chemical reaction is called the **stoichiometric yield** of a chemical reaction. By measuring the actual mass of a product produced in a chemical reaction and comparing it to the mass predicted from the stoichiometric yield it is possible to calculate the *percent yield.* This concept is illustrated by the following example.

Suppose that a water solution containing 25.0 g of $CaCl_2$ was mixed with a solution of sodium sulfate,

$$\text{CaCl}_2(aq) + \text{Na}_2\text{SO}_4(aq) \rightarrow \text{CaSO}_4(s) + 2\text{NaCl}(aq) \quad (4.10.2)$$

to produce a solid precipitate of $CaSO_4$, the desired product of the reaction. (Recall that a *precipitate* is a solid formed by the reaction of species in solution; such a solid is said to *precipitate* from the solution.) Removed by filtration and dried, the precipitate was found to have a mass of 28.3 g, the measured yield. What was the percent yield?

Using atomic masses Ca 40.0, Cl 35.5, Na 23.0, and O, 16.0 gives molar masses of 111 g/mol for $CaCl_2$ and 136 g/mol for $CaSO_4$. Furthermore, 1 mole of $CaCl_2$ yields 1 mol of $CaSO_4$. The stoichiometric yield of $CaSO_4$ is given by the following calculation:

$$\text{Mass CaSO}_4 = 25.0\text{ g CaCl}_2 \times \frac{1\text{ mol CaCl}_2}{111\text{ g CaCl}_2} \times \frac{1\text{ mol CaSO}_4}{1\text{ mol CaCl}_2} \times \frac{136\text{ g CaSO}_4}{1\text{ mol CaSO}_4}$$

$$\text{Mass CaSO}_4 = 30.6\text{ g CaSO}_4$$

The percent yield is calculated by the following:

$$\text{Percent yield} = \frac{\text{measured yield}}{\text{stoichiometric yield}} \times 100 \quad (4.10.4)$$

$$\text{Percent yield} = \frac{28.3\text{ g}}{30.6\text{ g}} \times 100 = 92.5\%$$

4.11. TITRATIONS: MEASURING MOLES BY VOLUMES OF SOLUTION

Masses are commonly measured with a laboratory balance that registers in grams. Masses of industrial chemicals are measured with much larger industrial scales that commonly give masses in kilograms or tons. In doing laboratory stoichiometric measurements with species in solution, it is often convenient to measure volumes of solution rather than masses of reactants. Solutions can be prepared that contain known numbers of moles per unit volume of solution. The volume of the reagent that must be added to another reagent to undergo a particular reaction can be measured with a device called a **buret**. A buret is shown in Figure 4.2. By measuring the volume of a solution of known concentration of solute required to react with another reactant, the number of moles of solute reacting can be calculated and stoichiometric calculations can be performed based upon the reaction. This procedure is commonly used in chemical analysis and is called **titration**.

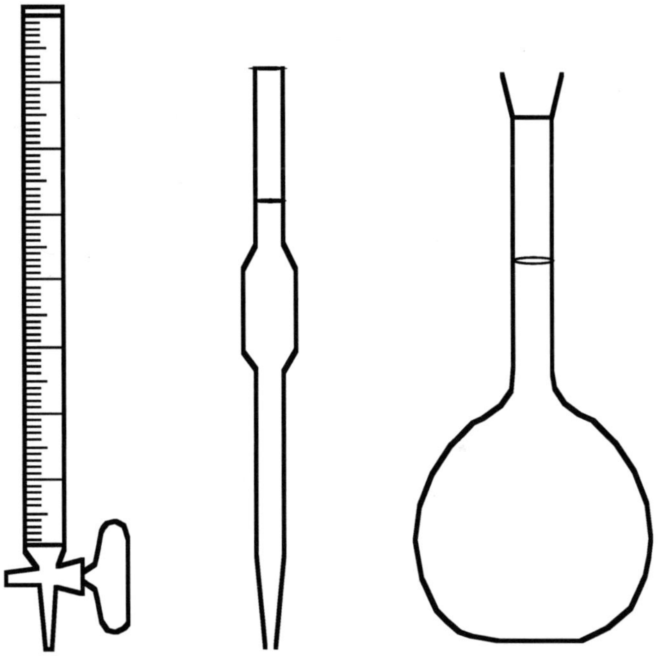

Figure 4.2. A buret consists of a narrow glass tube marked off in divisions of milliliters (mL) further subdivided into tenths of mL, usually with a total capacity of 50 mL. It enables accurate measurements of volumes of solution delivered through a stopcock estimated to the nearest 0.01 mL. A pipet delivers a fixed volume of solution. A volumetric flask contains a fixed volume of solution.

It is especially easy to relate volumes of solutions stoichiometrically when the solution concentrations are expressed as *molar concentration*, *M*. This concentration unit is defined as

$$M = \frac{\text{moles of solute}}{\text{number of liters of solution}} \tag{4.11.1}$$

The number of moles of a substance, in this case the moles of solute, is related to the mass of the substance by

$$\text{Moles of solute} = \frac{\text{mass of solute, g}}{\text{molar mass of solute, g/mol}} \tag{4.11.2}$$

These two relationships can be combined to give the following useful equation:

$$M = \frac{\text{mass of solute}}{(\text{molar mass of solute}) \times (\text{number of liters of solution})} \quad (4.11.3)$$

A solution of known concentration that is added to a reaction mixture during the procedure of titration is a **standard solution**. One of the most common of these is a standard base solution of sodium hydroxide, NaOH. Typically, the concentration of sodium hydroxide in such a standard solution is 0.100 mol/L. Suppose that it is desired to make exactly 2 liters of a solution of 0.100 mol/L sodium hydroxide. What mass of NaOH, molar mass 40.0 g/mol, is dissolved in this solution? To do this calculation, use Equation 4.11.3 rearranged to solve for mass of solute:

$$\text{Mass NaOH} = M \times (\text{molar mass NaOH}) \times (\text{liters NaOH}) \quad (4.11.4)$$

$$\text{Mass NaOH} = 0.100\ \text{mol/L} \times 40.0\ \text{g/mol} \times 2.00\ \text{L} = 8.00\ \text{g NaOH} \quad (4.11.5)$$

A common titration procedure is to use a standard solution of base to titrate an unknown solution of acid or to use standard acid to determine base. As an example consider a sample of water used to scrub exhaust gas from a hospital incinerator. The water is acidic because of the presence of hydrochloric acid produced by the scrubbing of HCl gas from the incinerator stack gas where the HCl was produced in the burning of polyvinylchloride in the incinerator. Suppose that a sample of 100 mL of the scrubber water was taken and that the volume of a 0.125 mol/L standard NaOH consumed was 11.7 mL. What was the molar concentration of HCl in the stack gas scrubber water? To solve this problem it is necessary to know that the reaction between NaOH and HCl is,

$$NaOH + HCl \rightarrow NaCl + H_2O \quad (4.11.6)$$

a neutralization reaction in which water and a salt, NaCl are produced. Examination of the reaction shows that 1 mole of HCl reacts for each mole of NaOH. Equation 4.11.1 applies to both the standard NaOH solution and the HCl solution being titrated leading to the following equations:

$$M_{HCl} = \frac{\text{moles}_{HCl}}{\text{Liters}_{HCl}} \text{ and } M_{NaOH} = \frac{\text{moles}_{NaOH}}{\text{Liters}_{NaOH}}$$

When exactly enough NaOH has been added to react with all the HCl present, the reaction is complete with no excess of either HCl or NaOH. In a titration this **end point** is normally shown by the change of color of a dye called an **indicator** dissolved in the solution being titrated. At the end point $\text{moles}_{HCl} = \text{moles}_{NaOH}$ and the two equations above can be solved to give,

$$M_{HCl} \times \text{Liters}_{HCl} = M_{NaOH} \times \text{Liters}_{NaOH} \quad (4.11.7)$$

which can be solved to give the molar concentration of HCl:

$$M_{HCl} = \frac{M_{NaOH} \times \text{Liters}_{NaOH}}{\text{Liters}_{HCl}} \quad (4.11.8)$$

Converting the volumes given from mL to liters and substituting into this equation gives the molar concentration of HCl in the incinerator scrubber water:

$$M_{HCl} = \frac{0.125\ \text{mol/L} \times 0.0117\ \text{L}}{0.100\ \text{L}} \quad (4.11.9)$$

$$M_{HCl} = 0.0146\ \text{mol/L}$$

Determining Percentage Composition by Titration

A useful application of titration, or **titrimetric analysis** as it is called, is to determine the percentage of a substance in a solid sample that will react with the titrant. To see how this is done, consider a sample consisting of basic lime, $Ca(OH)_2$, and dirt with a mass of 1.26 g. Using titration with a standard acid solution it is possible to determine the mass of basic $Ca(OH)_2$ in the solution and from that calculate the percentage of $Ca(OH)_2$ in the sample. Assume that the solid sample is placed in water and titrated with 0.112 mol/L standard HCl, a volume of 42.2 mL (0.0422 L) of the acid being required to reach the end point. The $Ca(OH)_2$ reacts with the HCl,

$$Ca(OH)_2 + 2HCl = CaCl_2 + 2H_2O \qquad (4.11.10)$$

whereas the dirt does not react. Examination of this reaction shows that at the end point the mole ratio

$$\frac{1 \text{ mol } Ca(OH)_2}{2 \text{ mol HCl}}$$

applies. At the end point, the number of moles of HCl can be calculated from

$$Mol_{HCl} = Liters_{HCl} \times M_{HCl} \quad (4.11.11)$$

and, since the molar mass of $Ca(OH)_2$ is 74.1 (given atomic masses 40.1, 16.0, and 1.0 for Ca, O, and H, respectively), the mass of $Ca(OH)_2$ is given by

$$Mass_{Ca(OH)_2} = moles_{Ca(OH)_2} \times \text{molar mass}_{Ca(OH)_2} \quad (4.11.12)$$

With this information it is now possible to calculate the mass of $Ca(OH)_2$:

$$Mass_{Ca(OH)_2} = mol_{Ca(OH)_2} \times \frac{74.1 \text{ g } Ca(OH)_2}{1 \text{ mol } Ca(OH)_2} \qquad (4.11.13)$$

$$Mass_{Ca(OH)_2} = \underbrace{Liters_{HCl} \times M_{HCl}}_{\text{Moles HCl reacting}} \times \underbrace{\frac{1 \text{ mol } Ca(OH)_2}{2 \text{ mol HCl}}}_{\substack{\text{Converts from moles} \\ \text{HCl to moles } Ca(OH)_2}} \times \underbrace{\frac{74.1 \text{ g } Ca(OH)_2}{1 \text{ mol } Ca(OH)_2}}_{\substack{\text{Gives mass } Ca(OH)_2 \\ \text{from moles } Ca(OH)_2}} \qquad (4.11.14)$$

$$Mass_{Ca(OH)_2} = 0.0422 \text{ L HCl} \times \frac{0.112 \text{ mol HCl}}{1 \text{ L HCl}} \times \frac{1 \text{ mol } Ca(OH)_2}{2 \text{ mol HCl}} \times \frac{74.1 \text{ g g } Ca(OH)_2}{1 \text{ mol } Ca(OH)_2} \qquad (4.11.15)$$

$$Mass_{Ca(OH)_2} = 0.175 \text{ g}$$

$$Percent_{Ca(OH)_2} = \frac{\text{mass } Ca(OH)_2}{\text{mass sample}} \times 100 = \frac{0.175 \text{ g}}{1.26 \text{ g}} \times 100 = 13.9\% \qquad (4.11.16)$$

Exercise: A 0.638 g sample consisting of oxalic acid, $H_2C_2O_4$, and sodium oxalate, $Na_2C_2O_4$ was dissolved and titrated with 0.116 mol/L sodium hydroxide, of which 47.6 mL (0.0476 L) was required. Each molecule of $H_2C_2O_4$ releases 2 H^+ ions. Calculate the percentage of oxalic acid in the sample.

Answer: 38.9%

4.12. INDUSTRIAL CHEMICAL REACTIONS: THE SOLVAY PROCESS

Literally thousands of chemical reactions are used to make important industrial products. Most of these involve organic chemicals, which are addressed in Chapter 9 and later chapters of this book. Some are used to make inorganic chemicals in large quantities. One such synthesis operation is the **Solvay process**, long used to make sodium bicarbonate and sodium carbonate, industrial chemicals required for glass making, cleaning formulations, and many other applications. The Solvay process is examined in some detail in this section because it illustrates some important inorganic chemical reactions and can be used for the discussion of green chemistry in industry.

The key reaction in Solvay synthesis is,

$$NaCl + NH_3 + CO_2 + H_2O \rightarrow NaHCO_3(s) + NH_4Cl \quad (4.12.1)$$

in which a sodium chloride solution (brine) is saturated with ammonia gas (NH_3), then with carbon dioxide, and finally cooled. This is a precipitation reaction in which solid sodium bicarbonate, $NaHCO_3$, comes out of solution. When heated, the solid sodium bicarbonate yields solid sodium carbonate, Na_2CO_3, water vapor, and carbon dioxide gas:

$$2NaHCO_3 + heat \rightarrow Na_2CO_3 + H_2O(g) + CO_2(g) \quad (4.12.2)$$

In keeping with the practice of green chemistry (although Solvay developed the process long before anyone ever thought of green chemistry), the CO_2 from Reaction 4.12.2 is recycled back into Reaction 4.12.1.

The raw materials for the Solvay process are cheap. The NaCl solution can be pumped from the ground from brine deposits in some locations, or fresh water can be pumped into a salt formation to dissolve NaCl and the resulting brine pumped to the surface. The most expensive raw material is ammonia, which is made by the reaction of elemental hydrogen and nitrogen over an iron-based catalyst,

$$3H_2 + N_2 \rightarrow 2NH_3 \quad (4.12.3)$$

a means of making ammonia developed by Haber and Bosch in Germany in 1913. However, as shown below, the ammonia is recycled, so only relatively small quantities of additional makeup NH_3 are required.

In addition to NaCl, the major consumable raw material in the Solvay process is calcium carbonate, $CaCO_3$, which is abundantly available from deposits of limestone. It is heated (calcined)

$$CaCO_3 + heat \rightarrow CaO + CO_2 \quad (4.12.4)$$

to produce calcium oxide and carbon dioxide gas. The carbon dioxide gas is used in Reaction 4.12.1, another green chemical aspect of the process. The calcium oxide is reacted with water (it is said to be slaked),

$$CaO + H_2O \rightarrow Ca(OH)_2 \quad (4.12.5)$$

to produce basic calcium hydroxide. This base is then reacted with the solution from which solid $NaHCO_3$ has been precipitated (Reaction 4.12.1) and that contains dissolved ammonium chloride,

$$Ca(OH)_2(s) + 2NH_4Cl(aq) \rightarrow 2NH_3(g) + CaCl_2(aq) + 2H_2O(l) \quad (4.12.6)$$

releasing ammonia gas that is recycled back into Reaction 4.12.1 for $NaHCO_3$ synthesis. This has the advantage of recycling ammonia, which is essential for the process to be economical. It has the disadvantage of generating a solution of calcium chloride, $CaCl_2$. The commercial demand for this salt is limited, although concentrated solutions of it are used for de-icing ice-covered roads. It has such a voracious appetite for water that it cannot be dried economically for storage in a dry form.

Does the Solvay process meet the criteria for a green chemical synthesis? There is not a simple answer to that question. There are two respects in which it does meet green chemical criteria:

1. It uses inexpensive, abundantly available raw materials in the form of NaCl brine and limestone ($CaCO_3$). A significant amount of NH_3 is required to initiate the process with relatively small quantities to keep it going.

2. It maximizes recycle of two major reactants, ammonia and carbon dioxide. The calcination of limestone (Reaction 4.12.4) provides ample carbon dioxide to make up for inevitable losses from the process, but some additional ammonia has to be added to compensate for any leakage.

What about the percent yield and atom economy of the Solvay process? The percent yield of reaction generating the product, Reaction 4.12.1, can be expected to be significantly less than 100% in large part because the stoichiometric amount of $NaHCO_3$ cannot be expected to precipitate from the reaction mixture. To calculate the maximum atom economy for Na_2CO_3 production, it must be assumed that all reactions go to completion without any losses. In such an ideal case, the overall reaction for the process is

$$CaCO_3 + 2NaCl \rightarrow Na_2CO_3 + CaCl_2 \qquad (4.12.7)$$

Using the atomic masses Ca 40.0, C 12.0, O 16.0, and Cl 35.5 gives the molar masses of $CaCO_3$, 100 g/mol; NaCl, 58.5 g/mol; Na_2CO_3, 106 g/mol; and $CaCl_2$, 111 g/mol. If the minimum whole number of moles of reactants were to react, 100 g of $CaCO_3$ would react with $2 \times 58.5 = 117$ g of NaCl to produce 106 g of Na_2CO_3 and 111 g of $CaCl_2$. Note that the mass of NaCl reacting is 2 times the molar mass because 2 moles of NaCl are reacting. So, for these amounts of materials in the reaction, a total mass of $100 + 117 = 217$ g of reactants produces 106 g of the Na_2CO_3 product. Therefore, the percent atom economy is

$$\text{Percent atom economy} = \frac{\text{mass of desired product}}{\text{total mass of reactants}} \times 100 = \frac{106\text{ g}}{217\text{ g}} \times 100 \qquad (4.12.8)$$

$$\text{Percent atom economy} = 48.8\%$$

This is the maximum possible assuming complete reactions and no losses. If the $CaCl_2$ byproduct is considered to be a useful product, the atom economy can be regarded as being higher.

Is the Solvay process green with respect to environmental impact? Again, the answer to this question is mixed. Extraction of the two major raw materials, limestone and NaCl, normally can be accomplished with minimal adverse effects on the environment. Quarrying of limestone in open pits results in dust production and blasting of the rock, which is usually carried out with an explosive mixture of fuel oil mixed with ammonium nitrate, $NaNO_3$, causes some disturbance. Open-pit limestone quarries can be unsightly, but can also serve as artificial lakes. In some places the underground spaces left from the underground quarrying of limestone have found excellent commercial use as low-cost warehouses that largely provide their own climate control. Truck transport of quarried lime definitely has negative environmental impacts. Extraction of liquid NaCl brine usually has minimal environmental impact. The Solvay process, itself, releases significant quantities of greenhouse gas carbon dioxide and some gaseous ammonia to the atmosphere. Solvay production of sodium carbonate requires significant amounts of energy.

There are numerous natural deposits of sodium bicarbonate and sodium carbonate, the most common of which is **trona**, for which the chemical formula is $Na_2CO_3 \cdot NaHCO_3 \cdot 2H_2O$. (This formula shows that a formula unit of trona mineral consists of 1 formula unit of ionic Na_2CO_3, 1 formula unit of ionic $NaHCO_3$ and 2 molecules of H_2O). The development of huge deposits of trona in the state of Wyoming and elsewhere in the world has lowered dependence on the Solvay process for sources of sodium bicarbonate and sodium carbonate and the process is no longer used in the United States.

QUESTIONS AND PROBLEMS

1. How do chemical equations relate to chemical reactions?

2. Summarize the information contained in the chemical equation below. How would this reaction be classified?

$$CaCl_2(aq) + Na_2CO_3(aq) \rightarrow CaCO_3(s) + 2NaCl(aq)$$

3. What are the meanings of (s), (l), (g), and (aq) after formulas in a chemical equation? What are the meanings of Δ and $\leftarrow\!\rightarrow$?

4. What is wrong with balancing the chemical equation $S + O_2 \rightarrow SO_3$ as $S + O_2 \rightarrow SO_2$?

5. From your knowledge of chemistry and chemical formulas write the balanced equation for heating magnesium carbonate to give magnesium oxide and carbon dioxide, indicating the physical states of the reactants and products.

6. Balance the equation $FeSO_4 + H_2SO_4 + O_2 \rightarrow Fe_2(SO_4)_3 + H_2O$, which is for a reaction involved in the formation of pollutant acid mine water.

7. Balance each of the following: (a) $C_2H_4 + O_2 \rightarrow CO_2 + H_2O$, (b) $KClO_4 \rightarrow KClO + O_2$, (c) $FeS + O_2 + H_2O \rightarrow FeSO_4 + H_2SO_4$ (d) $Fe_2O_3 + CO \rightarrow Fe + CO_2$, (e) $H_3PO_4 + H_2 \rightarrow PH_3$, (f) $P + Cl_2 \rightarrow PCl_5$

8. Explain how chemical equations fit in with the general scheme of chemistry as a language.

9. A chemical equation that describes the action of hydrogen sulfide, H_2S, dissolved in water is $H_2S \leftarrow\!\rightarrow H^+ + HS^-$. What does this equation say and how is it consistent with the fact that dissolved hydrogen sulfide is a weak acid?.

10. From the discussion of reactions of metals with sulfuric acid in Section 4.3 and your knowledge of the properties of silver jewelry, explain what is likely to happen when silver metal is placed in sulfuric acid.

11. Zinc is a very reactive metal. Explain with chemical equations what you would expect to happen if zinc metal were placed in sulfuric acid and what would happen if zinc oxide, ZnO, were placed in sulfuric acid.

12. Finely divided steel wool heated red hot and quickly placed into a bottle of oxygen burns vigorously undergoing the reaction $4Fe + 3O_2 \rightarrow 2Fe_2O_3$. Why is there no concern that a steel beam used in construction will burn in air? However, such a beam can be cut with an oxyacetylene torch by first heating a small portion of it with the torch, then turning off the acetylene and slowly running the torch across the beam. Explain why this procedure works to cut steel.

13. A water solution of hydrogen peroxide, H_2O_2, is relatively stable. But, if a small quantity of solid manganese oxide is placed in the solution of hydrogen peroxide, bubbles are given off near the surface of the manganese oxide, although it appears to remain intact. Explain what happens and the role of the manganese oxide.

14. The following reactions were given in connection with the Solvay process used to make sodium bicarbonate and sodium carbonate: (A) $NaCl + NH_3 + CO_2 + H_2O \rightarrow NaHCO_3(s) + NH_4Cl$, (B) $2NaHCO_3 + heat \rightarrow Na_2CO_3 + H_2O(g) + CO_2(g)$ (C) $3H_2 + N_2 \rightarrow 2NH_3$, (D) $CaCO_3 + heat \rightarrow CaO + CO_2$, (E) $CaO + H_2O \rightarrow Ca(OH)_2$. Classify each of these reactions in terms of the categories given in Section 4.6.

15. Given the chemical reaction $4CH_4 + 6NO_2 \rightarrow 4CO + 3N_2 + 8H_2O$, write all the possible mole ratios relating N2 to each of the other reaction participants.

16. Given the atomic masses N 14.0, H 1.0, and Cl 35.5 and the reaction below, calculate the mass of HCl produced when 12.7 g of NH_3 react.

    $2NH3 + 3Cl_2$ _ $N_2 + 6HCl$

17. Given the atomic masses C 12.0, H 1.0, and O 16.0 and the reaction below, calculate the mass of H_2O produced when 15.6 g of O_2 react.

    $C_2H_4 + 3O_2 \rightarrow 2CO_2 + 2H_2O$

18. Match the reaction type from the list on the left with the example reaction from the right, below. $PbSO_4$ is insoluble in water.

A. Decomposition	1. $HCl + NaOH \rightarrow H_2O + NaCl$
B. Neutralization	2. $Pb(NO_3)_2 + Na_2SO_4 \rightarrow PbSO_4 + 2NaCl$
C. Double displacement and precipitate formation	3. $2H_2O_2 \rightarrow 2H_2O + O_2$ 4. $CuSO_4(aq) + Fe(s) \rightarrow FeSO_4(aq) + Cu(s)$
D. Substitution	

19. Of the following, the untrue statement is

    A. The symbol $\longleftrightarrow$ is used to show that a reaction goes both ways.

    B. The notation (*l*) is used to show that a reactant or product is dissolved in water.

    C. A catalyst changes the rate of a reaction but is not itself consumed.

    D. The symbol, Δ, is used to show application of heat to a reaction.

    E. Simply bccause a chemical equation may be written and balanced does **not** indicate for certain that the chemical reaction it indicates will occur.

20. Given the reaction $2S + 3O_2 \rightarrow 2SO_2$ and atomic masses of 32.0 and 16.0 for S and O, respectively, calculate the mass of O_2 reacting with 15.0 g of S.

21. Given the reaction $CH_4 + 2H_2O \rightarrow CO_2 + 4H_2$ and atomic masses of C, 12.0, H, 1.0; and O 16.0, calculate the total mass of products formed when 24.0 g of CH_4 reacts.

22. Given the reaction $3CH_4 + 4Fe_2O_3 \rightarrow 3CO_2 + 6H_2O + 8Fe$ and atomic masses of C, 12.0; H, 1.0; Fe, 55.8; and O, 16.0, what is the mass of CO_2 produced by the reaction of 36.0 g of Fe_2O_3?

23. What is the basis of stoichiometry in respect to relative amounts of materials in reactions?

23. What are the major steps in doing a stoichiometric calculation?

24. What is a limiting reactant?

25. A solution of sulfuric acid was prepared by mixing 100 g of pure H_2SO_4 with water and putting it in contact with 50.0 g of iron metal. What reaction occurred? What masses of reaction products were generated and what were the masses of reactants, if any, left over? The atomic masses needed are H 1.0, Fe 55.8, S 32.0, and O 16.0.

26. What is the difference between the stoichiometric yield and the measured yield in a chemical reaction? How are they used to calculate percent yield?

27. How are titrations and stoichometry related?

28. A solid mineral sample consisting of calcium carbonate, $CaCO_3$, and nonreactive mineral matter weighing 0.485 g was stirred in some water to which 0.115 mol/L standard hydrochloric acid, HCl, was added from a buret. The reaction was $CaCO_3 + 2HCl \rightarrow CaCl_2 + CO_2 + H_2O$. If 48.6 milliliters (0.0486 L) of HCl was required to react with all the $CaCO_3$ in the sample, what was the percentage of $CaCO_3$ in the sample given that the molar mass of $CaCO_3$ is 100 g/mol?

29. A 250 mL sample of incinerator exhaust gas scrubber water contaminated with HCl was titrated with 0.104 mol/L standard NaOH, of which 11.3 mL were required to reach the end point. What was the molar concentration of HCl in the water sample?

30. What is made by the Solvay process? What is the overall chemical reaction that describes the Solvay process? What are the two major raw materials consumed and what are two major species that are recycled through the process?

31. What are major green aspects of the Solvay process? What are some aspects that are less green?

32. What is a major alternative to use of the Solvay process?

33. A total of 38.6 g of $AlCl_3$ contains (a) ____________ moles of the compound, whereas 217 g of methane, CH_4, contains (b) ______________ moles of CH_4. Use 27.0, 35.5, 12.0, and 1.0 for the atomic masses of Al, Cl, C, and H, respectively.

34. Why might you expect stoichiometric ratios to be used in industrial chemical reactions? If one of two reactants used in an industrial process is much more expensive than another, suggest why and in which way a stoichiometric ratio might not be used? Also, suppose that one of two reactants is quite toxic whereas the other reactant is not. Why might the practice of green chemistry suggest using a nonstoichometric ratio of reactants in such a case?

35. Given the reaction $2H_2 + O_2 \rightarrow 2H_2O$, identify which species is oxidized, which is reduced, which is the oxidizing agent, and which is the reducing agent.

36. Given the reaction that occurs when a direct electrical current is passed through liquid ionic NaCl, $2Na^+ + 2Cl^- \xrightarrow{\text{Electrolysis}} 2Na + Cl_2$, identify which species is oxidized and which is reduced. Justify the answer.

37. Identify which reactions given in Section 4.6 are oxidation-reduction reactions.

5. THE WONDERFUL WORLD OF CARBON: ORGANIC CHEMISTRY AND BIOCHEMICALS

5.1. RINGS AND CHAINS OF CARBON ATOMS

Most of the molecules of chemical compounds studied so far have been clusters of only a few atoms. Therefore, molecules of water, H_2O, exist as individual clusters of 2 H atoms bonded to 1 O atom and molecules of ammonia, NH_3, each consist of an atom of N to which are bonded 3 H atoms. In cases where atoms of a particular element in chemical compounds have a tendency to bond with atoms of the same element, the number of possible compounds is increased tremendously. This is the case with carbon, C. Groups of carbon atoms can bond together to form straight chains, branched chains, and rings, leading to a virtually limitless number of chemical compounds. Such carbon-containing compounds are **organic chemicals**, the study of which is **organic chemistry**. Adding to the enormous diversity of organic chemistry is the fact that two carbon atoms may be connected by **single bonds** consisting of 2 shared electrons, **double bonds** composed of 4 shared electrons, and even **triple bonds** that contain 6 shared electrons.

Organic chemicals comprise most of the substances with which chemists are involved. Petroleum, which serves as the raw material for vast polymer, plastics, rubber, and other industries consists of hundreds of compounds composed of hydrogen and carbon called **hydrocarbons**. Among organic chemicals are included the majority of important industrial compounds, synthetic polymers, agricultural chemicals, and most substances that are of concern because of their toxicities and other hazards. The carbohydrates, proteins, lipids (fats and oils), and nucleic acids (DNA) that make up the biomass of living organisms are organic chemicals made by biological processes. The feedstock chemicals needed to manufacture a wide range of chemical products are mostly organic chemicals, and their acquisition and processing are of great concern in the practice of green chemistry. The largest fraction of organic chemicals acquired from petroleum and natural gas sources are burned to fuel vehicles, airplanes, home furnaces, and power plants. Prior to burning, these substances may be processed to give them desired properties. This is particularly true of the constituents of gasoline, the molecules of which are processed and modified to give gasoline desired properties of smooth burning (good antiknock properties) and low air pollution potential. Pollution of the water, air, and soil environments by organic chemicals is an area of significant concern. Much of the effort put into green chemistry has involved the safe manufacture, recycling, and disposal of organic compounds.

A number of organic compounds are made by very sophisticated techniques to possess precisely tailored properties. This is especially true of pharmaceuticals, which must be customized to deliver the desired effects with minimum undesirable side effects. A single organic compound that is effective against one of the major health problems — usually one out of hundreds or even thousands tested — has the potential for hundreds of millions of dollars per year in profits.

Organic chemicals differ widely in their toxicities. Some compounds are made and used because of their toxicities to undesirable organisms. These are the **pesticides**, including, especially, **insecticides** used to kill unwanted insects and **herbicides** used to eradicate weeds that compete with desired crops. Green chemistry is very much involved with these kinds of applications. One of the more widely applied uses of genetically modified crops has been the development of crops that produce their own insecticides in the form of insecticidal proteins normally made by certain kinds of bacteria whose genes have been spliced into field crops. Another application of green chemistry through genetic engineering is the development of crops that resist the effects of specific organic molecules commonly used as herbicides. These herbicides may be applied directly to target crops, leaving them unscathed while competing weeds are killed.

It should be obvious from this brief discussion that organic chemistry is a vast, diverse, highly useful discipline based upon the unique bonding properties of the carbon atom. The remainder of this chapter discusses major aspects of organic chemistry. Many of the most interesting and important organic chemicals are made by biological processes. Indeed, until 1828, it was generally believed that only organisms could synthesize organic chemicals. In that year, Friedrich Wöhlers succeeded in making urea, an organic chemical that is found in urea, from ammonium cyanate, an inorganic material. Because of the important role of organisms in making organic chemicals, several of the most significant kinds of these chemicals made biologically are also discussed in this chapter. Additional details regarding the ways in which living organisms make and process chemicals are given in Chapters 9 and 13.

5.2. COMPOUNDS OF CARBON AND HYDROGEN: HYDROCARBONS

The tremendous variety and diversity of organic chemistry is due to the ability of carbon atoms to bond with each other in a variety of straight chains, branched chains, and rings and of adjacent carbon atoms to be joined by single, double, or triple bonds. This bonding ability can be illustrated with the simplest class of organic chemicals, the **hydrocarbons** consisting only of hydrogen and carbon. Figure 5.1 shows some hydrocarbons in various configurations.

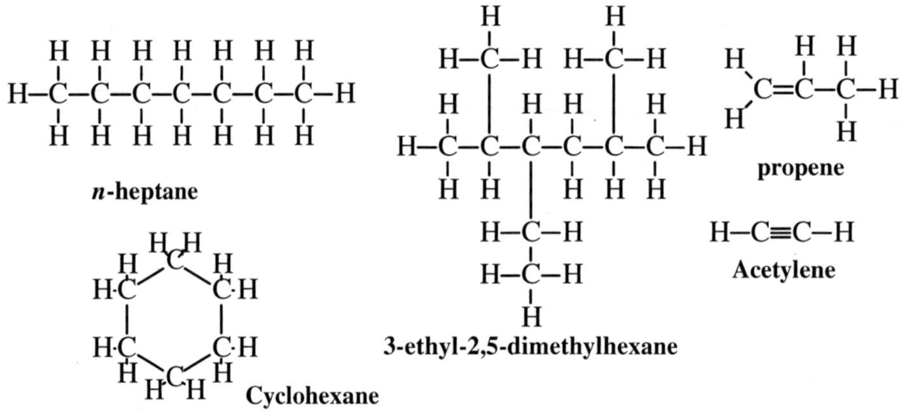

Figure 5.1. Some typical hydrocarbons. These formulas illustrate the bonding diversity of carbon which gives rise to an enormous variety of hydrocarbons and other organic compounds.

Hydrocarbons are the major ingredients of petroleum and are pumped from the ground as crude oil or extracted as natural gas. They have two major uses. The first of these is combustion as a source of fuel. The most abundant hydrocarbon in natural gas, methane, CH_4, is burned in home furnaces, electrical power plants, and even in vehicle engines,

$$CH_4 + 2O_2 \rightarrow CO_2 + 2H_2O + \text{heat energy} \qquad (5.2.1)$$

to provide energy. The second major use of hydrocarbons is as a raw material for making rubber, plastics, polymers, and many other kinds of materials. Given the value of hydrocarbons as a material, it is unfortunate that so much of hydrocarbon production is simply burned to provide energy, which could be generated by other means.

There are several major class of hydrocarbons, all consisting of only hydrogen and carbon. **Alkanes** have only single bonds between carbon atoms. Cyclohexane, *n*-heptane, and 3-ethyl-2,5-dimethylhexane in Figure 5.1 are alkanes; the cyclohexane is a cyclic hydrocarbon. **Alkenes**, such as propene shown in Figure 5.1, have at least one double bond consisting of 4 shared electrons between two of the carbon atoms in the molecule. **Alkynes** have at least one triple bond between carbon atoms in the molecule as shown for acetylene in Figure 5.1. Acetylene is an important fuel that is used with pure oxygen to produce very high temperature in welding and cutting torches. Otherwise, the alkynes are of relatively little importance and will not be addressed farther. A fourth class of hydrocarbon consists of **aromatic** compounds which have rings of carbon atoms with special bonding properties as discussed later in this chapter.

Alkanes

The molecular formulas of non-cyclic alkanes are C_nH_{2n+2}. By counting the numbers of carbon and hydrogen atoms in the molecules of alkanes shown in Figure 5.1, it is seen that the molecular formula of *n*-heptane is C_7H_{16} and that of 3-ethyl-2,5-dimethylhexane is $C_{10}H_{22}$, both of which fit the general formula given above. The general formula of cyclic alkanes is C_nH_{2n}; that of cyclohexane, the most common cyclic alkane, is C_6H_{12}. These formulas are **molecular formulas**, which give the number of carbon and hydrogen atoms in each molecule, but do not tell anything about the structure of the molecule. The formulas given in Figure 5.1 are **structural formulas** which show how the molecule is assembled. The structure of *n*-heptane is that of a straight chain of carbon atoms; each carbon atom in the middle of the chain is bound to 2 H atoms and the 2 carbon atoms at the ends of the chain are each bound to 3 H atoms. The prefix *hep* in the name denotes 7 carbon atoms and the *n*- indicates that the compound consists of a single straight chain. This compound can be represented by a **condensed structural formula** as $CH_3(CH_2)_5CH_3$ representing 7 carbon atoms in a straight chain. In addition to methane mentioned previously, the lower alkanes include the following:

Ethane: CH_3CH_3 Propane: $CH_3CH_2CH_3$ Butane: $CH_3(CH_2)_2CH_3$ *n*-Pentane: $CH_3(CH_2)_3CH_3$

For alkanes with 5 or more carbon atoms, the prefix (*pen* for 5, *hex* for 6, *hept* for 7, *oct* for 8, *non* for 9) shows the total number of carbon atoms in the compound and *n*- may be used to denote a straight-chain alkane. Condensed structural formulas may be used to represent branched chain alkanes as well. The condensed structural formula of 3-ethyl-2,5-dimethylhexane is

$$CH_3CH_2(CH_3)CH_2(C_2H_5)CH_2CH_2(CH_3)CH_3.$$

In this formula, the C atoms and their attached H atoms that are not in parentheses show carbons that are part of the main hydrocarbon chain. The (CH_3) after the second C in the chain shows a methyl group attached to it, the (C_2H_5) after the third carbon atom in the chain shows an ethyl group attached to it, and the (CH_3) after the fifth carbon atom in the chain shows a methyl group attached to it.

Compounds that have the same molecular formulas but different structural formulas are **structural isomers**. For example, the straight-chain alkane with the molecular formula $C_{10}H_{22}$ is *n*-decane,

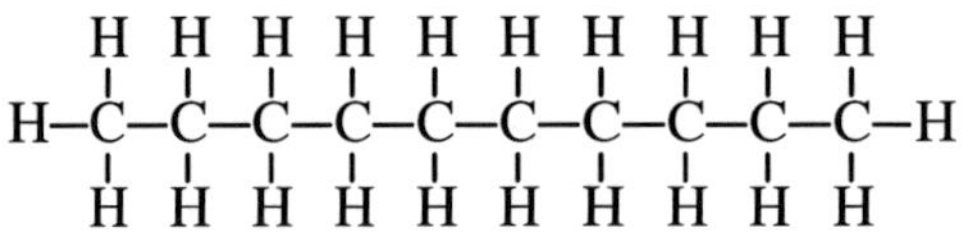

***n*-decane**

which is a structural isomer of 3-ethyl-2,5-dimethylhexane.

The names of organic compounds are commonly based upon the structure of the hydrocarbon from which they are derived using the longest continuous chain of carbon atoms in the compound as the basis for the name. For example, the longest continuous chain of carbon atoms in 3-ethyl-2,5-dimethylhexane shown in Figure 5.1 is 6 carbon atoms, so the name is based upon *hex*ane. The names of the chain branches are also based upon the alkanes from which they are derived. As shown below, the two shortest-chain alkanes are methane with 1 carbon atom and ethane with 2 carbon atoms. Removal of 1 of the H

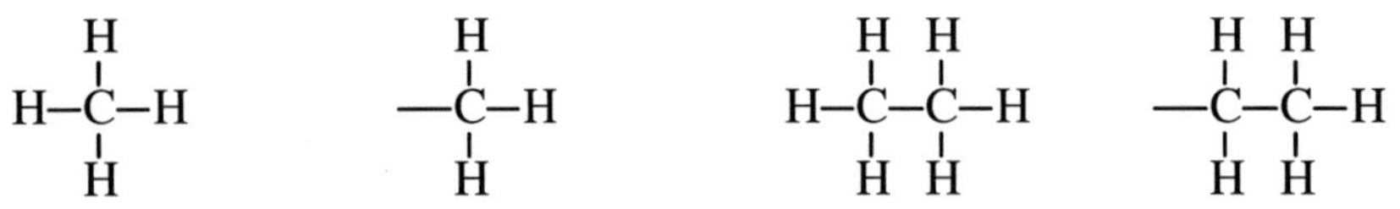

Methane (CH_4) Methyl group (CH_3) Ethane (C_2H_6) Ethyl group (C_2H_5)

atoms from methane gives the **methyl** group and removal of 1 of the H atoms from ethane gives the **ethyl** group. These terms are used in the name 3-ethyl-2,5-dimethylhexane to show groups attached to the basic hexane chain. The carbon atoms in this chain are numbered sequentially from left to right. An ethyl group is attached to the 3rd carbon atom, yielding the "3-ethyl" part of the name, and methyl groups are attached to the 2nd and 5th carbon atoms, which gives the "2,5-dimethyl" part of the name.

The names discussed above are **systematic names**, which are based upon the actual structural formulas of the molecules. In addition, there are **common names** of organic compounds that do not indicate the structural formulas. Naming organic compounds is a complex topic, and no attempt is made here to teach it to the reader. However, from the names of compounds given in this and later chapters, some appreciation of the rationale for organic compound names should be obtained.

Other than burning them for energy, the major kind of reaction with alkanes consists of **substitution reactions** such as,

$$C_2H_6 + 2Cl_2 \rightarrow C_2H_4Cl_2 + 2HCl \quad (5.2.2)$$

in which one or more H atoms are displaced by another kind of atom. This is normally the first step in converting alkanes to compounds containing elements other than carbon or hydrogen for use in synthesizing a wide variety of organic compounds.

Alkenes

Four common alkenes are shown in Figure 5.2. Alkenes have at least one C=C double bond per molecule and may have more. The first of the alkenes in Figure 5.2, ethylene, is a very widely produced hydrocarbon used to synthesize polyethylene plastic and other organic compounds. About 25 billion kilograms (kg) of ethylene are processed in the U.S. each year. About 14.5 billion kg of propylene are used in the U.S. each year to produce polypropylene plastic and other chemicals. The two 2-butene compounds illustrate an important aspect of alkenes, the possibility of ***cis-trans*** isomerism. Whereas carbon atoms and the groups substituted onto them joined by single bonds can freely rotate relative to each other as though they were joined by a single shaft, carbon atoms connected by a double bond behave as though they were attached by two parallel shafts and are not free to rotate. So, *cis*-2-butene in which the two end methyl (-CH_3) groups are on the same side of the molecule is a different compound from *trans*-2-butene in which they are on opposite sides. These two compounds are *cis-trans* isomers.

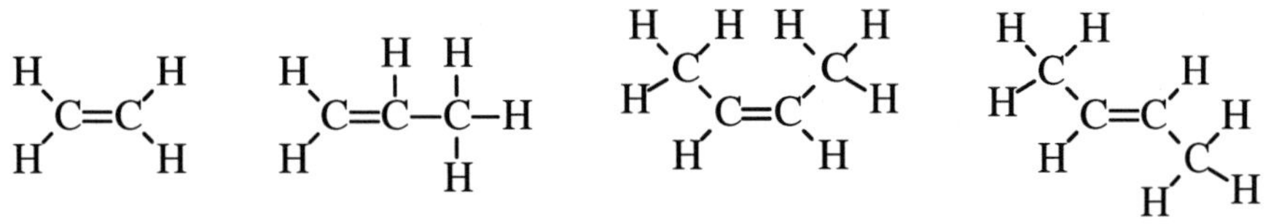

Ethylene **Propylene (propene)** ***Cis*-2-butene** ***Trans*-2-butene**

Figure 5.2. Examples of alkene hydrocarbons.

Alkenes are chemically much more active than alkanes. This is because the double bond is **unsaturated** and has electrons available to form additional bonds with other atoms. This leads to **addition reactions** in which a molecule is added across a double bond. For example, the addition of H_2O to ethylene,

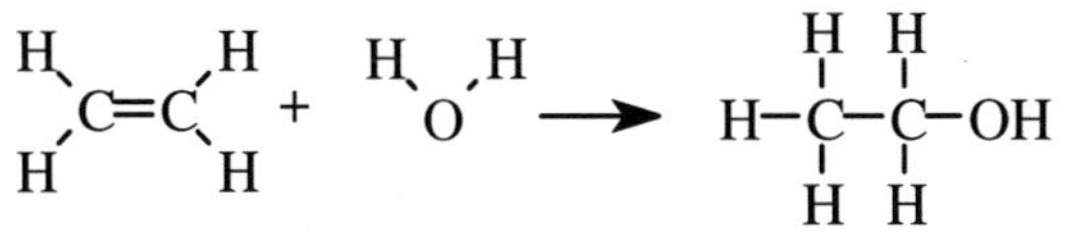

(5.2.3)

Ethylene **Water** **Ethanol**

yields ethanol, the same kind of alcohol that is in alcoholic beverages. In addition to adding immensely to the chemical versatility of alkenes, addition reactions make them quite reactive in the atmosphere during the formation of photochemical smog. The presence of double bonds also adds to the biochemical and toxicological activity of compounds in organisms.

Because of their double bonds, alkenes can undergo **polymerization** reactions in which large numbers of individual molecules add to each other to produce large molecules called **polymers**. For example, 3 ethylene molecules can add together as follows:

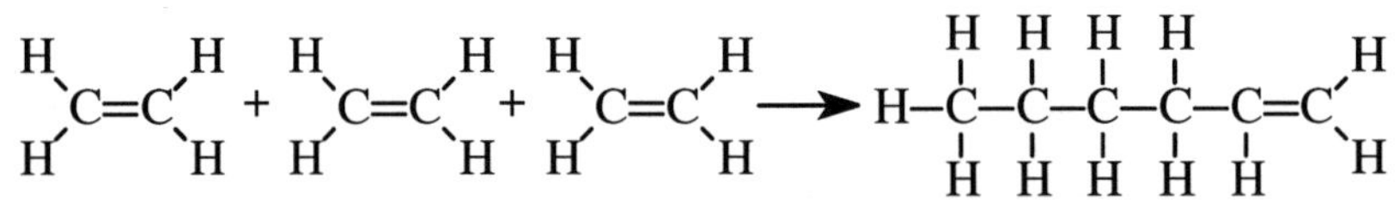

(5.2.4)

a process that can continue, forming longer and longer chains and resulting in the formation of the very large molecules that constitute polyethylene.

Aromatic Hydrocarbons

A special class of hydrocarbons consists of rings of carbon atoms, almost always containing 6 C atoms, which can be viewed as having alternating single and double bonds as shown below:

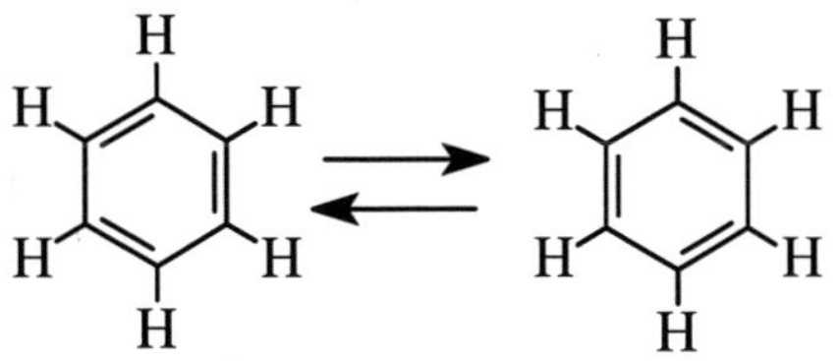

These structures show the simplest aromatic hydrocarbon, benzene, C_6H_6. Although the benzene molecule is represented as having 3 double bonds, chemically it behaves very differently from alkenes, for example undergoing substitution reactions rather than addition reactions. The properties of aromatic compounds are special properties called **aromaticity**. The two structures shown above are equivalent **resonance** structures, which can be viewed as having atoms that stay in the same places, but in which the bonds joining the atoms can shift positions with the movement of electrons composing the bonds. Since benzene has different chemical properties from those implied by either of the above structures, it is commonly represented as a hexagon with a circle in the middle:

Many aromatic hydrocarbons have two or more rings. The simplest of these is naphthalene,

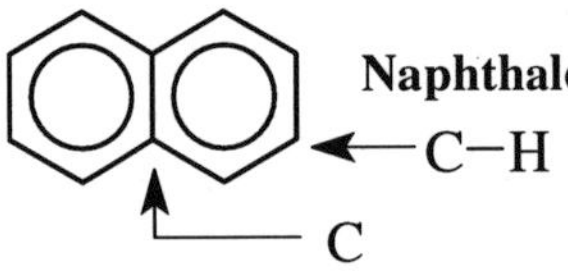

Naphthalene, $C_{10}H_8$

a two-ringed compound in which two benzene rings share the carbon atoms at which they are joined; these two carbon atoms do not have any H attached, each of the other 8 C atoms in the compound has 1 H attached. Aromatic hydrocarbons with multiple rings, called **polycyclic aromatic hydrocarbons**, PAH, are common and are often produced as byproducts of combustion. One of the most studied of these is benzo(a)pyrene,

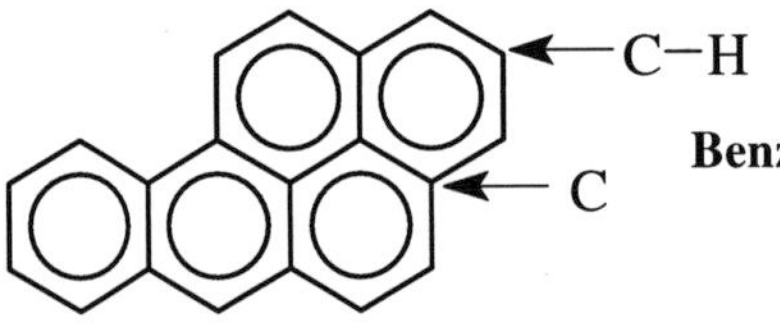

Benzo(a)pyrene, $C_{20}H_{12}$

found in tobacco smoke, diesel exhaust, and charbroiled meat. This compound is toxicologically significant because it is partially oxidized by enzymes in the body to produce a cancer-causing metabolite.

The presence of hydrocarbon groups and of elements other than carbon and hydrogen bonded to an aromatic hydrocarbon ring gives a variety of **aromatic compounds**. Three examples of common aromatic compounds are given below. Toluene is widely used for chemical synthesis and as a solvent. The practice of green chemistry now calls for substituting toluene for benzene wherever possible because benzene is suspected of causing leukemia, whereas the body is capable of metabolizing toluene to harmless metabolites (see Chapter 13). About 850 million kg of aniline are made in the U.S. each year as an intermediate in the synthesis of dyes and other organic chemicals. Phenol is a relatively toxic oxygen-containing aromatic compound which, despite its toxicity to humans, was the first antiseptic used in the 1800s.

Toluene

Aniline

Phenol

5.3. LINES SHOWING ORGANIC STRUCTURAL FORMULAS

The aromatic structures shown above use a hexagon with a circle in it to denote an aromatic benzene ring. Organic chemistry uses lines to show other kinds of structural formulas as well. The reader who may have occasion to look up organic formulas will probably run into this kind of notation, so it is important to be able to interpret these kinds of formulas. Some line formulas are shown in Figure 5.3.

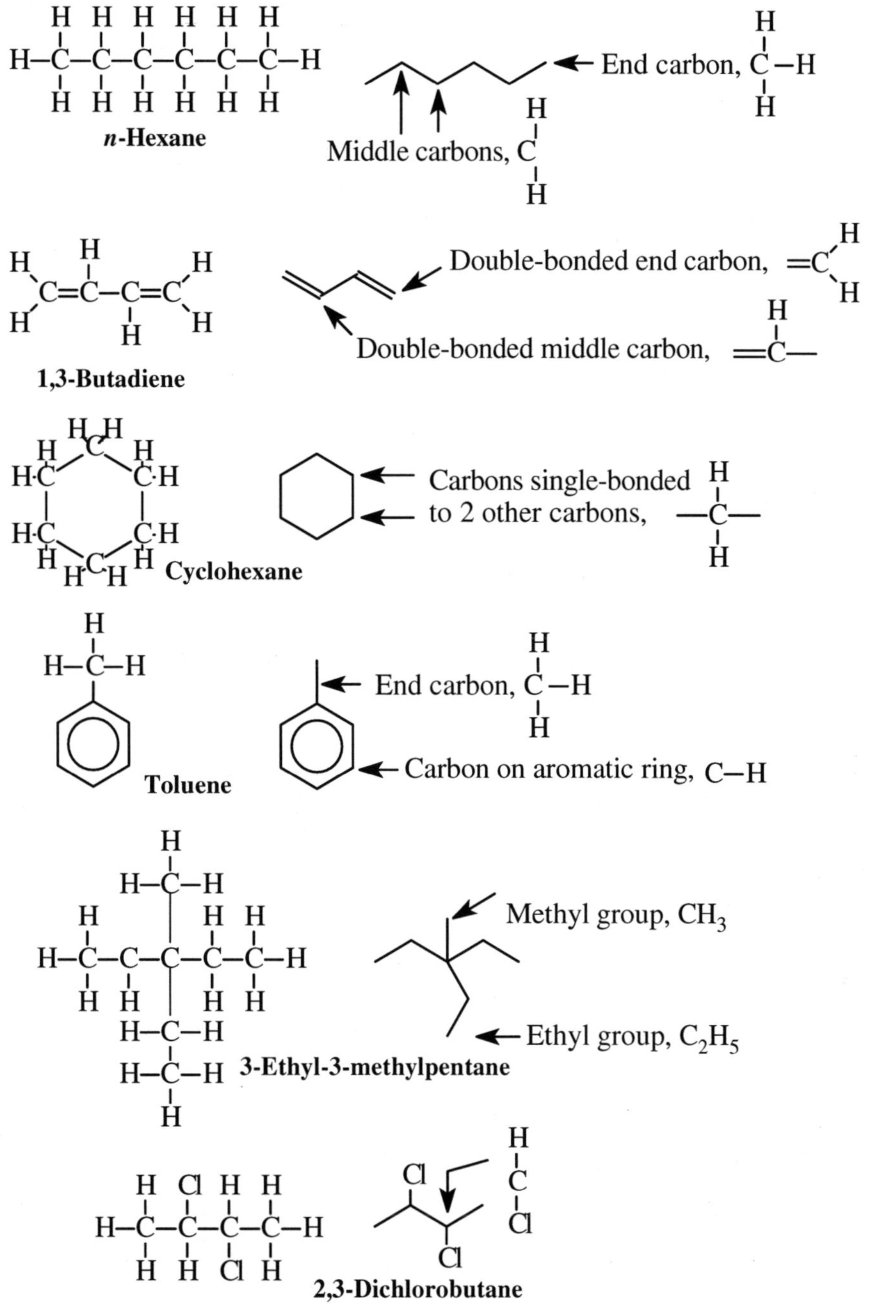

Figure 5.3. Representation of organic structural formulas with lines.

In using lines to represent organic structural formulas, the corners where lines intersect and the ends of lines represent C atoms, and each line stands for a covalent bond (2 shared electrons). It is understood that each C atom at the end of a single line has 3 H atoms attached, each C atom at the intersection of 2 lines has 2 C atoms attached, each C at the intersection of 3 lines has 1 H attached, and the intersection of 4 lines denotes a C atom with no H atoms attached. Multiple lines represent multiple bonds as shown for the double bonds in 1,3-butadiene. Substituent groups are shown by their symbols, or formulas of functional groups consisting of groups of atoms; it is understood that each such group substitutes for a hydrogen atom as shown in the formula of 2,3-dichlorobutane in Figure 5.2. The 6-carbon-atom aromatic ring is denoted by a hexagon with a circle in it.

Exercise: What is the structural formula of the compound represented on the right?

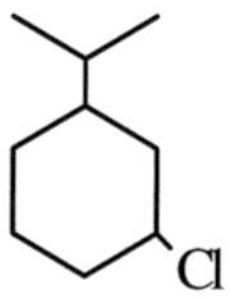

Answer:

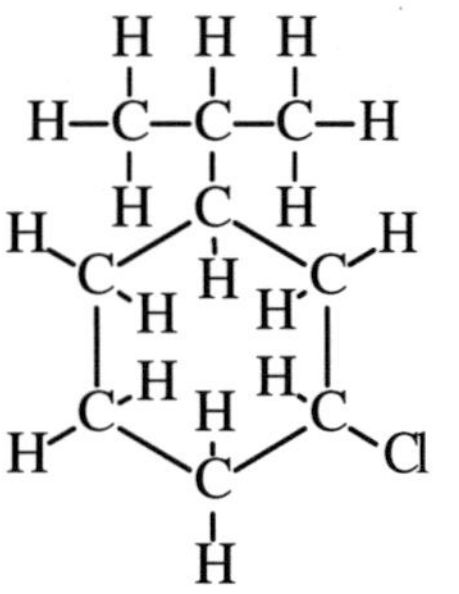

5.4. FUNCTIONAL GROUPS

Numerous elements in addition to carbon and hydrogen occur in organic compounds. These are contained in **functional groups**, which define various classes of organic compounds. The $-NH_2$ group in aniline and the -OH groups in phenol mentioned above are examples of functional groups. The same organic compound may contain two or more functional groups. Among the elements common in functional groups are O, N, Cl, S, and P. There is not space here to discuss all the possible functional groups and the classes of organic compounds that they define. Some important examples are given to provide an idea of the variety of organic compounds with various functional groups. Other examples are encountered later in the text.

Organooxygen Compounds

Figure 5.4 shows several important classes of organic compounds that contain oxygen. **Ethylene oxide** is a sweet-smelling, colorless, flammable, explosive gas. It is an **epoxide** characterized by an oxygen atom bridging two carbon atoms that are also bonded with each other. Ethylene oxide is toxic

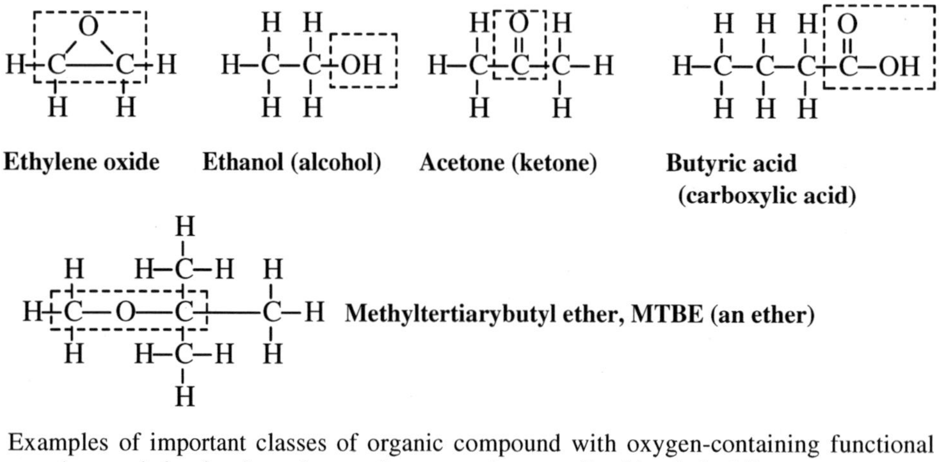

Figure 5.4. Examples of important classes of organic compound with oxygen-containing functional groups. The functional groups characteristic of various classes of compounds are outlined by dashed lines.

and is used as a sterilant and fumigant as well as a chemical intermediate. Because of the toxicity and flammability of this compound, the practice of green chemistry tries to avoid its generation and use. **Ethanol**, which occurs in alcoholic beverages, is an **alcohol**, a class of compound in which the -OH group is bonded to an alkane or alkene (attachment of the -OH group to an aromatic hydrocarbon molecule gives a **phenolic** compound). **Acetone** is a ketone, a class of compounds that has the C=O functional group in the middle of a hydrocarbon chain. Acetone is an excellent organic solvent and relatively safe. **Butyric** acid, which occurs in butter, is an organic carboxylic acid, all of which contain the functional group,

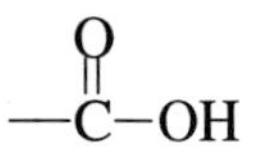

which can release the H^+ ion characteristic of acids. **Methyltertiary butyl ether**, MTBE, is an example of an ether in which an O atom connects 2 C atoms. When highly toxic tetraethyllead was phased out of gasoline as an octane booster, MTBE was chosen as a substitute. It was subsequently found to be a particularly noxious water pollutant, and a number of states have considered legislation to ban it.

The C=O group in the middle of an organic molecule is characteristic of ketones. When this group is located at the end of a molecule and the carbon is also bonded to H, the compound is an **aldehyde**. The two lowest aldehydes are formaldehyde and acetaldehyde,

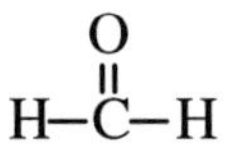
Formaldehyde

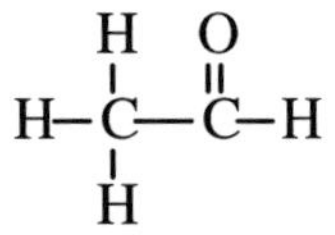
Acetaldehyde

of which formaldehyde is the most widely produced. Despite its many uses, formaldehyde lacks characteristics of green chemicals because it is a volatile, toxic, noxious substance. Formaldehyde tends to induce hypersensitivity (allergies) in people who inhale the vapor or whose skin is exposed to it.

The reaction of an alcohol and an organic acid,

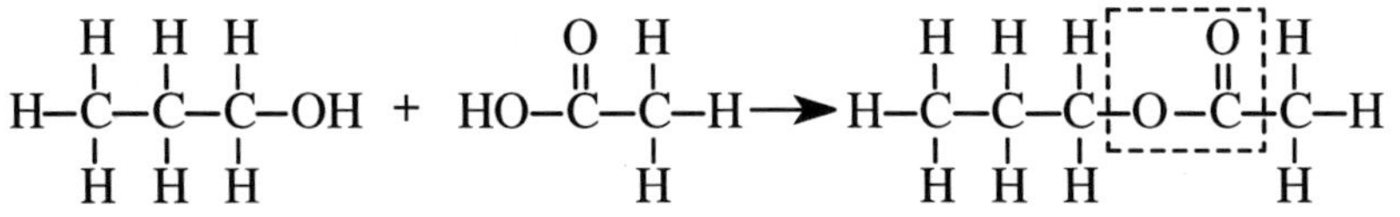
(5.4.1)

Propyl alcohol **Acetic acid** **Propyl acetate ester**

produces an important kind of organic compound called **esters**. The linkage characteristic of esters is outlined by the dashed box in the structure of propyl acetate above. A large number of the naturally-occurring esters made by plants are noted for their pleasant odors. Propyl acetate, for example gives pears their pleasant odor. Other fruit odors due to esters include methyl butyrate, apple; ethyl butyrate, pineapple; and methyl benzoate, ripe kiwi fruit.

Organonitrogen Compounds

Methylamine,

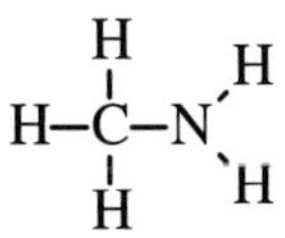
Methylamine

is the simplest of the amines in which an N atom is bonded to a hydrocarbon group. In an amine, the N atom may be bonded to 2 H atoms or one or both of these H atoms may be substituted by hydrocarbon groups as well. Although it is widely used in chemical synthesis because no suitable substitutes are available, methylamine is definitely not compatible with the practice of green chemistry. That is because it is highly flammable and toxic. It is a severe irritant to skin, eyes, and mucous membranes of the respiratory tract. It has a noxious odor and is a contributor to the odor of rotten fish (another amine, putrescine, gives decayed flesh its characteristic odor).

Many organonitrogen compounds contain oxygen as well. One such compound is nitromethane,

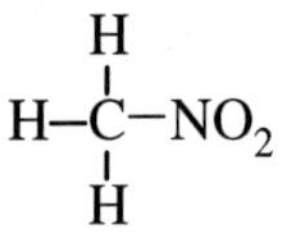
Nitromethane

used in chemical synthesis and as a fuel in some race cars. The nitro group, -NO_2, is the functional group in this compound. Another class of organonitrogen compounds also containing oxygen consists of the **nitrosamines**, or N-nitroso compounds, which have figured prominently in the history of green chemistry before it was defined as such. These are compounds that have the N-N=O functional group, which are of concern because several are known carcinogens (cancer-causing agents). The most well known of these is **dimethylnitrosamine** shown below:

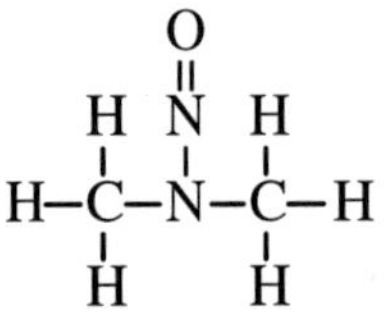
Dimethylnitrosamine

This compound used to be employed as an industrial solvent and was used in cutting oils. However, workers exposed to it suffered liver damage and developed jaundice, and the compound as well as other nitrosamines was found to be a carcinogen. A number of other nitrosamines were later found in industrial materials and as byproducts of food processing and preservation.

Organohalide Compounds

Organohalides exemplified by those shown in Figure 5.5 are organic compounds that contain halogens — F, Cl, Br, or I — but usually chlorine, on alkane, alkene, or aromatic molecules. Organohalides have been widely produced and distributed for a variety of applications, including industrial solvents, chemical intermediates, coolant fluids, pesticides, and other applications. They are for the most part environmentally persistent and, because of their tendency to accumulate in adipose (fat) tissue, they tend to undergo bioaccumulation and biomagnification in organisms.

Carbon tetrachloride is produced when all four H atoms on methane, CH_4, are substituted by Cl. This compound was once widely used and was even sold to the public as a solvent to remove stains and in fire extinguishers, where the heavy CCl_4 vapor smothers fires. It was subsequently found to be very toxic, causing severe liver damage, and its uses are severely restricted. **Dichlorodifluoromethane** is a prominent member of the **chlorofluorocarbon** class of compounds, popularly known as Freons. Developed as refrigerant fluids, these compounds are notably unreactive and nontoxic. However, as discussed in Chapter 8, they were found to be indestructible in the lower atmosphere, persisting to very high altitudes in the stratosphere where chlorine split from them by ultraviolet radiation destroys stratospheric ozone. So the manufacture of chlorofluorocarbons is now prohibited. **Vinyl chloride**, an alkene-based organohalide compound, is widely used to make polyvinylchloride polymers and pipe. Unfortunately, it is a known human carcinogen, so human exposure to it is severely limited. **Trichloroethylene** is an excellent organic solvent that is nonflammable. It is used as a drycleaning solvent and for degreasing manufactured parts, and was formerly used for food extraction, particularly to decaffeinate coffee). **Chlorobenzene** is the simplest aromatic organochloride. In addition to its uses in making other chemicals, it serves as a solvent and as a fluid for heat transfer. It is extremely stable, and its destruction is a common test for the effectiveness of hazardous waste incinerators. The **polychlorinated biphenyl (PCB)** compound shown is one of 209 PCB compounds that can be formed by substituting from 1 to 10 Cl atoms onto the basic biphenyl (two-benzene-ring) carbon skeleton. These compounds are notably stable and persistent, leading to their uses in electrical equipment, particularly as coolants in transformers and in industrial capacitors, as hydraulic fluids, and other applications. Their extreme environmental persistence has led to their being banned. Sediments in New York's Hudson River are badly contaminated with PCBs that were (at the time, legally) dumped or leaked into the river from electrical equipment manufacture from the 1950s into the 1970s.

Carbon tetrachloride **Dichlorodifluoromethane**
(Both of these compounds are alkyl halides.)

Vinyl chloride **Trichloroethylene**
(These compounds are alkenyl halides.)

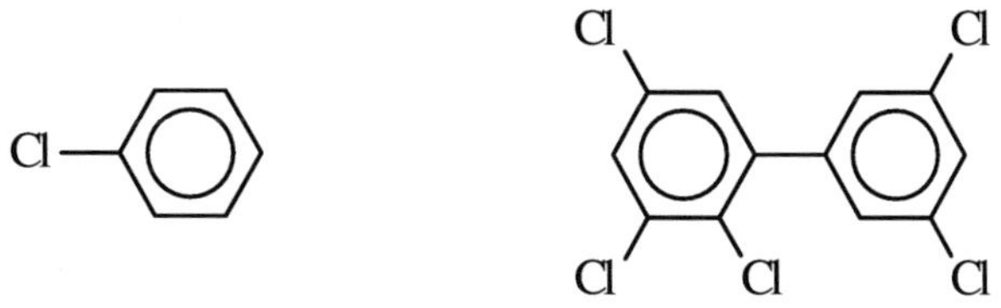

Chlorobenzene **A polychlorinated biphenyl (PCB)**
(These compounds are aromatic halides)

Figure 5.5. Examples of important organohalide compounds including alkyl halides based upon alkanes, alkenyl halides based upon alkenes, and aromatic halides.

From the discussion above, it is obvious that many organohalide compounds are definitely not green because of their persistence and biological effects. A lot of the effort in the development of green chemistry has been devoted to finding substitutes for organohalide compounds. Indeed, international agreements have been made aimed at the elimination of the use of several organohalide compounds contained in a list of "dirty dozen" compounds that are deemed particularly undesirable from an environmental viewpoint.

Organosulfur and Organophosphorus Compounds

A number of organosulfur and organophosphorus compounds have been synthesized for various purposes including pesticidal applications. A common class of organosulfur compounds consists of the thiols, the simplest of which is methanethiol:

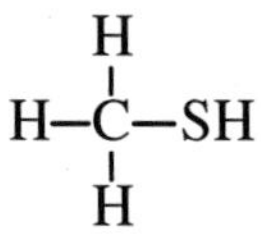 **Methanethiol** 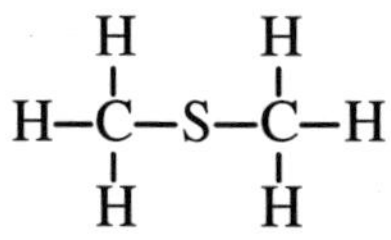 **Dimethylsulfide**

As with other thiols, which contain the -SH group, this compound is noted for its foul odor. Thiols are added to natural gas so that their odor can warn of gas leaks. Dimethylsulfide, also shown above, is a volatile compound released by ocean-dwelling microorganisms to the atmosphere in such quantities that it constitutes the largest flux of sulfur-containing vapors from Earth to the atmosphere.

Among the most prominent organophosphorus compounds are the organophosphates as shown by the two compounds below. These compounds are both insecticides and contain sulfur as well as phosphorus. Parathion was developed during the 1940s and was once widely used as an insecticide in place of DDT because parathion is very biodegradable, whereas DDT is not and undergoes bioaccumulation and biomagnification in ecosystems. Unfortunately, parathion has a high toxicity to humans and other animals and some human fatalities have resulted from exposure to it. Like other organophosphates, it inhibits acetylcholinesterase, an enzyme essential for nerve function (the same mode of action as its deadly cousins, the "nerve gas" military poisons, such as Sarin). Because of its toxicity, parathion is now banned from general use. Malathion is used in its place and is only about 1/100 as toxic as parathion to mammals because they — though not insects — have enzyme systems that can break it down.

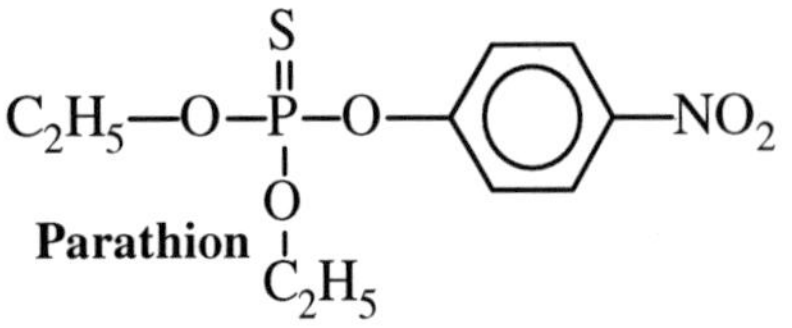
Parathion

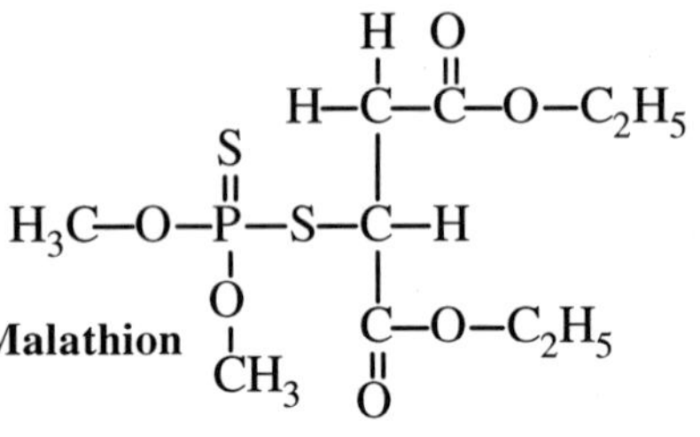
Malathion

5.5. GIANT MOLECULES FROM SMALL ORGANIC MOLECULES

Reaction 5.2.4 showed the bonding together of molecules of ethylene to form larger molecules. This process, widely practiced in the chemical and petrochemical industries, is called *polymerization* and the products are *polymers*. Many other unsaturated molecules, usually based upon alkenes, undergo polymerization to produce synthetic polymers used as plastics, rubber, and fabrics. As an example, tetrafluoroethylene polymerizes as shown in Figure 5.6 to produce a polymer (Teflon) that is exceptionally heat- and chemical-resistant and that can be used to form coatings to which other materials will not stick (for example, frying pan surfaces).

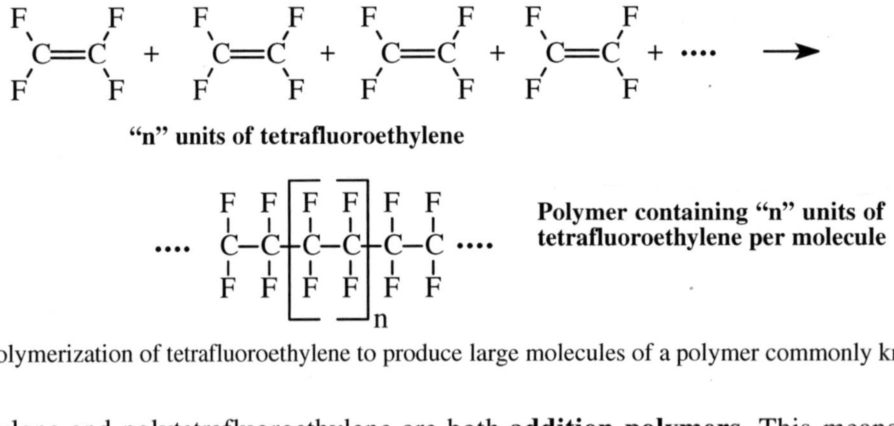

Figure 5.6. Polymerization of tetrafluoroethylene to produce large molecules of a polymer commonly known as Teflon.

Polyethylene and polytetrafluoroethylene are both **addition polymers**. This means that they are formed by the chemical addition together of the small-molecule monomers that make up the large polymer molecules. Another kind of polymers consists of **condensation polymers** that join together with the elimination of a molecule of water for each monomer unit joined. A common condensation polymer is **nylon**, which is formed by the bonding together of two different kinds of molecules. There are several forms of nylon, the original form of which is nylon 66 discovered by Wallace Carothers, a DuPont chemist, in 1937 and made by the polymerization of adipic acid and 1,6-hexanediamine as shown in Reaction 5.5.1

There are many different kinds of synthetic polymers that are used for a variety of purposes. Some examples in addition to the ones already discussed in this chapter are given in Table 5.1.

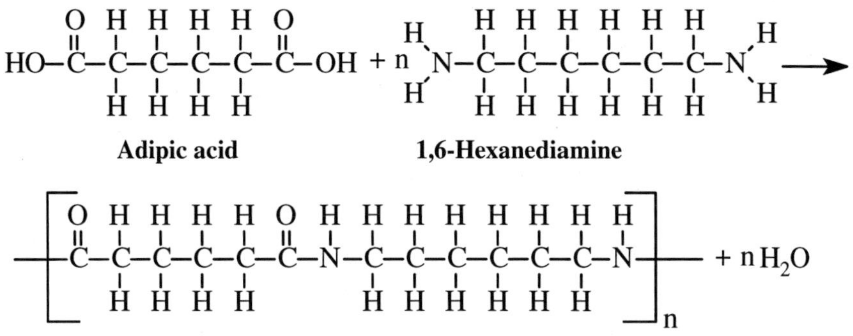

Nylon 66 polymer (5.5.1)

Table 5.1. Some Typical Polymers and the Monomers from which they are Formed

Monomer (polymer formed)	Monomer formula	Polymer formula	Polymer uses
Propylene (polypropylene)	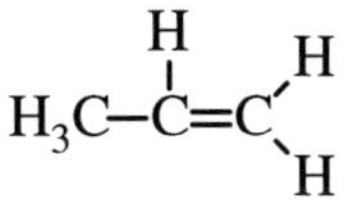	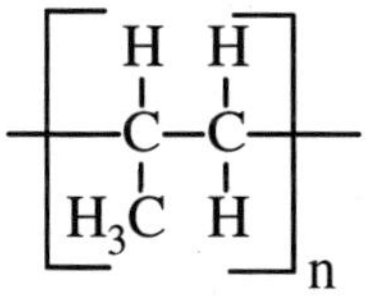	Applications requiring harder plastic, luggage, bottles, outdoor carpet
Vinyl chloride (polyvinyl chloride)	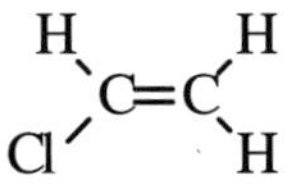	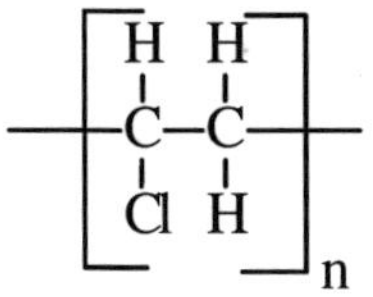	Thin plastic wrap, hose, flooring, PVC pipe
Styrene (polystyrene)	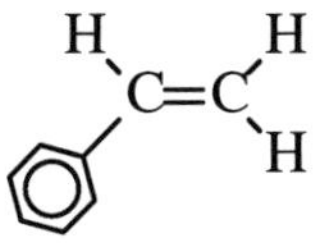	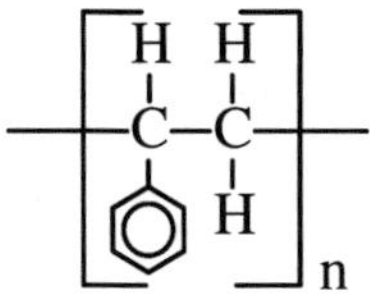	Plastic furniture, plastic cups and dishes, blown to produce Styrofoam plastic products
Acrylonitrile (polyacrylonitrile)	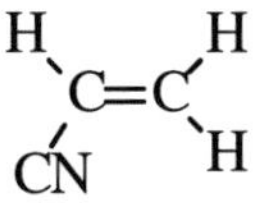	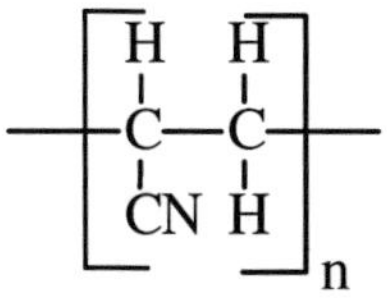	Synthetic fabrics (Orlon, Acrilan, Creslan), acrylic paints
Isoprene (polyisoprene)	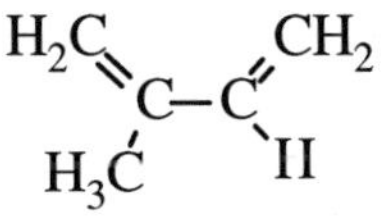	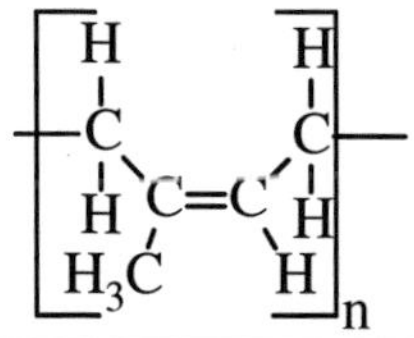	Natural rubber

Polymers and the polymer industry are of particular concern in the practice of green chemistry. One reason is the huge quantities of materials consumed in the manufacture of polymers. In addition to the enormous quantities of ethylene and propylene previously cited in this chapter, the U.S. processes about 1.5 billion kg of acrylonitrile, 5.4 billion kg of styrene, 2.0 billion kg of butadiene, and 1.9 kg of adipic acid (for nylon 66) each year to make polymers containing these monomers. These and similarly large quantities of monomers used to make other polymers place significant demands upon petroleum resources and the energy, materials, and facilities required to make the monomers.

There is a significant potential for the production of pollutants and wastes from monomer processing and polymer manufacture. Some of the materials contained in documented hazardous waste sites are byproducts of polymer manufacture. Monomers are generally volatile organic compounds with a tendency to evaporate into the atmosphere, and this characteristic combined with the presence of reactive C=C bonds tends to make monomer emissions active in the formation of photochemical smog (see Chapter 8). Polymers, including plastics and rubber, pose problems for waste disposal, as well as opportunities and challenges for recycling. On the positive side, improved polymers can provide long-lasting materials that reduce material use and have special applications, such as liners in waste disposal sites that prevent waste leachate migration and liners in lagoons and ditches that prevent water loss. Strong, lightweight polymers are key components of the blades and other structural components of huge wind generators that are making an increased contribution to renewable energy supplies around the world (see Section 6.8).

Some of the environmental and toxicological problems with polymers have arisen from the use of additives to improve polymer performance, durability, degradation resistance and other characteristics. The most notable of these are **plasticizers**, normally blended with plastics to improve flexibility, such as to give polyvinylchloride the flexible characteristics of leather. The plasticizers are not chemically bound as part of the polymer and they leak from the polymer over a period of time, which can result in human exposure and environmental contamination. The most widely used plasticizers are phthalates, esters of phthalic acid as shown by the example of di(2-ethylhexyl) phthalate below. Though not particularly toxic, these compounds are

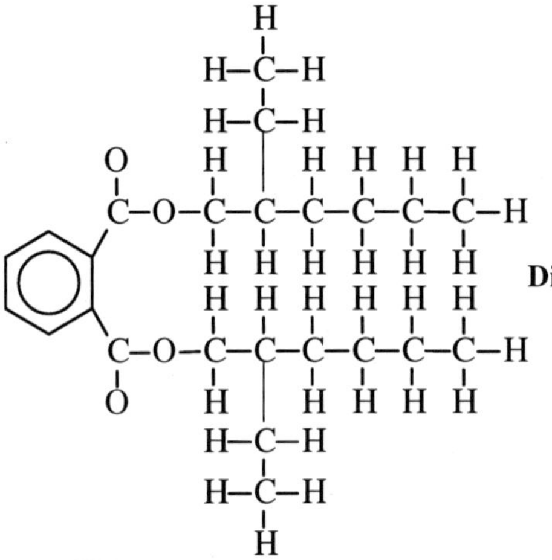

Di(2-ethylhexyl) phthalate

environmentally persistent, resistant to treatment processes, and prone to undergo bioaccumulation. They are found throughout the environment and have been implicated by some toxicologists as possible estrogenic agents that mimic the action of female sex hormone and cause premature sexual development in young female children.

Alternative means of making monomers by green processes have consumed significant effort in the practice of green chemistry and will continue to be an area of high future activity in the green chemical arena. Significant progress has been made in the green synthesis of adipic acid, one of the two monomeric molecules used to make nylon 66 (see Reaction 5.5.1). The conventional synthesis of adipic acid as it has been practiced since the 1940s begins with the addition of H_2 to benzene (see discussion of aromatic compounds in Section 5.2) to produce cyclohexane, (Figure 5.1). Air oxidation over metal catalysts attaches an -OH group to the cyclohexane to make the alcohol, cyclohexanol. This compound is then oxidized with 60% nitric acid, a very severe oxidizing agent, to adipic acid in a process that releases air pollutant nitrous oxide. As a green alternative to the severe chemical conditions required by this synthesis, laboratory studies have shown that genetically engineered *Escherichia coli* bacteria can convert glucose sugar to *cis,cis*-muconic acid, which requires only mild treatment with hydrogen gas to give adipic acid.

5.6. LIFE CHEMICALS

As noted at the beginning of this chapter, living organisms produce a variety of organic chemicals or **biochemicals**. These are considered under the topic of **biochemistry**, the chemistry of life processes. The topic of biochemistry and its relationship to green chemistry is addressed in more detail in later chapters, especially Chapters 9 and 13. At this point, however, it is useful to introduce several classes of the most important kinds of chemicals produced by organisms.

Biochemicals are governed by the same laws of chemistry as are other kinds of organic chemicals. For example, many fats, oils, and waxes produced by organisms are esters, a class of organooxygen compounds described in Section 5.4 and shown in Reaction 5.4.1. Many biochemicals are polymers, which may consist of huge macromolecules. One such material is DNA, the basic molecule of genetic material that may contain billions of atoms per molecule. Proteins are polymers of amino acids. Starch and cellulose are polymers of sugars. DNA, proteins, starch, and cellulose are condensation polymers which release a molecule of H_2O for every monomer molecule bonded to the polymer.

There are four major general groups of kinds of chemical species that are made and used by living organisms. These are proteins, carbohydrates, lipids, and nucleic acids, which are addressed in the sections that follow.

5.7. CARBOHYDRATES

Carbohydrates are biomolecules consisting of carbon, hydrogen, and oxygen having the approximate simple formula CH_2O. One of the most common carbohydrates is the simple sugar glucose:

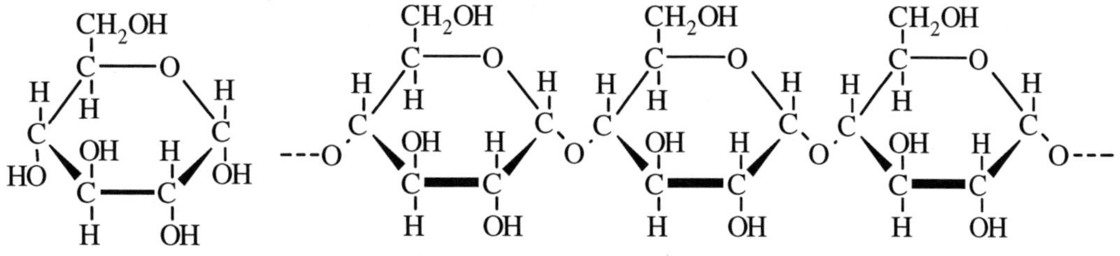

Glucose **Three units of the starch macromolecule**

Figure 5.7. Glucose, a monosaccharide, or simple sugar, and a segment of the starch molecule, which is formed when glucose molecules polymerize with the elimination of one H_2O molecule per glucose monomer.

Units of glucose and other simple sugars called **monosaccharides** join together in chains with the loss of a water molecule for each linkage to produce macromolecular **polysaccharides**. These include **starch** and **cellulose** in plants and starch-like glycogen in animals.

Glucose carbohydrate is the biological material generated from water and carbon dioxide when solar energy in sunlight is utilized in photosynthesis. The overall reaction is

$$6CO_2 + 6H_2O \rightarrow C_6H_{12}O_6 + 6O_2 \tag{5.7.1}$$

This is obviously an extremely important reaction because it is the one by which inorganic molecules are converted to high-energy carbohydrate molecules that are in turn converted to the vast number of biomolecules that comprise living systems. There are other simple sugars, including fructose, mannose, and galactose, that have the same simple formula as glucose, $C_6H_{12}O_6$, but which must be converted to glucose before being utilized by organisms for energy. Common table sugar, sucrose, $C_{12}H_{22}O_{11}$, consists of a molecule of glucose and one of fructose linked together (with the loss of a water molecule) and, because of these two simple sugars of which it is composed, is called a **disaccharide**.

Starch molecules, which may consist of several hundred glucose units joined together, are readily broken down by organisms to produce simple sugars used for energy and to produce biomass. For example, humans readily digest starch in potatoes or bread to produce glucose used for energy (or to make fat tissue). Another chemically very similar polysaccharide consisting of even more glucose units is cellulose, which comprises much of the biomass of plant cells. We and other animals cannot digest cellulose directly to use as a food source but some bacteria and fungi do so readily. Such bacteria living in the stomachs of termites and ruminant animals (cattle, sheep, moose) break down cellulose to small molecules that are converted to molecules that can be absorbed through the digestive systems of animals and utilized as food.

Carbohydrates are potentially very important in green chemistry. For one thing, they are a concentrated form of organic energy that enables capture of solar energy by photosynthetic processes. Carbohydrates can be utilized directly for energy or fermented to produce ethanol, C_2H_6O, a combustible alcohol that is added to gasoline or can even be used in place of gasoline. Secondly, carbohydrates represent a rich source of organic raw material that can be converted to other organic molecules to make plastics and other useful materials. The potential for the biosynthesis from glucose of adipic acid used to make nylon was discussed in Section 5.5.

5.8. PROTEINS

Proteins are macromolecules that are composed of nitrogen, carbon, hydrogen, and oxygen along with smaller quantities of sulfur. The small molecules of which proteins are made are composed of 20 naturally occurring **amino acids**. The simplest of these, glycine, is shown in the first structure in Figure 5.8, along with two other amino acids. As shown in Figure 5.8, amino acids join together with the loss of a molecule of H_2O for each linkage formed. The three amino acids in Figure 5.8 are shown linked together as they would be in a protein in the bottom structure in the figure. Many hundreds of amino acid units may be present in a protein molecule.

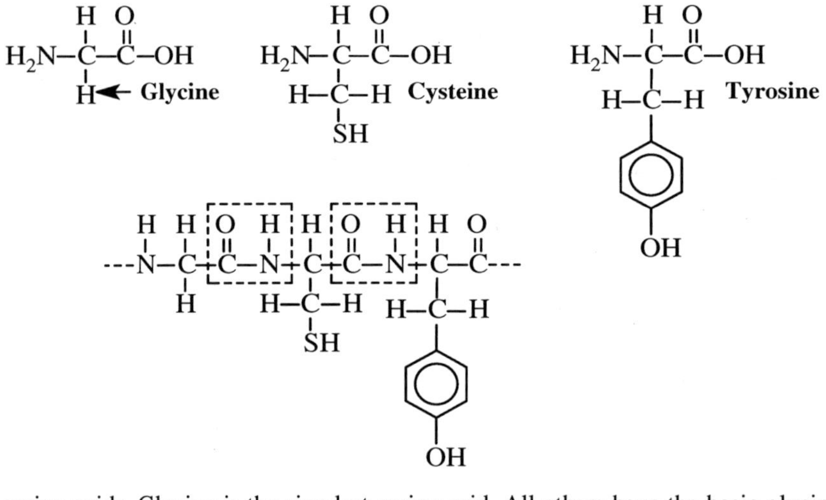

Figure 5.8. Three amino acids. Glycine is the simplest amino acid. All others have the basic glycine structure except that different groups are substituted for the H designated in glycine by an arrow. The lower structure shows these three amino acids linked together in a macromolecule chain composing a protein. For each linkage, one molecule of H_2O is lost. The peptide linkages holding amino acids together in proteins are outlined by dashed rectangles.

The three-dimensional structures of protein molecules are of the utmost importance and largely determine what the proteins do in living systems and how they are recognized by other biomolecules. Enzymes, special proteins that act as catalysts to enable biochemical reactions to occur, recognize the substrates upon which they act by the complementary shapes of the enzyme molecules and substrate molecule. There are several levels of protein structure. The first of these is determined by the order of amino acids in the protein macromolecule. Folding of protein molecules and pairing of two different protein molecules further determine structure. The loss of protein structure, called **denaturation**, can be very damaging to proteins and to the organism in which they are contained.

Two major kinds of proteins are tough **fibrous proteins** that compose hair, tendons, muscles, feathers, and silk, and spherical or oblong-shaped **globular proteins**, such as hemoglobin in blood or the proteins that comprise enzymes. Proteins serve many functions. These include **nutrient proteins**, such as casein in milk, **structural proteins**, such as collagen in tendons, **contractile proteins**, such as those in muscle, and **regulatory proteins**, such as insulin, that regulate biochemical processes.

5.9. LIPIDS: FATS, OILS, AND HORMONES

Lipids differ from most other kinds of biomolecules in that they are repelled by water. Lipids can be extracted from biological matter by organic solvents, such as diethyl ether or toluene. Recall that proteins and carbohydrates are distinguished largely by chemical structures (chemically similar amino acids for proteins and chemically similar simple sugar units for carbohydrates). However, lipids have a variety of chemical structures that share the common physical characteristic of solubility in organic solvents. Many of the commonly encountered lipid fats and oils are esters of glycerol alcohol, $CH_2(OH)CH(OH)\ CH_2(OH)$, and long-chain carboxylic acids (fatty acids), such as stearic acid, $CH_3(CH_2)_{16}CO_2H$. The glycerol molecule has three -OH groups to each of which a fatty acid molecule may be joined through the carboxylic acid group with the loss of a water molecule for each linkage that is formed. Figure 5.9 shows a fat molecule formed from three stearic acid molecules and a glycerol molecule. Such a molecule is one of many possible **triglycerides**. Also shown in this figure is cetyl palmitate, the major ingredient of spermaceti wax extracted sperm whale blubber and used in some cosmetics and pharmaceutical preparations. Cholesterol shown in Figure 5.9 is one of several important lipid **steroids**, which share the ring structure composed of rings of 5 and 6 carbon atoms shown in the figure for cholesterol. Steroids act as **hormones**, chemical messengers that convey information from one part of an organism to another.

Although the structures shown in Figure 5.9 are diverse, they all share a common characteristic. This is the preponderance of hydrocarbon chains and rings, so that lipid molecules largely resemble hydrocarbons. This is the characteristic that makes lipids soluble in organic solvents.

Lipids are important in green chemistry for several reasons. Lipids are very much involved with toxic substances, the generation and use of which are always important in green chemistry. Poorly biodegradable substances, particularly organochlorine compounds, that are always an essential consideration in green chemistry, tend to accumulate in lipids in living organisms, a process called bioaccumulation. Lipids can be valuable raw materials and fuels. Therefore, the development and cultivation of plants that produce oils and other lipids is a major possible route to the production of renewable resources.

5.10. NUCLEIC ACIDS

Nucleic acids are biological macromolecules that store and pass on the genetic information that organisms need to reproduce and synthesize proteins. The two major kinds of nucleic acids are **deoxyribonucleic acid, DNA**, which basically stays in place in the cell nucleus of an organism and **ribonucleic acid, RNA**, which is spun off from DNA and functions throughout a cell. Molecules of nucleic acids contain three basic kinds of materials. The first of these is a simple sugar, 2-deoxy-β-D-ribofuranose (deoxyribose) contained in DNA and β-D-ribofuranose (ribose) contained in RNA. The second major kind of ingredient in nucleic acids consists of nitrogen-containing bases: cytosine, adenine,and guanine, which occur in both DNA and RNA, thymine, which occurs only in DNA, and uracil, which occurs only in RNA. The third consituent of both DNA and RNA is inorganic phosphate, PO_4^{3-}.These three kinds of substances occur as repeating units called **nucleotides** each consisting of one of the simple sugars, one of the nitrogen-containing bases, and phosphate joined together in astoundingly long chains in the nucleic acid polymer as shown in Figure 5.10.

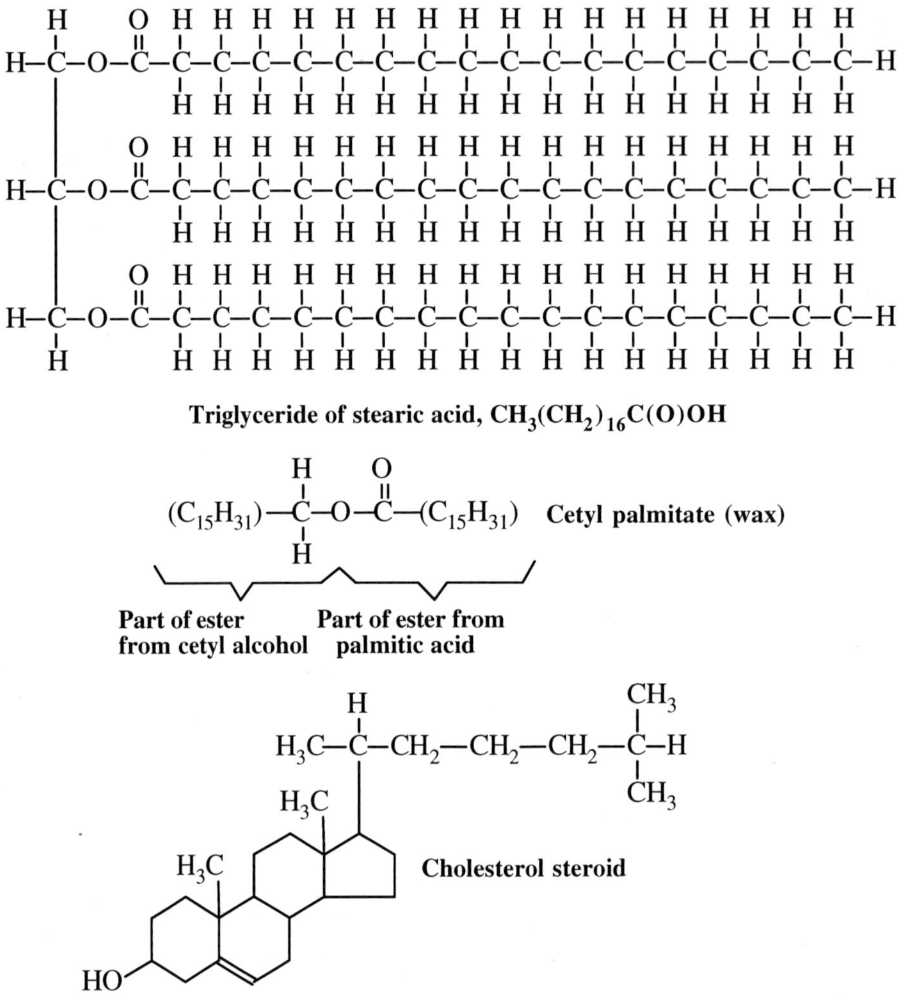

Figure 5.9. Three examples of lipids formed in biological systems. Note that a line structure is used to show the ring structure of cholesterol. The hydrocarbon-like nature of these compounds which makes them soluble in organic compounds is obvious.

The remarkable way in which DNA operates to pass on genetic information and perform other functions essential for life is the result of the structure of the DNA molecule. In 1953, James D. Watson, and Francis Crick deduced that DNA consisted of two strands of material counterwound around each other in a structure known as an α-helix, an amazing bit of insight that earned Watson and Crick the Nobel Prize in 1962. These strands are held together by hydrogen bonds between complementary nitrogenous bases. Taken apart, the two strands resynthesize complementary strands, a process that occurs during reproduction of cells in living organisms. In directing protein synthesis, DNA becomes partially unravelled and generates a complementary strand of material in the form of RNA, which in turn directs protein synthesis in the cell.

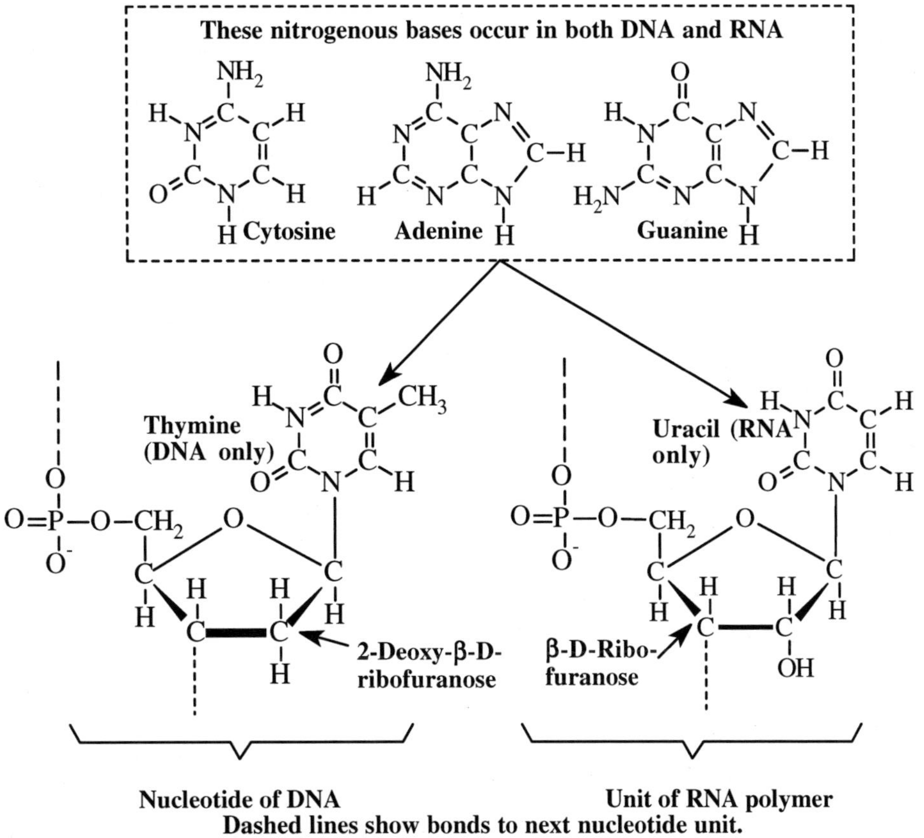

Figure 5.10. Basic units of nucleic acid polymers. These units act as a code in directing reproduction and other activities of organisms.

Nucleic acids have an enormous, as of yet largely unrealized, potential in the development of green chemistry. Much of the hazard of many chemical substances results from potential effects of these substances upon DNA. Of most concern is the ability of some substances to alter DNA and cause uncontrolled cell replication characteristic of cancer.

In recent years humans have developed the ability to alter DNA so that organisms synthesize proteins and perform other metabolic feats that would otherwise be impossible. Such alteration of DNA is commonly known as **genetic engineering** and **recombinant DNA** technology. Organisms produced by recombinant DNA techniques that contain DNA from other organisms are called **transgenic organisms**. The potential of this technology to produce crops with unique characteristics, to synthesize pharmaceuticals, and to make a variety of useful raw materials as renewable feedstocks is discussed in later chapters.

QUESTIONS AND PROBLEMS

1. What are two major reactions of alkanes?

2. What is the difference between molecular formulas and structural formulas of organic compounds?

3. What is the difference between ethane and the ethyl group?

4. What is the structural formula of 3-ethyl-2,3-dimethylpentane?

5. What is a type of reaction that is possible with alkenes, but not with alkanes?

6. What is represented by the structure below?

7. Suggest a name for the compound below, which is derived from the hydrocarbon toluene:

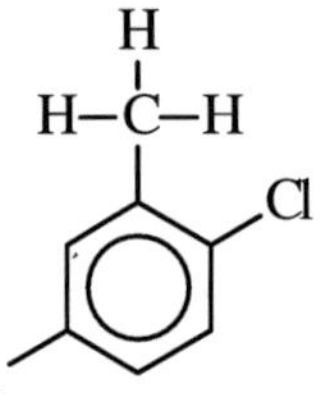

8. What is a health concern with the aromatic compound below?

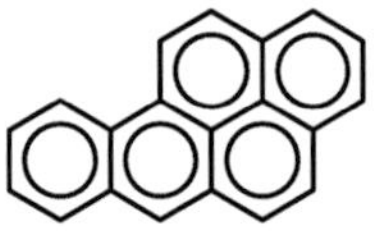

9. What do the groups of atoms outlined by dashed lines represent in the structure below?

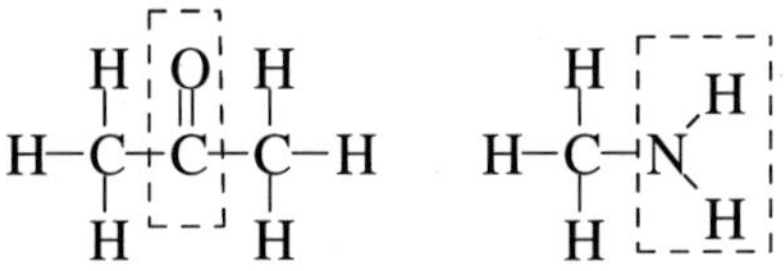

10. Based upon the structures shown in Figure 5.4, what are the similarities and differences between organic oxides and ethers.

11. What are 3 separate kinds of groups characteristic of organonitrogen compounds?

12. What is a class of organochlorine compounds consisting of many different kinds of molecules that is noted for environmental persistence?

13. What is a notable characteristic of organosulfur thiols?

14. What is a particularly toxic organophosphorus compound? What is a biochemical molecule containing phosphorus?

15. What are polymers and why are they important?

16. What are the basic building blocks of proteins, and how do they determine the primary structure of proteins?

17. Write the complete structural formulas correponding to each of the line structures on the right.

(A) 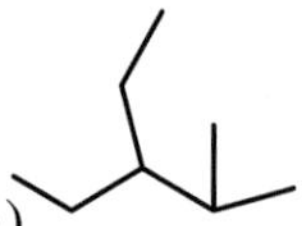 (B) 

18. What is meant by denaturation of proteins? Is it bad?

19. What are some major kinds of proteins?

20. What is the approximate simple formula of carbohydrates?

21. Fill in the blanks of the following pertaining to carbohydrates: Glucose is an example of a ________________, sucrose is a ____________________________, and starch and cellulose are both ____________________________.

22. How are lipids defined and how does this definition differ from that of other biomolecules?

23. What does DNA stand for? What are 6 specific ingredients of DNA?

24. Although lipids are defined by a physical property that they all share, what is a common characteristic of lipid structure?

25. From the structures given in Figure 5.9, what kind of functional group seems to be common in lipids?

26. How does the compatibility of lipids with organic substances, such as organochlorine compounds, influence the environmental behavior of such compounds?

27. What distinguishes RNA from DNA? How are they similar?

28. What are the three constituents of all basic units of nucleic acids?

6. ENERGY RELATIONSHIPS

6.1. ENERGY

Energy is the capacity to do work or to transfer **heat** (the form of energy that flows from a warmer to a colder object). A farm tractor working in a field illustrates the definition of energy and several forms of energy. **Chemical energy** in the form of petroleum hydrocarbons is used to fuel the tractor's diesel engine. In the engine the hydrocarbons combine with oxygen from air,

$$2C_{16}H_{34} + 49O_2 \rightarrow 32CO_2 + 34H_2O + \text{heat energy} \qquad (6.1.1)$$

to produce **heat energy**. As the hot gases in the engine's cylinders push the pistons down, some of this heat energy is converted to **mechanical energy**, which is transferred by the engine crankshaft, gears, and axle to propel the tractor forward. A plow or other implement attached to the tractor moves soil.

Figure 6.1. A farm tractor using energy to do work in tilling soil. Chemical energy in the diesel fuel used to run the tractor is converted to heat energy and then to mechanical energy in the tractor's engine and the mechanical energy is used to move the tractor and till soil.

The standard unit of energy is the **joule**, abbreviated **J**. A total of 4.184 J of heat energy will raise the temperature of 1 g of liquid water by 1˚C. This amount of heat is equal to 1 **calorie** of energy (1 cal = 4.184 J), the unit of energy formerly used in scientific work. A joule is a small unit, and the kilojoule, kJ, equal to 1000 J is widely used in describing chemical processes. The "calorie" commonly used to express the energy value of food (and its potential to produce fat) is actually a kilocalorie, kcal, equal to 1000 cal.

The science that deals with energy in its various forms and with work is **thermodynamics**. There are some important laws of thermodynamics. The **first law of thermodynamics** states that energy is neither created nor destroyed. This law is also known as the **law of conservation of energy**. As an example of the application of this law, consider Figure 1. The energy associated with cultivating the land enters the system as chemical energy in the form of diesel fuel, and the oxygen from the air required for its combustion. This is a valuable form of concentrated chemical energy that can be used to propel a tractor or truck, in a turbine attached to a generator for the generation of electrical energy, or as a fuel to generate heat in an oil-fired furnace. The fuel is burned in the tractor's engine, and more than half of its energy is dissipated as heat to the surroundings. The rest is used to move the tractor and dirt. The energy originally contained in a concentrated useful form in the diesel fuel is not destroyed, but it is dissipated in a dilute form, mostly to warm the surroundings very slightly. The energy that was originally present as a very useful form in the diesel fuel has not been destroyed, but it has been dispersed in a form that is no longer of practical use.

The first law of thermodynamics must always be kept in mind in the practice of green chemistry. The best practice of green chemistry and, indeed, of all environmental science, requires the most efficient use of energy as it goes through a system. The availability of energy is often the limiting factor in using and recycling materials efficiently. If enough energy is available, almost anything can be done. For example, many water-deficient areas of the world, such as the northern coast of Africa, Israel, and Saudi Arabia are adjacent to limitless supplies of ocean water. If sufficient energy were available, seawater could simply be distilled to provide an abundant supply. But the consumption of energy would be prohibitive. It is not only a question of availability of energy, but rather of energy that can be used in a way that does not do unacceptable environmental harm. Many countries, including the U.S. have vast deposits of coal. But to rely on coal for all the energy that various societies perceive that they need in the future would place huge amounts of carbon dioxide in Earth's atmosphere almost certainly resulting in intolerable global warming.

6.2. RADIANT ENERGY FROM THE SUN

The sun is the ultimate source of most of the energy that we use. How much energy does Earth receive from the sun? If the sun were to abruptly "go out" (not to worry, it won't happen for another billion years or so) we would quickly find out, because within hours Earth would become a frozen rock in space. The rate of energy transfer from the sun is expressed by the **solar flux**, which is the rate at which solar energy is transmitted through space at Earth's distance from the sun. This rate of energy transfer (power) is 1.34×10^3 watts/m^2. What this means, as illustrated in Figure 6.2, is that a 1 square meter area (a square just over 3 feet to the side) with the sun shining perpendicular to it just above Earth's atmosphere receives energy at a rate of 1,340 watts. A watt is a measure of power, that is, energy per unit time. A power level of 1,340 watts is a lot of power. In the form of electrical energy, it would easily power an electric iron or toaster and would provide the energy equivalent to 13 100-watt incandescent bulbs plus a 40-watt bulb.

Where does the sun get all this energy? It gets it by consuming itself in a gigantic thermonuclear fire, the same basic process of converting mass to energy that gives a "hydrogen bomb" its enormous destructive force. The fuel for the sun is ordinary hydrogen. But the energy-yielding reaction is not an ordinary chemical reaction. Instead, it is a **nuclear reaction** in which the nuclei of 4 hydrogen atoms fuse together to produce the nucleus of a helium atom of mass number 4, plus 2 positrons, subatomic particles with the same mass as the electron, but with a positive, instead of a negative, charge. There is a net loss of mass in the process (in nuclear reactions mass **can** change) and this loss translates into an enormous amount of energy. The fusion of only 1 gram of hydrogen releases as much energy as the heat from burning about 20 tons of coal. Using superscripts to express mass number and subscripts for charge, the thermonuclear fusion of hydrogen in the sun may be expressed as follows:

$$4\,{}^{1}_{1}\text{H} \rightarrow {}^{4}_{2}\text{He} + 2\,{}^{0}_{+1}\text{e} + \text{energy} \qquad (6.2.1)$$

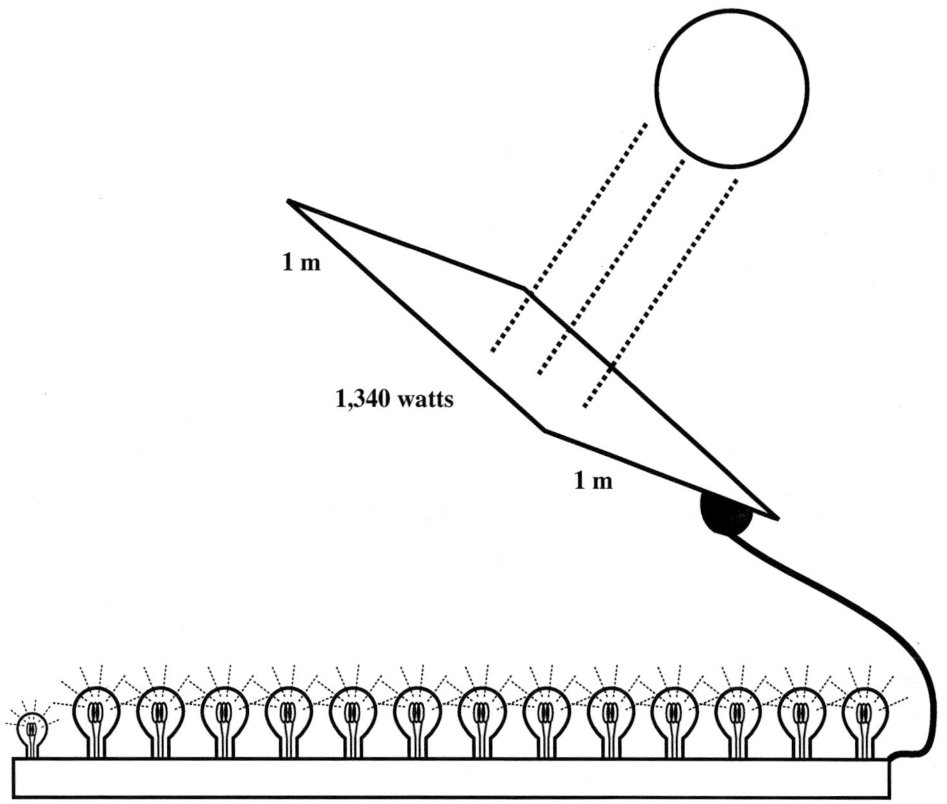

Figure 6.2. The solar flux at the distance of Earth from the sun is 1,340 watts/m^2. This amount of power is equivalent to that consumed by 13 100-watt incandescent light bulbs plus a 40-watt bulb.

Electromagnetic Radiation

The space between the sun and Earth is mostly just that — empty space. So how does the enormous amount of energy that the sun generates get carried through space to Earth? It does so as **electromagnetic radiation**, which includes ultraviolet radiation, visible light, infrared radiation, microwaves, and radio waves. Most of the energy that actually reaches Earth's surface does so as visible light and infrared radiation.

All electromagnetic radiation moves at the same speed through the vacuum of space, a very fast 3.00×10^8 m/s (meters per second). As shown in Figure 6.3, the waves of electromagnetic radiation have characteristics of **wavelength** (λ, Greek lambda), amplitude, and **frequency** (ν, Greek "nu"). These two parameters are related by the equation,

$$\nu\lambda = c \qquad (6.2.2)$$

where ν is in units of cycles per second (s^{-1}, a unit called the **hertz**, Hz), λ is in meters (m), and c is the speed of light in m s^{-1}. The wavelength is the distance required for one complete cycle and ν is the number of cycles per unit time.

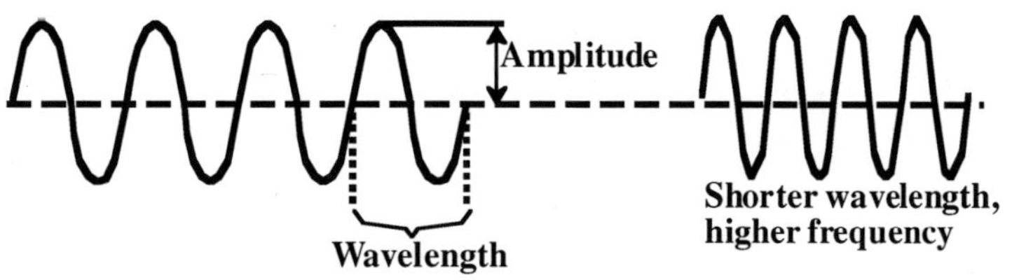

Figure 6.3. Important characteristics of electromagnetic radiation.

Energy is associated with electromagnetic radiation. In addition to its wave character, such radiation can be regarded as particles or **quanta**. According to the **quantum theory** of electromagnetic radiation, electromagnetic radiation can be absorbed or emitted only in discrete quanta, also called **photons**. A specific energy, E, is associated with each photon and is related to the frequency, ν, of the associated electromagnetic radiation by

$$E = h\nu \quad (6.2.3)$$

where h is Planck's constant, 6.63×10^{-34} J-s (joule × second). An important consequence of this relationship is that *the higher the frequency (the shorter the wavelength) of electromagnetic radiation, the more energetic the photon associated with it.*

There are some important environmental consequences associated with the energy of photons, as well as some implications for green chemistry. As noted above, the sun's energy reaches Earth largely in the visible wavelength region, a very narrow band of electromagnetic radiation having λ between 400 and 700 nanometers (nm). A significant amount of the incoming energy is in the **infrared region** above 700 nm. Some of the inbound electromagnetic radiation is reflected directly from the atmosphere, clouds in the atmosphere, and even Earth's surface (snow is especially effective at reflecting visible light). But most of the outbound energy is in the longer wavelength infrared region. As discussed in Chapter 8, so-called greenhouse gases, such as carbon dioxide and methane in the atmosphere, reabsorb this radiation and delay its eventual exit from Earth. This is a good thing because it is what keeps the atmosphere tolerably warm. But if it happens to excess, it may result in global warming, probably the most challenging environmental problem of our time.

Electromagnetic radiation with wavelengths shorter than 400 nm cannot be seen and is called **ultraviolet radiation** (the wavelengths of X-rays and gamma rays are even shorter than those of ultraviolet). Infrared and visible radiation impinging on matter mainly serve to warm it up. But, ultraviolet photons are sufficiently energetic that they can "excite" the valence electrons of molecules to higher levels. This can result in effects such as splitting molecules apart and cause chemical reactions to occur. Such reactions are **photochemical reactions**. They are responsible for effects such as the formation and destruction of ozone in the stratosphere and the formation of noxious photochemical smog at ground level.

Exposed to the energetic ultraviolet radiation from the sun, we would quickly perish because of its adverse effects on flesh. Fortunately, the atmosphere above us, thin though it is, absorbs the most damaging ultraviolet radiation, although a little bit gets through, and it is a good idea to minimize direct exposure to sunlight.

Direct and Indirect Solar Energy

From the discussion above, it is seen that a lot of energy comes from the sun. Most of it is absorbed by the atmosphere, but a significant fraction reaches Earth's surface directly. We certainly use that energy because it keeps us and other living organisms warm enough to sustain life. Photovoltaic cells that convert solar energy directly to electricity, enable use of solar energy as a power source.

Living organisms use solar energy. Chlorophyll in plants capture the energy of photons of visible light and use it to perform photosynthesis,

$$6CO_2 + 6H_2O \xrightarrow{h\nu} C_6H_{12}O_6 + 6O_2 \quad (6.2.4)$$

in which carbon dioxide from the atmosphere is combined with water to produce glucose carbohydrate, $C_6H_{12}O_6$, and molecular oxygen. The glucose is converted by plants and other organisms into cellulose and other biomolecules making up biomass. Whereas there is no usable energy to be obtained from the carbon dioxide and water reactants of the photosynthesis reaction, there is a lot of chemical energy to be had from the glucose and the biomolecules made from it. **Aerobic respiration**, essentially the reverse of photosynthesis,

$$C_6H_{12}O_6 + 6O_2 \rightarrow 6CO_2 + 6H_2O + \text{energy} \qquad (6.2.5)$$

is the process that we and most other organisms use to obtain energy. So photosynthesis enables the capture of energy from photons of visible light, a form of electromagnetic energy, and its conversion to chemical energy.

Even the tractor shown in Figure 6.1 runs indirectly on solar energy. That is because in eons past, plants and microscopic photosynthetic organisms (phytoplankton) performed photosynthesis to produce large quantities of biomass. Subjected to high temperatures and pressures in the absence of air underground, this biomass got converted to oils (petroleum), coal, and organic matter closely associated with rocks called kerogen, the source of shale oil. It is these fossil fuels upon which the world depends for most of its energy today.

7.3. STORAGE AND RELEASE OF ENERGY BY CHEMICALS

Consider the burner on a kitchen range fueled by natural gas. The flame is obviously hot; something is going on that is releasing heat energy. The flame is also giving off light energy, probably as a light blue glow. A chemical reaction is taking place as the methane in the natural gas combines with oxygen in the air,

$$CH_4 + 2O_2 \rightarrow CO_2 + 2H_2O + \text{energy} \qquad (6.3.1)$$

to produce carbon dioxide and water. Most of the energy released during this chemical reaction is released as heat, and a little bit as light. It is reasonable to assume that the methane and oxygen contain stored energy as **chemical potential energy** and that it is released in producing carbon dioxide and water. Common sense tells us that it would be hard to get heat energy out of either of the products. They certainly won't burn! Water is used to put out fires, and carbon dioxide is even used in fire extinguishers.

The potential energy contained in chemical species is contained in the chemical bonds of the molecules that are involved in the chemical reaction. Figure 6.4 shows the kinds of bonds involved in methane, elemental oxygen, carbon dioxide, and water and the energy contained in each. The bond energies are in units of the number of kilojoules (kJ) required to break a mole (6.02×0^{23}) of the bonds (kJ/mol). The same amount of energy is released when a mole of a bond is formed. By convention, energy put into a system is given a positive sign and energy released is denoted by a negative sign.

To calculate the energy change when a mole of methane reacts with oxygen as shown in Reaction 6.3.1, the difference is taken between the sum of the energies of the bonds in the products and the sum of the energies of the bonds in the reactants. Examination of Reaction 6.3.1 and Figure 6.4 shows the following total bond energies in the products:

$$1 \text{ mol } CO_2 \times \frac{2 \text{ mol C=O}}{\text{mol } CO_2} \times \frac{799 \text{ kJ}}{\text{mol C=O}} = 1598 \text{ kJ}$$

$$2 \text{ mol } H_2O \times \frac{2 \text{ mol O-H}}{\text{mol } H_2O} \times \frac{459 \text{ kJ}}{\text{mol O-H}} = 1836 \text{ kJ}$$

Total bond energy in products = 1598 kJ + 1836 kJ = 3434 kJ

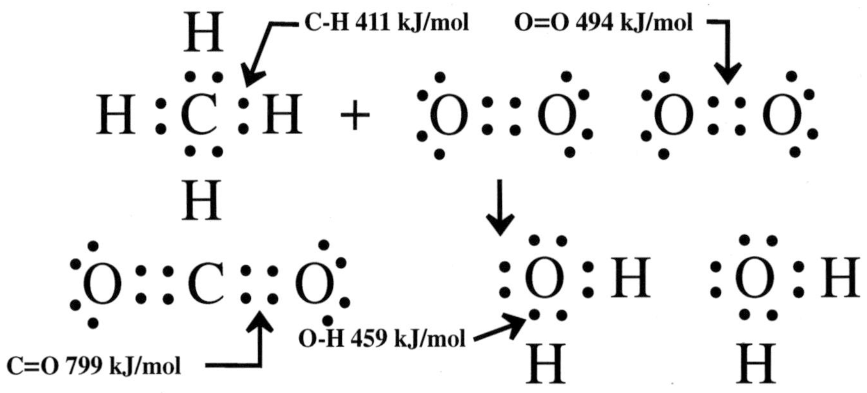

Figure 6.4. Bonds and their energies in the chemical species involved when methane burn in oxygen to produce carbon dioxide and water.

A similar calculation gives the total bond energies in the reactants:

$$1 \text{ mol } CH_4 \times \frac{4 \text{ mol C-H}}{\text{mol } CH_4} \times \frac{411 \text{ kJ}}{\text{mol C-H}} = 1644 \text{ kJ}$$

$$2 \text{ mol } O_2 \times \frac{1 \text{ mol O=O}}{\text{mol } O_2} \times \frac{494 \text{ kJ}}{\text{mol O=O}} = 988 \text{ kJ}$$

Total bond energy in reactants = 1644 kJ + 988 kJ = 2632

The difference in bond energies between products and reactants is

3434 kJ - 2632 kJ = 802 kJ

This calculation states that, based upon considerations of bond energy, alone, the energy released when 1 mole of CH_4 reacts with 2 moles of O_2 to produce 1 mole of CO_2 and 2 moles of H_2O, is 802 kJ. This is an **exothermic** reaction in which heat energy is *released*, so it is denoted as -802 kJ. This value is close to the value that would be obtained by experimentally measuring the heat energy released by the reaction, assuming all the reactants and products were in the gas phase (a significant amount of heat energy is released when vapor-phase water condenses to liquid). For the most part, therefore, the amount of heat energy released in a chemical reaction, and the amount of potential chemical energy contained in the reactants is equal to the difference between the total bond energies of the products and those of the reactants.

6.4. ENERGY SOURCES

Two centuries ago, before the steam engine became commonplace, virtually all the energy used in the world was from biomass sources. Energy for heating came from burning wood. Other than walking, people and goods moved on land mostly by means of horses and oxen. Cultivation of soil and other agricultural activities were powered largely by animals with a significant contribution from people, themselves. The energy required for people and for beasts of burden was provided by food, a form of biomass. A significant amount of energy for transportation was from wind, which drove sailing boats and ships. In any case, most of the energy was from renewable sources.

The use of coal for energy grew spectacularly during the 1800s and by the end of that century coal had become the predominant source of energy in the United States, England, Europe, and other countries that had readily accessible coal resources. So there was a major shift from renewable biomass energy sources to coal, a depletable resource that had to be dug from the ground. By 1900, petroleum had become a significant source of energy and by 1950 had surpassed coal as a source of energy in the U. S. Also by 1950 natural gas had become a major energy source lagging behind petroleum in its rate of development. Hydroelectric power had become a significant source of energy worldwide by 1900, and retained a significant share of energy production through the 1900s. By around 1975, nuclear energy had become a significant source of electricity and has maintained a share of several percent worldwide until the present. Although just a "blip" in the total energy picture, miscellaneous sources including geothermal and, more recently, solar and wind energy now make contributions to total energy supply. Biomass still contributes a little to the total of the sources of energy used.

Figure 6.5 shows U.S. and world energy sources used annually as of the year 2000. The predominance of petroleum and natural gas, both in the U.S. and globally, are obvious. With coal, these **fossil fuel** sources account for the vast majority of energy used. There are two major problems with this tremendous reliance on fossil fuels. One of these problems is that fossil fuel sources are running out. There is a lot of debate about how fast these resources are being depleted and when they will be gone (or become so costly that they are impossible to afford). Indications are that world petroleum production has peaked around the year 2000, give or take several years. Resources of coal are considerably more abundant, but this gets into the second major problem with fossil fuels, their contribution to greenhouse gas atmospheric carbon dioxide. As discussed in Chapter 8, atmospheric carbon dioxide levels have grown by about 50% during the last 150 years. Now at a level of about 370 parts per million by volume in the atmosphere, these levels are increasing by about 1 part per million per year. Most reputable models for the effect this will have upon global temperatures project an increase of several degrees C within the next several decades as atmospheric CO_2 levels increase. Although such a temperature increase seems small, its effects upon global climate would be profound and perhaps even catastrophic.

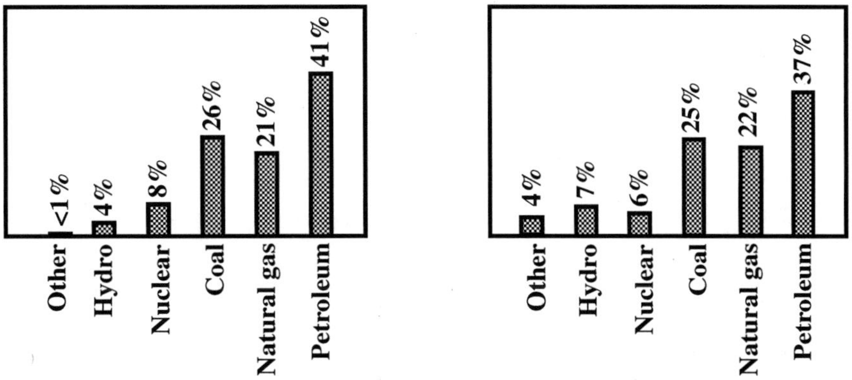

Figure 6.5. U.S. (left) and world (right) sources of energy. Percentages of total rounded to the nearest 1%.

Fossil fuels vary in their contribution to greenhouse gas carbon dioxide; the lower the content of H, the greater the contribution to atmospheric carbon dioxide. Consider the combustion of an atom of carbon in natural gas, liquid petroleum, and coal. Natural gas consists of molecules of methane, CH_4, so the combustion of an atom of carbon in methane can be represented by the chemical reaction

$$CH_4 + 2O_2 \rightarrow CO_2 + 2H_2O + \text{energy} \tag{6.4.1}$$

This reaction produces 1 molecule of CO_2, but it also involves the combustion of 2 hydrocarbon-bound H atoms, which also produces a lot of energy. In the case of liquid petroleum, such as gasoline or diesel fuel, the approximate ratio of C to H atoms is 1:2, rather than 1:4 as in methane. So the combustion of a petroleum hydrocarbon containing 1 C atom can be represented as

$$CH_2 + {}^3/_2O_2 \rightarrow CO_2 + H_2O + \text{energy} \quad (6.4.2)$$

In this case only half as much hydrocarbon-bound H is burned, so significantly less energy is produced per C atom than in the combustion of natural gas. Coal is even worse. Despite its being a black solid that would lead one to believe coal is pure carbon, it is in fact a hydrocarbon with an approximate simple formula of $CH_{0.8}$. So the combustion of an atom of carbon in coal can be represented in a somewhat cumbersome manner as

$$CH_{0.8} + 1.2O_2 \rightarrow CO_2 + 0.4H_2O + \text{energy} \quad (6.4.3)$$

Much less hydrocarbon-bound hydrogen is available to burn, so the amount of carbon dioxide emitted to the atmosphere per unit energy produced from coal is higher than with petroleum and much higher than with natural gas.

The problem with world energy supplies is clear; the solution is not so obvious. Industrialized societies have become dependent upon fossil carbon sources for energy and will have to develop alternatives. It will not be an easy transition. Several of the possible alternatives are discussed later in this chapter.

6.5. CONVERSIONS BETWEEN FORMS OF ENERGY

The most abundant sources of energy are usually not directly useful and must be converted to other forms. Therefore, much of what is done with energy involves converting it from one form to another. As an example, the nuclear energy that can be extracted from a few pounds of natural uranium is enormous. But in order to get any benefit from it, the uranium must first be enriched in the isotope whose nucleus can undergo fission (split) to release the energy, the enriched uranium must be placed in a nuclear reactor where fission occurs, converting the nuclear energy to heat, this heat must be used to produce steam, the steam must be run through a turbine to produce mechanical energy, and the turbine must be coupled to a generator to convert its mechanical energy to electrical energy. The various conversions of energy from one form to another occur with different efficiencies. The successful practice of industrial ecology tries to maximize the efficiencies of energy conversion.

Figure 6.6 illustrates major forms of energy and conversions between them. Examination of the different percentage efficiencies for energy conversion given in Figure 6.6 shows differences ranging from very low to almost 100%. But they point to areas in which improvements may be sought. For example, photosynthesis is less than about 0.5% efficient in converting light energy to chemical energy. Despite this dismal figure, photosynthesis has generated the fossil fuels from which industrialized societies now get their energy and provides a significant fraction of energy in areas where wood and agricultural wastes are used. The intriguing possibility is suggested that genetically modified plants (see Chapters 10, 11, and 12) may be developed with much higher photochemical efficiencies, leading to greatly increased use of renewable biomass as an energy source. The poor efficiency of conversion of electricity to light in the incandescent light bulb points to the need to replace these wasteful devices with fluorescent bulbs that are 5 or 6 times more efficient.

The most common kind of energy conversion carried out in the anthrosphere is the conversion of heat, produced by chemical combustion processes, to mechanical energy used to propel a vehicle or run an electrical generator. This occurs, for example, when gasoline in a gasoline engine burns, generating hot gases that move pistons in the engine connected to a crankshaft that converts the up-and-down movement of the piston to rotary motion that drives a vehicle's wheels. It also occurs when hot steam generated at high pressure in a boiler flows through a turbine connected directly to an electrical generator. Unfortunately, the laws of thermodynamics dictate that the conversion of heat to mechanical energy is always much less than 100% efficient. The Carnot equation,

$$\text{Percent efficiency} = \frac{T_1 - T_2}{T_1} \times 100 \qquad (6.5.1)$$

states that the percent efficiency is a function of the inlet temperature (for example, of steam), T_1, and the outlet temperature, T_2, both expressed in Kelvin (°C + 273). Consider a steam turbine as shown in Figure 6.7. If the inlet temperature is 850 K and the outlet temperature is 330 K, substitution into the Carnot equation gives a maximum theoretical efficiency of 61%. An inability to introduce all the steam at the highest temperature combined with friction losses of energy reduce the energy conversion efficiency of most modern steam turbines to just below 50%. Since only about 80% of the chemical

To / From	Radiant (light)	Kinetic	Mechanical	Electrical	Thermal	Chemical	Nuclear
Radiant (light)	—	—	—	Incandescent bulb (4%)	Hot filament (variable)	Chemiluminescence (variable)	—
Kinetic	—	—	—	—	Aircraft jet engine (35%)	—	—
Mechanical	—	Flywheel storage generator	—	Electric motor (60-90%)	Steam turbine (46%)	—	—
Electrical	Photovoltaic cell (15%)	—	Electric generator (98%)	—	Thermocouple (7%)	Storage battery, dry cell	—
Thermal	Solar water heater (can be high)	—	—	Hot water heater (100%)	—	Boiler, furnace (90%)	Nuclear reactor (100%)
Chemical	Photosynthesis (<0.5%)	—	—	Storage battery (variable)	—	—	—
Nuclear	—	—	—	—	—	—	—

Figure 6.6. Devices for the conversion of energy from one form to another (efficiencies in parentheses).

energy used to raise steam by combustion of fossil fuel in a boiler is actually transferred to water to produce steam, the net efficiency for conversion of chemical energy in fossil fuels to mechanical energy to produce electricity is about 40%. Fortunately, essentially all the mechanical energy in a rotating turbine can be converted to electricity in the generator to which it is connected, so the overall efficiency of conversion of fossil fuel chemical energy to electricity is about 40%. The conversion of nuclear energy to mechanical energy in a reactor-powered steam turbine is only about 30% because reactor peak temperatures are limited for safety reasons.

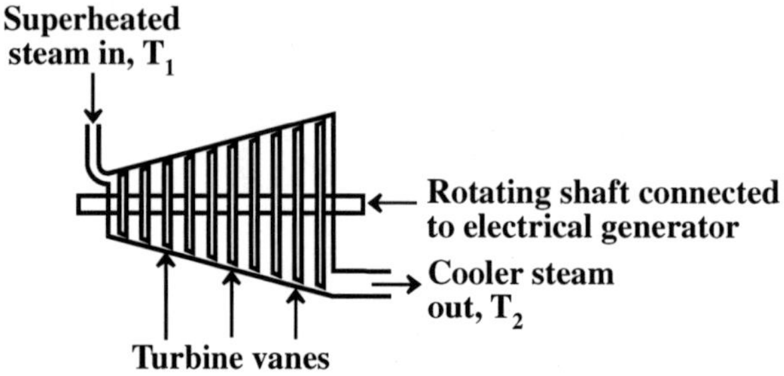

Figure 6.7. A steam turbine in which superheated steam drives vanes attached to a shaft, thus converting heat energy in steam to mechanical energy. The rapidly rotating shaft is usually attached to an electrical generator to generate electricity.

Another example of the application of the Carnot equation is provided by the internal combustion piston engine shown in Figure 6.8. A complete cycle of this engine consists of two downward and two upward strokes of the piston. As the piston moves downward during the first stroke, air or an air/fuel mixture is drawn into the cylinder through the open intake valve. This gas is then compressed to a very small volume by the upward movement of the piston during the next compression stroke, during which both valves are closed. Near the top of the compression stroke, fuel is injected, if it were not taken in as a mixture with air by the intake stroke. The fuel/air mixture is ignited at the end of the compression stroke, and the sudden formation of very hot, high-pressure exhaust gas drives the piston downward in the power stroke, which is the one shown in Figure 6.8. For the last of the four strokes, the exhaust valve is opened, and the exhaust gases are expelled as the piston moves upward, at which point the entire 4-step cycle is ready to start again. An amazing number of these cycles occur each second for each piston in an engine. If you have a tachometer on your car you may notice that it registers around 3000 rpm (revolutions per minute) at cruising speed. That means that each piston performs an up and down motion 50 times each *second* and a total of 25 complete cycles each second!

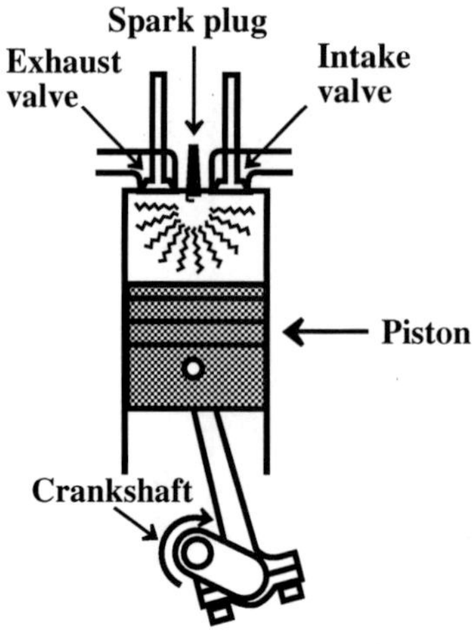

Figure 6.8. An internal combustion piston engine in which a very rapidly burning mixture of air and fuel drives a piston downward during the power stroke and this motion is converted to rotary mechanical motion by the crankshaft.

The efficiency of an internal combustion engine in converting heat from burning fuel into mechanical motion is described by the Carnot equation, where T_1 is the peak temperature of the highly compressed combustion gas at the beginning of the power stroke and T_2 is the lower temperature at the bottom of the stroke. Overall, a typical gasoline engine converts about 25% of the energy in the gasoline to mechanical energy. You may know that diesel engines get much better mileage than gasoline engines, and U.S. tourists are thankful for the efficient diesel automobiles that they can rent for European travel after seeing the exorbitant prices of fuel in Europe. The reason that the diesel engine is about 50% more efficient than a comparable gasoline engine is that a diesel engine compresses the air in the cylinder to a very small volume; the diesel engine has a compression *ratio* of about 16:1. Fuel is injected into the cylinder at the top of the compression stroke, and the compressed gases are so hot that they ignite without requiring a spark plug for ignition. This extreme compression results in a very high peak temperature, the engine follows Carnot's equation, and the fuel efficiency is relatively much higher than that of a gasoline engine.

6.6. GREEN ENGINEERING AND ENERGY CONVERSION EFFICIENCY

All agree that increased efficiency of energy conversion and utilization is desirable. During the last century enormous advances have been made in the efficiency of energy utilization. The early fossil-fueled electrical power generating plants from around 1900 were only about 4% efficient in converting chemical energy to electrical energy; modern ones exceed 40%. The change from picturesque steam locomotives to bland, but efficient diesel locomotives that took place during the 1940s and 1950s resulted in an approximately four-fold increase in the energy efficiency of rail transport.

The improvements described above were due in part to improved materials that can tolerate higher peak temperatures. They were also due to advances in engineering. Engineers of a century ago had never heard of green chemistry or green engineering, and probably would not have cared had they known about it. But they did understand costs of fuel (which on the basis of constant value currency were often higher then than they are now) and they welcomed the greater efficiencies they achieved on the basis of costs.

Improved efficiencies of energy utilization are to be expected in the future. In modern times engineers and the concerns that employ them are very much aware of the virtues of green engineering, and the inevitable rise of energy prices in the future will continue to add economic necessity to green scientific and engineering virtue. Materials continue to improve, and that can lead to increased energy efficiency. Remarkable advances in computers and their application to energy conversion processes are making possible the precision control that leads to energy efficiency.

One area in which substantial increases in energy efficiency can be made by simply integrating known technology is by using fossil fuels and the heat that they produce in several steps. This can be done with **combined** power **cycles** as outlined in Figure 6.9. Typically, in combined power cycle installations gas or fuel oil is burned in a turbine engine that is much like the engine of a turboprop airplane, and the rotating shaft of this engine is coupled to a generator to produce electricity. The hot exhaust gases from the combustion turbine can be exhausted to a boiler where their heat turns liquid water to steam. This steam can be run through a steam turbine coupled to a generator to produce more electricity. Steam leaving the steam turbine still contains a lot of heat, and can be conveyed to homes and other structures for heating. The water condensed from this steam is pure and is recycled to the boiler, thus minimizing the amount of makeup boiler feedwater, which requires expensive treatment to make it suitable for use in boilers. Such a system as the one described is in keeping with the best practice of industrial ecology. The use of steam leaving a steam turbine for heating, a concept known as **district heating,** is commonly practiced in Europe (and many university campuses in the U.S.) and can save large amounts of fuel otherwise required for heating.

6.7. CONVERSION OF CHEMICAL ENERGY

In some cases a need exists to convert chemical energy from one form to another so that it can be used in a desired fashion. The generation of hydrogen gas from fossil fuels is an important chemical energy conversion process that may become much more widely practiced as fuel cells, which use elemental hydrogen as a fuel, come into more common use. Hydrogen can be obtained from a number of sources, but the cheapest and most abundant raw material for hydrogen generation is coal. When coal is used to generate hydrogen, the hydrogen actually comes from steam. In this process, known as **coal gasification** part of the coal is burned in an oxygen stream,

$$C(coal) + O_2 \rightarrow CO_2 + \text{heat} \qquad (6.7.1)$$

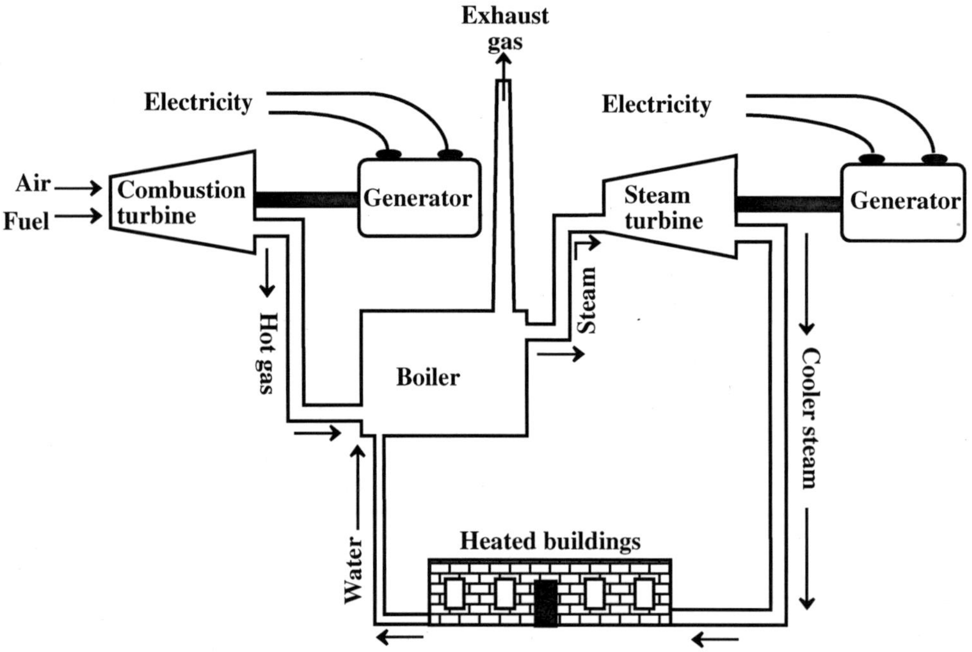

Figure 6.9. A combined power cycle in which natural gas or oil is first used to fire a gas turbine connected to an electrical generator. The hot exhaust gases from this turbine are fed to a boiler to produce steam, which drives a steam turbine, also connected to a generator. The still hot exhaust steam from the steam turbine is distributed to residential and commercial buildings for heating. The water condensed from the steam used to heat buildings is returned to the power plant to generate more steam, which conserves water and also prevents the necessity of treating more water to the high standards required for a boiler.

leaving a solid residue of very hot carbon from the unburned coal. This material reacts with water in steam,

$$C(hot) + H_2O \rightarrow H_2 + CO \qquad (6.7.2)$$

to generate elemental H_2 and CO in a reaction that absorbs heat. The CO can be reacted with more steam over an appropriate catalyst,

$$CO + H_2O \rightarrow H_2 + CO_2 \qquad (6.7.3)$$

The reactions shown above for the generation of elemental hydrogen from coal and water have been used for well more than a century in the coal gasification industry. Before natural gas came into common use, steam blown over heated carbon was used to generate a synthesis gas mixture of H_2 and CO that was piped into homes and burned for lighting and cooking. The mixture burned well, but, in addition to forming treacherous explosive mixtures with air, it was lethal to inhale because of the toxic carbon monoxide. But the process may have a future for the generation of elemental hydrogen for use in fuel cells. By using pure oxygen as an oxidant, it raises the possibility of producing greenhouse gas carbon dioxide in a concentrated form that can be pumped underground or otherwise prevented from getting into the atmosphere. Retention of carbon dioxide in this manner is called carbon sequestration and is the subject of some intense research.

The synthesis gas mixture of H_2 and CO_2 is a good raw material for making other chemicals, including hydrocarbons that can be used as gasoline or diesel fuel. Combined in the correct ratios over a suitable catalyst, these two gases can be used to make methanol:

$$CO + H_2 \rightarrow H_3COH + CO_2 \qquad (6.7.4)$$

Methanol is used as an additive to gasoline. The Mercedes-Benz automobile company is advocating methanol as a liquid fuel that can be transported in the fuel tank of an automobile and reacted over a catalyst to generate elemental hydrogen (Equation 6.7.5) to use in a fuel cell to power an electric motor in the car. This is the fuel of choice for Mercedes-Benz's experimental NECAR 4.

$$2H_3COH + O_2 \rightarrow 4H_2 + 2CO_2 \qquad (6.7.5)$$

6.8 RENEWABLE ENERGY SOURCES

Ideal energy sources are those that do not pollute and never run out. Such sources are commonly called **renewable energy resources**. There are several practical renewable energy resources that are discussed briefly in this section.

Solar Energy is The Best — When The Sun Shines

Sunshine comes close to meeting the criteria of an ideal energy source, including widespread availability, an unlimited supply, and zero cost up to the point of collection. The utilization of solar energy does not cause air, heat, or water pollution. Sunshine is intense and widely available in many parts of the world. If it were possible to collect solar energy with a collection efficiency of 10%, approximately one-tenth of the area of Arizona would suffice to meet U.S. energy needs, and at 30% collection efficency, only about one-thirtieth of the area of that generally sunny state would suffice. But, keep in mind that such an area is still enormous and the implications of covering it with solar collectors would be profound.

There are several ways in which solar energy can be utilized. The simplest of these is simply to capture the sun's heat for heating, and solar-heated houses and solar water heaters have been developed and used successfully. At a somewhat more sophisticated level, solar boilers have been developed that are located on towers and receive concentrated sunlight from an array of parabolic mirrors, thus generating steam to make electricity. Some years ago a serious proposal was even made to use solar collectors in Earth orbit and convert the energy to a beam of microwave radiation focused on a receiver on Earth's surface. Visions of this beam deviating from its aiming point or hapless birds or even aircraft straying into it and being instantly cooked by an extraterrestrial microwave oven have prevented this plan from coming to fruition. Photosynthesis is another way of utilizing solar energy and the potential of photosynthetic biomass for energy is discussed in a later section of this chapter.

Other than low-grade building and water heating, the most promising way to utilize solar energy is by its direct conversion to electricity in photovoltaic cells (see Figure 6.10). Originally just a laboratory curiosity, these devices became practical sources of electricity for satellites and space vehicles where their high cost was of little concern. But over the years they have become more efficient and cheaper, and it is now common to see arrays of these cells used to power data processors and signalling devices in remote locations. And some houses even have banks of photovoltaic cells.

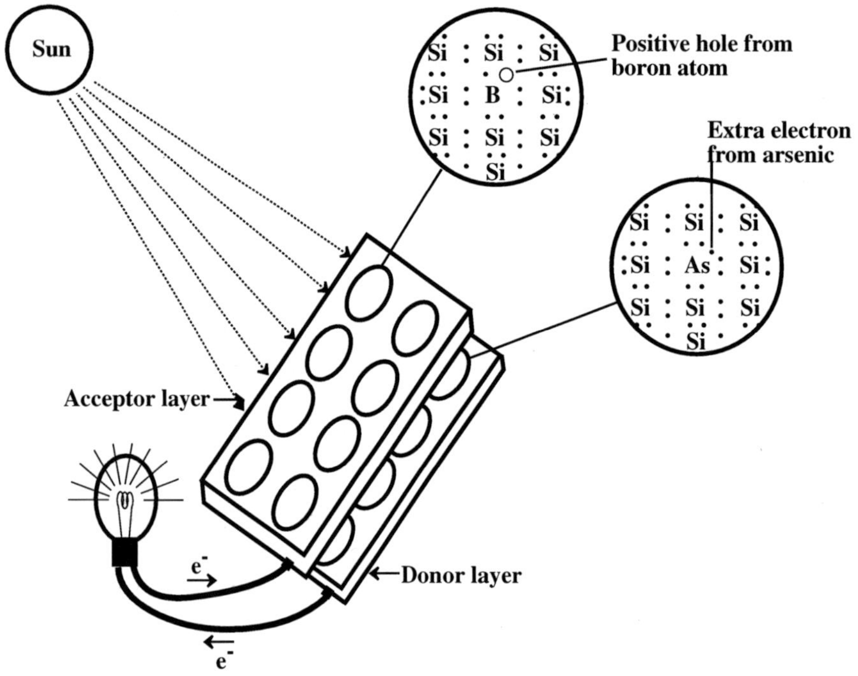

Figure 6.10. Photovoltaic cell

Photovoltaic cells depend upon the special electronic properties of silicon atoms containing low levels of other elements. The cell consists of two layers of silicon, a donor layer that is **doped** with about 1 part per million of arsenic atoms and an acceptor layer doped with about 1 part per million of boron. Examination of the Lewis symbols of these three elements,

$$\cdot\dot{\mathrm{Si}}: \qquad \cdot\dot{\underset{\cdot}{\mathrm{As}}}: \qquad \dot{\mathrm{B}}:$$

shows that substitution of an arsenic atom with its 5 valence electrons for a silicon atom with its 4 valence electrons in the donor layer gives a site with an excess of 1 electron whereas substitution of a boron atom with only 3 electrons for a silicon atom in the acceptor layer gives a site "hole" that is deficient in one electron. The surface of a donor layer in contact with an acceptor layer contains electrons that are attracted to the acceptor layer. When light shines on this area, the energy of the photons of light can push these electrons back onto the donor layer, from which they can go through an external circuit back to the acceptor layer. This flow of electrons constitutes an electrical current that can be used for energy.

Current photovoltaic cells are around 12–15% efficient in converting radiant solar energy to electricity at a cost 4–5 times that of electricity generated in fossil fuel power plants. However, advances are continually being made in solar cell technology and it can be anticipated that efficiencies will continue to increase as costs decrease.

The obvious major disadvantage of solar energy is that it does not work in darkness, and variable atmospheric conditions affect its output. Flexibility in electrical power grids allows such intermittent sources for up to 15% of power without using special devices for energy storage. Furthermore, there are means of storing energy, such as by extremely high-temperature/high-pressure supercritical water stored deep underground or mechanical energy stored in the extremely rapid rotation of flywheels.

A very attractive energy storage option for solar energy given the growing use of fuel cells is hydrogen gas. Electrolysis of water,

$$2H_2O + \text{electrical energy} \rightarrow 2H_2(g) + O_2(g) \quad (6.8.1)$$

with solar-generated electricity provides elemental hydrogen and oxygen, which are exactly the fuels used by fuel cells. The overall efficiency of this process can be increased significantly by the development of direct means for splitting water molecules into hydrogen and oxygen using the energy of light photons.

Favorable Winds

As the sun heats air masses unevenly, winds are generated that can be tapped as an indirect form of solar energy. Wind power is undergoing rapid growth in some areas as an energy source and has become competitive in cost with more conventional sources in some areas. In parts of Europe, California, Wyoming, and other areas, the sight of wind-powered generators mounted on towers has become common (Figure 6.11). Denmark now leads the world in wind-generated electricity. Using 3-bladed turbines spanning a huge 56-meter diameter and mounted on 64-m towers, National Wind Power, Ltd, operates a large wind farm in Carno, Wales. This facility produces 33.6 megawatts of power and is a serious contributor to the area's electrical power supply. The U.S. has committed to increasing its wind power capacity. In January, 2001, FPL Energy announced plans for the world's largest wind farm consisting of 450 windmills, each almost 80 meters tall, located along the border of Walla Walla County, Washington, and Umatilla County, Oregon. The electricity generated by this wind farm is sufficient for 70,000 homes. Many areas of the world, including parts of Alaska, Canada, Scandinavia, and Russia have strong, consistent winds favorable to the development of wind energy.

Biomass Energy

As noted earlier in this chapter, photosynthesis

$$6CO_2 + 6H_2O \xrightarrow{h\nu} C_6H_{12}O_6 + 6O_2 \quad (6.8.2)$$

Figure 6.11. Wind-powered electrical generators mounted on towers are becoming increasingly common sights in the world in areas where consistent wind makes this nonpolluting source of renewable energy practical.

enables the conversion of solar energy to chemical energy in the form of biomass. Photosynthetically generated biomass is the source of food energy for essentially all organisms and, until about 200 years ago, was the source of most fuel, which it still is in parts of the world. Today, Finland gets about 15% of its energy from wood and wood products. A large fraction of this energy comes from burning the black liquor byproduct from pulp and paper manufacture, which uses wood as a raw material. As shown in Figure 6.6, photosynthesis suffers from the disadvantage of having less than 0.5% efficiency in the conversion of solar energy to chemical energy. However, some plants, most notably sugarcane, convert solar energy to biomass energy with an efficiency of around 0.6%.

As shown in Reaction 6.8.2, carbohydrates, such as glucose, $C_6H_{12}O_6$, are produced by photosynthesis. They can be burned directly, converted chemically to other fuels, or fermented to produce ethyl alcohol fuel. Hydrocarbons are more desirable as fuels, and some plants produce them directly. One example is the Philippine plant, *Pittsosporum reiniferum*, the fruits of which contain such a high content of hydrocarbon terpenes, primarily α-pinene and myrcene, that they can be burned to provide illumination. Rubber trees and other plants, such as *Euphorbia lathyrus* (gopher plant), a small bush growing wild in California, produce hydrocarbon emulsions. Seed oils, such as those produced by sunflowers and peanuts, and more exotic sources including buffalo gourd, cucurbits, and Chinese tallow tree, can be used for fuel, especially in diesel engines.

There are several major objections to the use of biomass as an energy source. One of these is the perception that it cannot produce enough fuel to provide for energy needs and that use of land to grow biomass fuel detracts from its use for growing food. However, it should be noted that about 150 billion metric tons of biomass are produced in the world each year by photosynthesis, mostly from uncontrolled plant growth. A typical farm crop can produce 10-20 metric tons of dry biomass annually, a figure that reaches 50 metric tons per acre per year for some algae and grasses (there are 640 acres in a square mile of land). About 6% of the biomass generated globally each year would be equivalent to the world's demand for fossil fuels. Cultivation for fuel biomass of 6-8% of the land area of the 48 contiguous states would provide energy equivalent to annual U.S. consumption of petroleum and natural gas. Furthermore, only a small fraction of widely grown grain crops goes into grain; the rest is plant biomass, much of which could be used for energy production. And the U.S. has vast areas of underutilized land that could be devoted to the cultivation of energy-yielding plants. Much of this neglected, erosion-prone land would benefit from the cultivation of perennial plants that could be harvested for energy and regrow from roots left in the ground, thus lowering water and wind erosion.
Properly utilized, biomass is a largely nonpolluting source of energy. Since it is produced by photosynthesis, there is no net addition to global atmospheric carbon dioxide. Although the heating value of dried biomass is only about half that of coal, biomass combustion produces very little sulfur dioxide, and the ash residue can be returned to soil without adding harmful heavy metals, which can be a problem with coal ash.

Biological fermentation can be used to produce fuels from biomass. Yeasts act upon carbohydrates,

$$C_6H_{12}O_6 \rightarrow 2C_2H_5OH + 2CO_2 \qquad (6.8.3)$$

to produce ethanol, C_2H_5OH. This liquid alcohol can be used alone as a fuel, but is usually added to gasoline at levels of about 10% to produce **gasohol**, which burns more cleanly and with less CO output than ordinary gasoline. The source of carbohydrate for ethanol production is usually corn grain or sugar produced by sugarcane. Another biomass fermentation occurs with methane-forming bacteria,

$$C_6H_{12}O_6 \rightarrow 3CH_4 + 3CO_2 \qquad (6.8.4)$$

to produce methane gas, CH_4. The gas mixture produced by this reaction can be burned directly, or the carbon dioxide can be removed to produce pure methane gas. Anaerobic (air-free) methane digesters used to degrade the biomass in sewage sludge, the residue from biological treatment of wastewater can generate enough power to provide for the pumping and electrical needs of a large sewage treatment plant. Small methane digesters running on crop and food residues and human and animal wastes are used in rural areas of China to provide methane for cooking and lighting.

Geothermal Energy

Underground heat reaching the surface from Earth's hot, molten rock interior has been tapped as a source of energy for approximately a century. This **geothermal energy** was first used to generate electricity at Larderello, Italy, in 1904. It has since been developed in Iceland, Japan, Russia, New Zealand, the Phillipines, and at the Geysers in northern California. The most desirable form of geothermal energy coming from the ground is in the form of dry steam. Such steam is relatively rare and superheated water mixed with steam is the most common source of geothermal energy. The water byproduct is sometimes relatively pure, but often contains salts and gases that can cause problems. Air pollutant hydrogen sulfide is also evolved with some geothermal steam. (The odor of hydrogen sulfide is very noticeable around steam vents in Yellowstone National Park. If the park were an industrial plant, it would be closed for violating safety standards because of the toxic H_2S gas emitted by geothermal features in the park.) Experiments have been conducted with injecting water into hot, dry underground water formations to produce steam. Success with this technology could increase resources of geothermal energy ten-fold.

6.9 NUCLEAR ENERGY: WILL IT RISE AGAIN?

The only source of nuclear energy currently available depends upon the neutron-induced fissioning of heavy atomic nuclei, most commonly those of the uranium isotope with a mass number of 235, to produce radioactive fission products, an average of 2.5 more neutrons and an astounding amount of energy compared to an ordinary chemical reaction. A typical example of such a fission reaction is

$$^{235}_{92}U + ^{1}_{0}n \rightarrow ^{133}_{51}Sb + ^{99}_{41}Nb + 4^{1}_{0}n \qquad (6.9.1)$$

A nuclear reactor operating at a constant power level is controlled such that on average 1 neutron from each fission reaction is absorbed to cause another fission reaction, thus sustaining a **chain reaction**. The excess neutrons are absorbed by nonfissionable material. In order to cause the desired fission, the neutrons, initially released as rapidly-moving, high energy particles must be slowed down, which is done by a **moderator**, such as water, in the reactor.

The basic function of a nuclear power reactor is as a heat source to produce steam to generate mechanical energy. In a nuclear power reactor (Figure 6.11) pressurized superheated water circulates through the hot reactor core in an enclosed loop (to prevent escape of radioactive contaminants). Heat from this water is used to convert water to steam in a heat exchanger. The rest of the power plant is like a conventional fossil-fueled plant with a steam turbine coupled to a generator and the steam from the steam turbine being cooled to provide water for the heat exchanger.

There is an adequate supply of uranium. Only 0.71% of natural uranium is fissionable uranium-235, and uranium to be used for fission must be enriched in this isotope. In principle, the remaining 99.28% of uranium that consists of uranium-238 could be converted to fissionable plutonium by absorption of neutrons in **breeder reactors**. Plutonium is actually generated by uranium-238 absorbing neutrons in a conventional nuclear power reactor, and after the reactor has operated for a few months after refueling, a large fraction of its energy output comes from plutonium generated in the reactor.

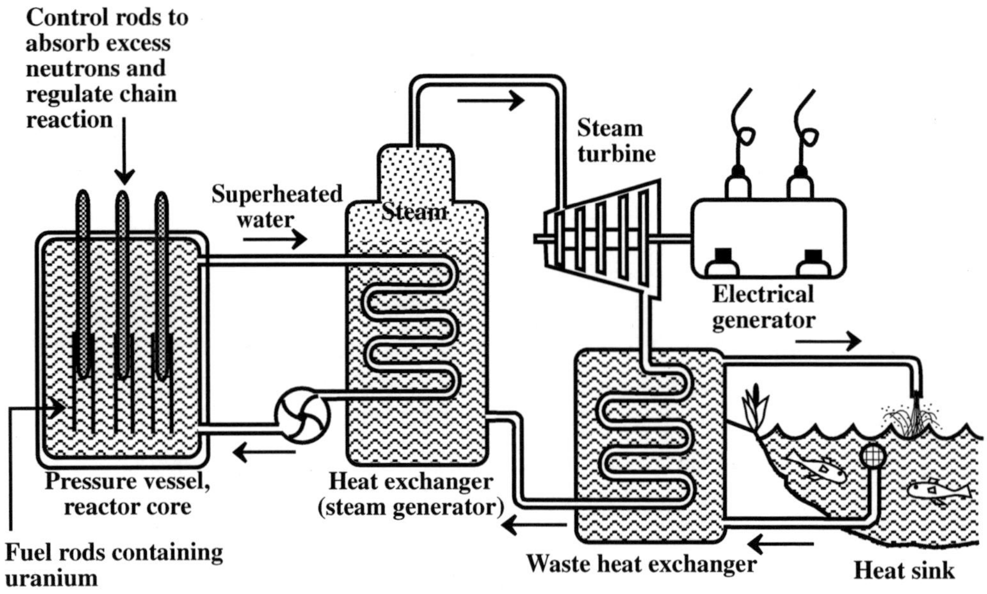

Figure 6.11. A typical nuclear fission power plant.

A big problem with nuclear power reactors is the radioactive fission products generated when the uranium nucleus splits apart. These remain lethal for thousands of years, so the spent fuel, or the fission products isolated from it, must be put in a secure location. So far, efforts to settle upon an appropriate nuclear waste repository have met with such opposition that a permanent site is not yet in operation. In the meantime, spent fuel is stored temporarily under water in containers located on the reactors' premises. This is actually a good thing because the short-lived wastes that are responsible for most of the radioactivity in nuclear fuel freshly removed from a reactor decay rapidly, and after a few years of storage only a small fraction of the original activity is present.

Another problem with nuclear reactors is their decommissioning. One option is to dismantle the reactor soon after it is shut down using apparatus operated by remote control. The radioactive reactor parts are then disposed. Another approach is to allow the reactor to stand for 30–100 years before dismantling, by which time most of the radioactivity has decayed (and the people responsible for the reactor initially have died). A third option is to entomb the reactor in a concrete structure.

Two accidents have dealt a strong blow to the future of nuclear energy. The first, and much lesser of these, occurred on March 28, 1979, when Metropolitan Edison Company's nuclear reactor located on Three Mile Island in the Susquehanna River, 28 miles outside of Harrisburg, Pennsylvania, lost much of its coolant resulting in overheating, and partial disintegration of the reactor core. Some radioactive xenon and krypton gases were released to the atmosphere and some radioactive water entered the river. The problem was remediated and the reactor building sealed. Then in April of 1986 a reactor of inherently dangerous Soviet design blew up in Chernobyl, which is now part of Ukraine. Officially, 31 people were killed, but the death toll was probably many more, especially when delayed effects of exposure to radioactive materials are considered. Food, including reindeer meat in Lapland, was contaminated as far away as Scandinavia, thousands of people were evacuated, and the entire reactor building was entombed in a massive concrete structure. The reactor that blew up was one of four units, the last of which was not shut down permanently until the end of 2000!

Given the horrors described above, why would reputable scientists even advocate development of nuclear energy? The answer is, simply, carbon dioxide. With massive world resources of coal and other nonpetroleum fossil fuels, the world has at least enough readily available fossil fuel to last for a century. But, as discussed in Chapter 8, evidence is mounting that the carbon dioxide from fossil fuel combustion will lead to global warming accompanied by effects such as rising sea levels that will inundate many coastal cities. Other alternatives, such as the renewable energy resources discussed in Section 6.8, are generally intermittent, disperse, and accompanied with environmental problems of their own. Humans do know how to design and operate nuclear reactors safely and reliably; indeed, France has done so for years and gets most of its electricity from nuclear fission. So, it may be that nuclear energy is far from dead and that humankind, reluctantly and with great care, will have to rely on it as the major source of energy in the future. A new generation of nuclear power plants is waiting to be built that have the desirable characteristics of **passive stability**. This means that measures such as gravity feeding of coolant, evaporation of water, or convection flow of fluids operating automatically provide for safe operation of the reactor and automatic shutdown of the reactor if something goes wrong. New designs are also much more reliable with only about half as many pumps, pipes, and heat exchangers as are contained in older power reactors.

Nuclear Fusion

The fusion of a deuterium nucleus and a tritium nucleus releases a lot of energy as shown below, where Mev stands for million electron volts, a unit of energy:

$$^{2}_{1}H + ^{3}_{1}H \rightarrow ^{4}_{2}He + ^{1}_{0}n + 17.6 \text{ Mev (energy released per fusion)} \qquad (6.9.2)$$

This reaction is responsible for the enormous explosive power of the "hydrogen bomb." So far it has eluded efforts at containment for a practical continuous source of energy. And since physicists have been trying to make it work on a practical basis for the last approximately 50 years, it will probably never be done. (Within about 15 years after the discovery of the phenomenon of nuclear fission, it was being used in a power reactor to power a nuclear submarine.) However, the tantalizing possibility of using the essentially limitless supply of deuterium, an isotope of hydrogen, from Earth's oceans for nuclear fusion still give some investigators hope of a practical nuclear fusion reactor.

Nuclear fusion was the subject of one of the greatest scientific embarrassments of modern times when investigators at the University of Utah in 1989 announced that they had accomplished so-called cold fusion of deuterium during the electrolysis of deuterium oxide (heavy water). This resulted in an astonishing flurry of activity as scientists throughout the world sought to repeat the results, whereas others ridiculed the idea. Unfortunately, for the attainment of a cheap and abundant source of energy, the skeptics were right, and the whole story of cold fusion stands as a lesson in the (temporary) triumph of wishful technological thinking over scientific good sense.

QUESTIONS AND PROBLEMS

1. The tail of a firefly glows, although it is not hot. Explain the kind of energy transformation that is most likely involved in the firefly's producing light.
2. What is the standard unit of energy? What unit did it replace? What is the relationship between these two units?
3. Which law states that energy is neither created nor destroyed?
4. What is the special significance of 1,340 watts?
5. Despite the unfavorable status of controlled thermonuclear fusion as an energy source discussed at the end of this chapter, how can the statement be justified that "thermonuclear fusion is responsible for most of the energy used in the world today?"

6. How is energy carried through empty space?

7. What happens to the energy of electromagnetic radiation as the wavelength becomes shorter?

8. What is electromagnetic radiation with a wavelength (A) just shorter and (B) just longer than that of visible light called?

9. What are photochemical reactions?

10. What is the reaction in nature by which solar energy is converted to chemical energy?

11. Given that the energies of H-H, O=O, and O-H bonds are 432, 494, and 459 kJ/mol, respectively, calculate the heat energy involved when 2 moles of H_2, react with 1 mole of O_2 to produce 2 moles of H_2O.

12. Given that the energies of H-H, N≡N, and N-H bonds are 432, 942, and 386 kJ/mol, respectively, calculate the heat energy involved when 1 mole of N_2 undergoes the reaction $N_2 + 3H_2 \rightarrow 2NH_3$. To which bond may this low energy release be attributed?

13. In what respects is wind both one of the oldest, as well as one of the newest, sources of energy?

14. What are two major problems with reliance upon coal and petroleum for energy?

15. Why does natural gas contribute much less to greenhouse warming than does coal?

16. How might coal be utilized for energy without producing greenhouse gas carbon dioxide?

17. What is a large limiting factor in growing biomass for fuel, and in what respect does this limit hold hope for the eventual use of biomass fuel?

18. What relationship describes the limit to which heat energy can be converted to mechanical energy?

19. Why does a diesel-powered vehicle have significantly better fuel economy than a gasoline-powered vehicle of similar size?

20. Why is a nuclear power plant less efficient in converting heat energy to electricity than is a fossil-fueled power plant?

21. Instead of having a spark plug that ignites the fuel, a diesel engine has a glow plug that heats up only during engine startup. Explain.

22. Cite two examples of vastly increased efficiency of energy utilization that have taken place during the last century.

23. Describe a combined power cycle. How may it be tied with district heating?

24. What are three reactions used in coal gasification?

25. What is a major proposed use of liquid methanol as a fuel for the future?

26. Describe a direct and an indirect way to produce electricity from solar energy.

27. What is the distinction between donor and acceptor layers in photovoltaic cells?

28. What are some possible means for storing energy generated from solar radiation?

29. How is wind power utilized as an energy source?

30. What are the advantages of *Pittsosporum reiniferum* and *Euphorbia lathyrus* for the production of biomass energy?

31. Corn produces biomass in large quantities during its growing season.What are two potential sources of biomass fuel from corn, one that depends upon the corn grain and the other that does not.

32. Does biomass contribute to greenhouse gas carbon dioxide? Explain.

33. What fermentation process is used to generate a fuel from wastes, such as animal wastes?

34. What are two potential pollution problems that accompany the use of geothermal energy to generate electricity?

35. What basic phenomenon is responsible for nuclear energy? What keeps the process going?

36. What is the biggest problem with nuclear energy? Why is it not such a bad idea to store spent nuclear fuel at a reactor site for a number of years before moving it?

37. What is meant by passive stability in nuclear reactor design?

38.What is the status of thermonuclear fusion for power production?

7. WATER, THE ULTIMATE GREEN SOLVENT: ITS USES AND ENVIRONMENTAL CHEMISTRY

7.1. H_2O: SIMPLE FORMULA, COMPLEX MOLECULE

Water, H_2O, is an amazing chemical compound. It is the true medium of life. Early life forms developed in water and only much later in their evolution ventured out of water, but never very far. Our own bodies are largely water. Our blood is a water solution of sodium chloride and other essential salts, in which are suspended colloidal-sized red blood cells that carry oxygen from the lungs throughout the body. Even those organisms that dare to live in water-deficient areas — camels, gila monsters, cactus, and some humans — must have elaborate mechanisms to conserve, store, and obtain the small quantities of water in their surroundings.

The thing that makes water so special is its molecular structure represented in Figure 7.1. Recall that the oxygen atom in the water molecule has a stable octet of valence electrons. These 8 electrons are grouped into 4 pairs. Two of these pairs are shared with the two H atoms in the water molecule and the other two are unshared pairs. The pairs repel each other by arranging themselves around the spherical surface of the water molecule in a way that enables them to be as far apart as possible. Try to visualize a sphere with 4 things arranged around its surface as far apart as possible. They would be at angles of somewhat more than 100 degrees. In fact, the angle formed by two lines, each joining the center of an H atom to the center of the O atom in the water molecule is 105°

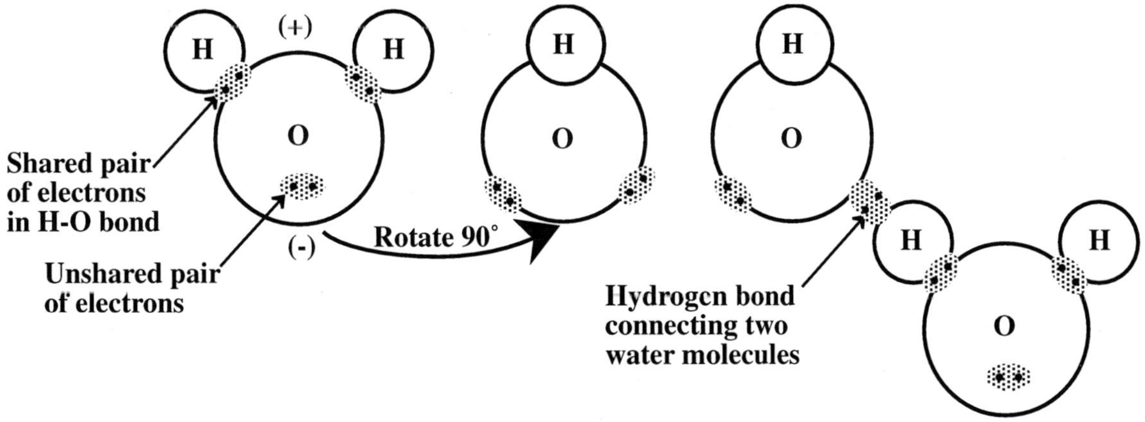

Figure 7.1. The water molecule (left) is electrically polar and has 2 pairs of valence electrons in H-O bonds and 2 unshared pairs, both of which are shown when the molecule is rotated (middle). The unshared pairs of electrons can form hydrogen bonds to other water molecules (right).

So, the first significant characteristic of the water molecule is that it has a bent structure with the two H atoms not on opposite sides of the O atom, but located at an angle somewhat greater than a right angle. A second characteristic of the water molecule is it is a *dipole* in which the O atom side of the molecule is relatively more negatively charged than the side with the two H atoms as shown on the left of Figure 7.1.

Now let us rotate the water molecule 90° around its vertical axis as shown in the second structure in Figure 7.1. We see one H atom, but it obscures the other H atom. We can see the two unshared pairs of electrons, but why bother? They are important because they can bond to H atoms on *other* water molecules with a special kind of bond called a **hydrogen bond**, as shown on the right in Figure 7.1. Through this characteristic of hydrogen bonding, water molecules in liquid water and in ice are connected to each other, which has profound effects upon water's properties. Furthermore, water molecules can form hydrogen bonds to molecules dissolved in water, greatly enhancing water's solvent properties for some kinds of substances, including many of biological importance.

Because of their polar nature, water molecules surround cations and anions from ionic compounds dissolved in water. The negative ends of the water molecules are oriented toward the positive cations and the positive ends of the water molecules toward the negative anions. This makes water a good solvent for ionic compounds.

7.2. IMPORTANT PROPERTIES OF WATER

Water has several important properties that are of crucial importance for its role as a solvent, life medium, environmental behavior, and industrial uses, based upon the following characteristics of the water molecule:

- Unsymmetrical molecule
- Polar nature of the water molecule
- Ability to form hydrogen bonds

Water is an excellent solvent for a number of materials, including salts, acids, bases, and substances that have H, O, and N atoms capable of forming hydrogen bonds. As a result, water is the solvent in biological fluids, such as blood or urine; it acts to weather minerals and transport dissolved minerals in the geosphere; it transports nutrients to plant roots in soil; and it has many industrial uses. Water has a very high surface tension and liquid water, such as that in rain droplets, acts like it is covered with a thin membrane. Ducks appreciate this characteristic of water because it enables them to float on a water surface. A duck will sink in water to which a detergent has been added to lower the surface tension, which the birds find to be very distressing. Water is transparent to visible light and to the longer-wavelength fraction of ultraviolet radiation. This enables photosynthesis to occur in algae suspended under the surface of water. Water freezes at 0°C and solid ice is less dense than liquid water. Liquid water has its maximum density as a liquid at 4°C. This causes bodies of water to become stratified with colder, denser layers on the bottom.

Arguably the most important physical property of water is its behavior with heat. Liquid water has a **heat capacity** of 4.184 joules per gram per °C (J/g°C), which means that 4.184 joules of heat energy are required to raise the temperature of 1 gram of liquid water by 1°C. This very high heat capacity stabilizes the temperatures of regions near bodies of water. The high heat capacity of water is due to the fact that the water molecules are strongly tied together by hydrogen bonds. A lot of heat energy is required to break these bonds and enable the water molecules to move more rapidly at higher temperatures.

Water has a very high **heat of fusion** of 334 joules per gram (J/g). This means that a very large amount of heat is required to break apart water molecules held in fixed positions by hydrogen bonding in ice in order to convert the water to a liquid state. As ice is melted, so long as both solid ice and liquid water are present, the temperature remains at the melting temperature, a constant 0˚C. Heat added to the system is used to break molecules apart in the solid ice, rather than raising the temperature.

The **heat of vaporization** of water is 2,259 J/g, so 2,259 joules of heat energy are required to vaporize a gram of liquid water. This is the highest heat of vaporization of any common liquid. It has an enormous influence on bodies of water and on the weather. By absorbing so much heat in changing from liquid to vapor, water stabilizes atmospheric temperatures. However, the **latent heat** contained in water vapor is released when the vapor condenses, which is what happens during rainfall. This heat release warms masses of air causing them to rise, and it is the driving force behind thunderstorms and hurricanes. Latent heat in the form of water vapor evaporated from oceans near the equator is carried away from the equator in masses of air and released when the water vapor condenses to form rain.

7.3. WATER DISTRIBUTION AND SUPPLY

Even more so than energy, the availability of affordable water will determine the development of civilization in future decades. Much of the world suffers from a chronic shortage of water. An idea of problems with water distribution may be had by examining Figure 7.2 showing distribution of precipitation in the 48 contiguous states of the U.S. The map shows distinct water shortages in Southern California, Arizona, Nevada, Utah, and Colorado, areas of the country experiencing exceptionally high population growth that also happen to be water deficient.

Throughout the first 3/4 of the 1900s, water use in the U.S. climbed steadily, but use became level around 1980 and in the last two decades of the century actually decreased somewhat, although population increased. This encouraging trend has been the result of water conservation efforts, especially in industry and agriculture. Industrial consumers of water have become very clever in recycling water, including uses through several levels requiring progressively lower water quality. Irrigation used to employ large numbers of spray irrigators that spray the water up into the air, where much of it evaporates or is carried away by wind. These have been largely replaced by systems that apply water directly to soil. The ultimate in irrigation efficiency is trickle irrigation, which drips the exact amount of water required directly onto plant roots. Exact computer control of water usage has helped conserve water in the industrial sector and is even helping in irrigation.

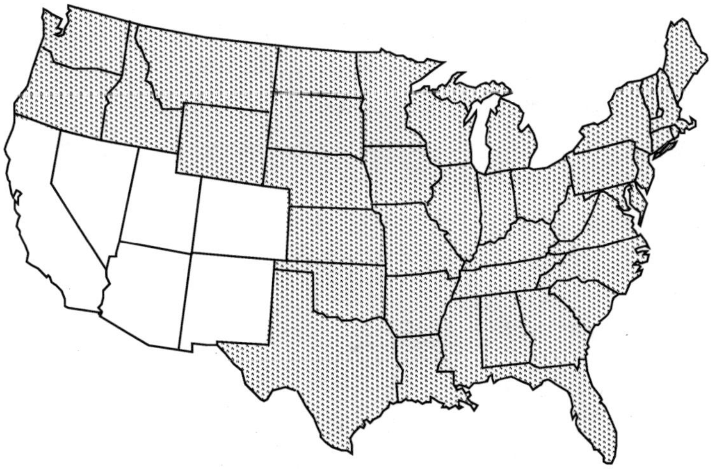

Figure 7.2. The six southwestern contiguous continental United States share characteristics of rapid population growth and water deficiency leading to serious concerns regarding water supply.

The Hydrologic Cycle

One of nature's major cycles is the **hydrologic cycle** (Figure 7.3). The oceans constitute a vast reservoir of water with about 97% of Earth's water. Most of the remaining water is in the form of solid snow and ice, predominantly in Greenland and at both poles. This leaves considerably less than 1% of Earth's water as water vapor and clouds in the atmosphere, as surface water in lakes, streams, and reservoirs, and as groundwater in underground aquifers.

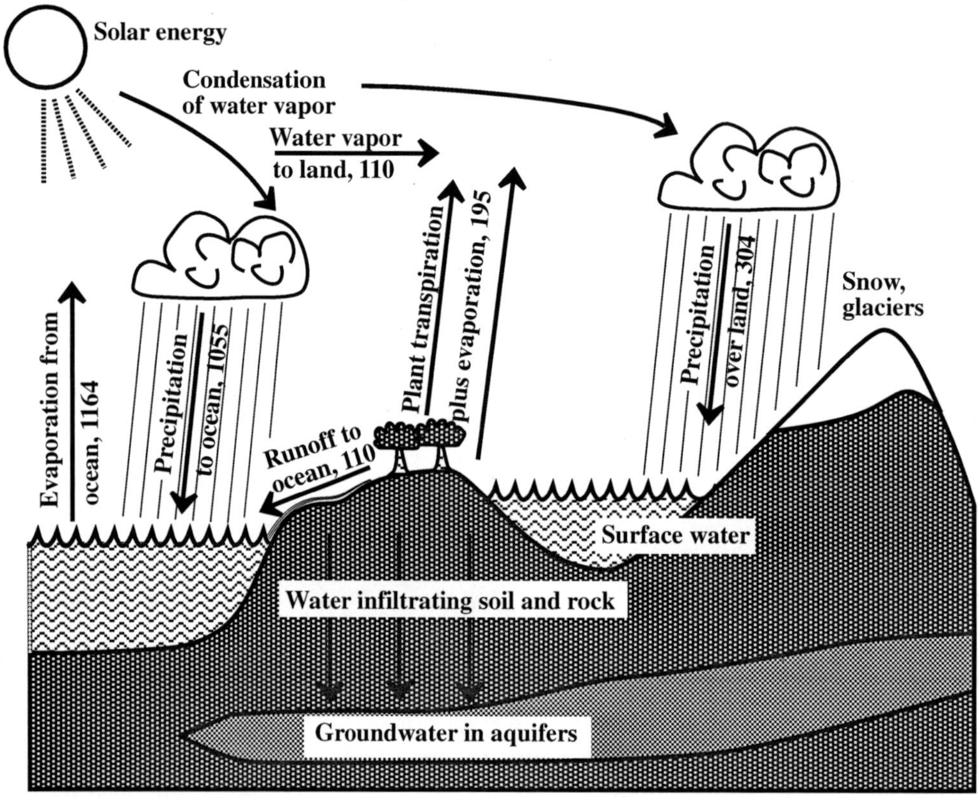

Figure 7.3. The hydrologic cycle, quantities of water in trillions of liters per day.

7.4. BODIES OF WATER AND LIFE IN WATER

Thermal stratification of bodies of water into a warmer, less dense, oxygen-rich upper **epilimnion** and a cooler, more dense, oxygen-poor **hypolimnion** (Figure 7.4) strongly influences water chemistry and biology. The epiliminion often supports a high growth of photosynthetic algae. The oxygen-rich epilimnion generally contains a high concentration of **oxidized** chemical species, which normally have high oxygen contents. These include CO_2 and HCO_3^- for carbon, NO_3^- for nitrogen, and SO_4^{2-} for sulfur. The hypolimnion is often deficient in O_2 and tends to contain chemically **reduced** species with low oxygen contents and high H contents, including CH_4, NH_3 (or NH_4^+), and H_2S for carbon, nitrogen, and sulfur, respectively. Under the reducing conditions of the hypolimnion, iron may be present as soluble Fe, whereas in the epilimnion it exists as solid $Fe(OH)_3$ or Fe_2O_3.

Living Organisms in Water

A normal body of water provides a habitat for large numbers of organisms ranging from single-celled algae to fish. Such a body of water can be regarded as an **ecosystem,** usually based upon a food supply consisting of the biomass produced photosynthetically by the algae and plants living in it:

$$6CO_2 + 6H_2O \xrightarrow{\text{Light energy}} C_6H_{12}O_6 + 6O_2 \qquad (7.4.1)$$

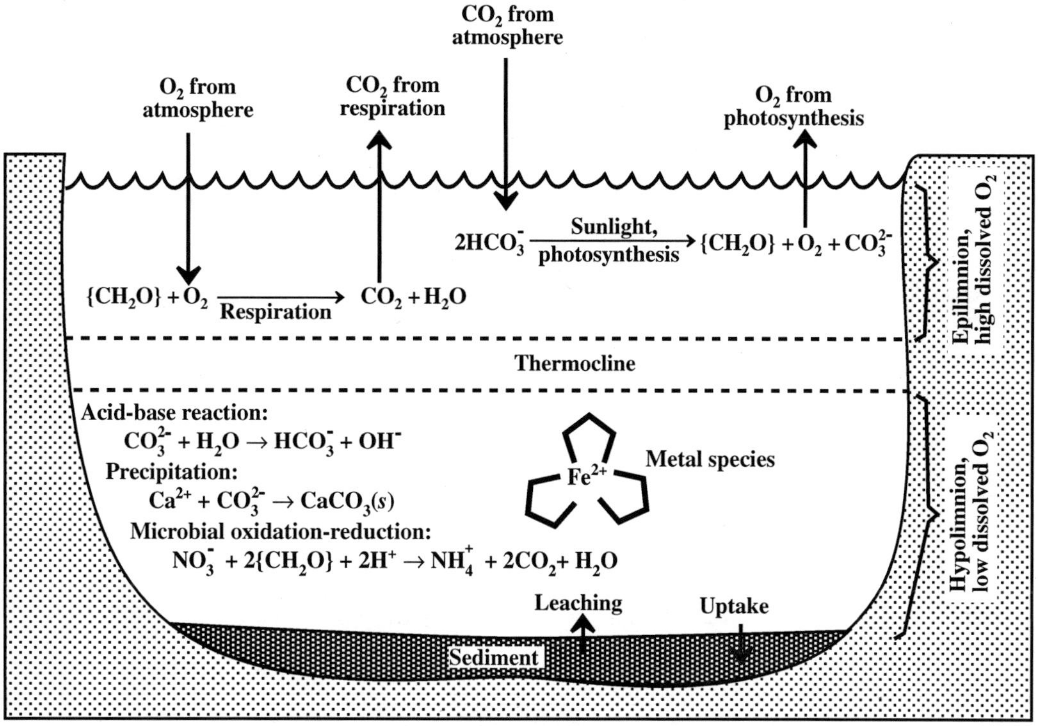

Figure 7.4. Stratification of a lake affects its physical characteristics, which in turn influences chemical phenomena in the body of water. The formula $\{CH_2O\}$is a generic expression representing degradable biomass.

The photosynthesis produces glucose, $C_6H_{12}O_6$, which is converted to other forms of biomass. This reaction produces elemental oxygen and, during daytime when photosynthesis occurs, can contribute substantially to the dissolved oxygen content of the epilimnion. Whereas algae and plants are **producers** that generate biomass, other **heterotrophic** organisms, usually bacteria in water, metabolize biomass,

$$C_6H_{12}O_6 + 6O_2 \rightarrow 6CO_2 + 6H_2O \qquad (7.4.2)$$

consuming oxygen. Normally, biomass is produced by photosynthesis near the surface of the water in the epilimnion. The biomass sinks to the hypolimnion and is consumed by heterotrophic organisms, consuming oxygen and lowering the levels of dissolved oxygen in the hypolimnion, often to around zero. Specialized bacteria in water can utilize *oxidized* chemical species with high oxygen contents other than molecular O_2 for oxygen sources. One such source is nitrate ion, NO_3^-, which acts as an *oxidizing agent* in the bacterially-mediated biodegradation of biomass,

$$C_6H_{12}O_6 + 3NO_3^- + 6H^+ \rightarrow 6CO_2 + 3H_2O + 3NH_4^+ \qquad (7.4.3)$$

producing ammonium ion (or ammonia, NH_3)

By mediating chemical reactions, such as the one above, microorganisms, particularly bacteria, largely determine the chemistry that occurs in water. As examples, bacteria in the oxygen-deficient hypolimnion convert *oxidized* NO_3^- and SO_4^{2-} to *reduced* NH_4^+ and H_2S, whereas bacteria in the oxygen-rich epilimnion convert reduced NH_4^+ and H_2S to oxidized NO_3^- and SO_4^{2-}.

A key species in determining the biological and chemical processes in water is **dissolved oxygen**. Fish cannot live in water where the dissolved oxygen is too low. Organic pollutants in the form of biodegradable organic matter may not be directly toxic, but result in fish kills because of their high **biochemical oxygen demand, BOD**, a measure of the amount of oxygen consumed in the biodegradation of the pollutant biomass.

7.5. CHEMICAL PROCESSES IN WATER

Many chemical and biochemical reactions occur in water in the environment. These are discussed here on the basis of their chemical classification. Several of these were shown by example reactions in Figure 7.4

The photosynthesis reaction, which utilizes sunlight energy to produce biomass,

$$2HCO_3^- \xrightarrow[\text{energy}]{\text{Sunlight}} \{CH_2O\} + O_2 + CO_3^{2-} \quad (7.5.1)$$

is shown here for the conversion of inorganic carbon from dissolved HCO_3^- ion to organic carbon (biomass) abbreviated as $\{CH_2O\}$. This reaction produces biomass that can be acted upon biochemically by other organisms to form the basis of a number of important biochemical processes in water.

The carbonate ion, CO_3^{2-}, generated by photosynthesis reacts with water,

$$CO_3^{2-} + H_2O \rightarrow HCO_3^- + OH^- \quad (7.5.2)$$

removing a hydrogen ion, H^+, from the water molecule and producing OH^- ion. Reactions involving the exchange of H^+ or the generation or consumption of OH^- are **acid-base reactions**. This reaction generates OH^- ion, so it makes the water more basic.

The carbonate ion generated by photosynthesis may become involved in another kind of reaction as exemplified by its reaction with dissolved calcium ion, Ca^{2+}, in water,

$$Ca^{2+} + CO_3^{2-} \rightarrow CaCO_3(s) \quad (7.5.3)$$

to produce solid $CaCO_3$. This is a **precipitation** reaction. $CaCO_3$ is limestone, and it is this kind of reaction, beginning with the CO_3^{2-} generated by photosynthesis, that is responsible for large formations of limestone rock throughout the world.

Oxidation-reduction reactions (see Section 4.7), usually carried out by bacteria, are common in natural waters. The bacterially-mediated reaction of sulfate ion, SO_4^{2-}, acting as an oxidizing agent in the O_2-deficient bottom regions of a body of water to oxidize biodegradable organic matter, $\{CH_2O\}$,

$$2\{CH_2O\} + SO_4 + 2H^+ \rightarrow H_2S(g) + 2H_2O + 2CO_2(g) \quad (7.5.4)$$

is one in which the sulfate ion loses oxygen (is reduced). As the H_2S gas bubbles up through the water, it may **contact** molecular oxygen and other kinds of bacteria that cause it to undergo the following reaction in which the sulfur is *oxidized* with the addition of oxygen atoms to produce SO_4^{2-} ion:

$$H_2S + 2O_2 \rightarrow SO_4^{2-} + 2H^+ \quad (7.5.5)$$

7.6. FIZZY WATER FROM UNDERGROUND

Natural waters contain dissolved gases. The oxygen that fish utilize, for example, is contained as dissolved oxygen from the air. Pressurized carbon dioxide, CO_2, dissolved in groundwater in some

areas causes the water to "fizz" when it is pumped to the surface and some of the dissolved gases escape. Entrepreneurs bottle this water and sell it as mineral water "with gas." It may not taste so great, but since it is "natural," some people assume it must be good for them and pay a relatively high price for the privilege of drinking it. Dissolved CO_2 can be dangerous as was the case with dissolved volcanic carbon dioxide in Lake Nyos in the African country of Cameroon in 1986 when CO_2 evolved from the lake asphyxiated about 1,700 people. (In an effort to prevent this tragedy from being repeated, wells have now been drilled in the bottom of the lake to release CO_2, to the atmosphere.)
Dissolved gases are often in equilibrium with the gas phase. A common example is dissolved oxygen, which is in equilibrium with oxygen in air. The concentration of dissolved gas is related to the pressure of the gas in the gas phase by **Henry's Law**, which states that *the solubility of a gas in a liquid is proportional to the partial pressure of that gas in contact with the liquid.* The solubility of a gas in water *decreases* with *increasing* temperature.

Oxygen in Water

The most important dissolved gas in water is molecular oxygen, O_2. At 25°C the concentration of oxygen dissolved in water in equilibrium with atmospheric air is only about 8 milligrams per liter of water (mg/L). It is this very low concentration of oxygen upon which fish depend for their oxygen supply. This much oxygen is readily lost or consumed and, without some means of replenishing it, water can rapidly become too oxygen-deficient to support fish life. The main way in which this happens is by the consumption of biomass ($\{CH_2O\}$) by oxygen-utilizing bacteria:

$$\{CH_2O\} + O_2 \rightarrow CO_2 + H_2O \qquad (7.6.1)$$

A simple stoichiometric calculation shows that only about 8 mg of biodegradable organic matter (a few flakes of goldfish food, for example) is a sufficient mass to consume all the oxygen in a liter of water initially saturated with air at 25°C.

Gas solubilities decrease with increasing temperature. You can see evidence of this by heating water in a pan and observing the air bubbles that come out of the water while it is warming, but still substantially below the boiling temperature.

7.7. (WEAK) ACID FROM THE SKY

Acids and bases and their reactions are of the utmost importance in water. An *acid* is a substance that contains or produces H^+ ion in water, whereas a *base* is a substance that accepts H^+ ion in water or contains or produces hydroxide ion, OH^-. Whether water is acidic or basic is expressed by a parameter called **pH**.It is defined as

$$pH = -\log [H^+] \qquad (7.7.1)$$

where $[H^+]$ is the molar concentration of H^+ in water, that is, the number of moles of this ion per liter of water. Several pH values corresponding to various concentrations of H^+ are given below:

$[H^+]$, mol/L	$\log[H^+]$	pH
0.100	-1.00	1.00
1.00×10^{-3}	-3.00	3.00
1.00×10^{-5}	-5.00	5.00
1.00×10^{-9}	-9.00	9.00

The value of $[H^+]$ in pure water at 25°C is 1.00×10^{-7} mol/L and the pH is 7.00. Such water is said to be **neutral**, neither acidic nor basic. Water with a pH *less than* 7.00 is acidic, whereas water with a pH greater than 7.00 is basic.

The average global concentration of CO_2 gas in air in the early 2000s is about 370 parts per million by volume, and going up by about 1 ppm per year. The concentration of dissolved carbon dioxide, $[CO_2(aq)]$, in water in equilibrium with 370 ppm atmospheric air at 25°C is 1.21×10^{-5} mol/L. This low concentration is enough to make the water slightly acidic because the dissolved carbon dioxide reacts with water,

$$CO_2 + H_2O \leftrightarrow H^+ + HCO_3^- \qquad (7.7.2)$$

to a sufficient extent to make $[H^+] = 2.3 \times 10^{-6}$ mol/L. This corrresponds to a pH of 5.6. This pH is lower than the neutral pH of 7, and the water is slightly acidic. Since unpolluted rainfall formed in the atmosphere is in equilibrium with atmospheric CO_2, it is slightly acidic.

7.7. WHY NATURAL WATERS CONTAIN ALKALINITY AND CALCIUM

Most surface water and groundwater contains **water alkalinity**, defined as the ability of water to react with and neutralize acid (H^+). This is because such water contains bicarbonate ion, HCO_3^-, which can react as follows with H^+ ion:

$$HCO_3^- + H^+ \leftrightarrow CO_2(aq) + H_2O \qquad (7.8.1)$$

This is a useful reaction because, among other reasons, it acts to neutralize pollutant acids, such as those from acid rain, that get into water. This water obtains alkalinity when dissolved carbon dioxide in the water reacts with calcium carbonate mineral (Figure 7.5), the same process by which the water acquires **water hardness** in the form of dissolved Ca^{2+} ion:

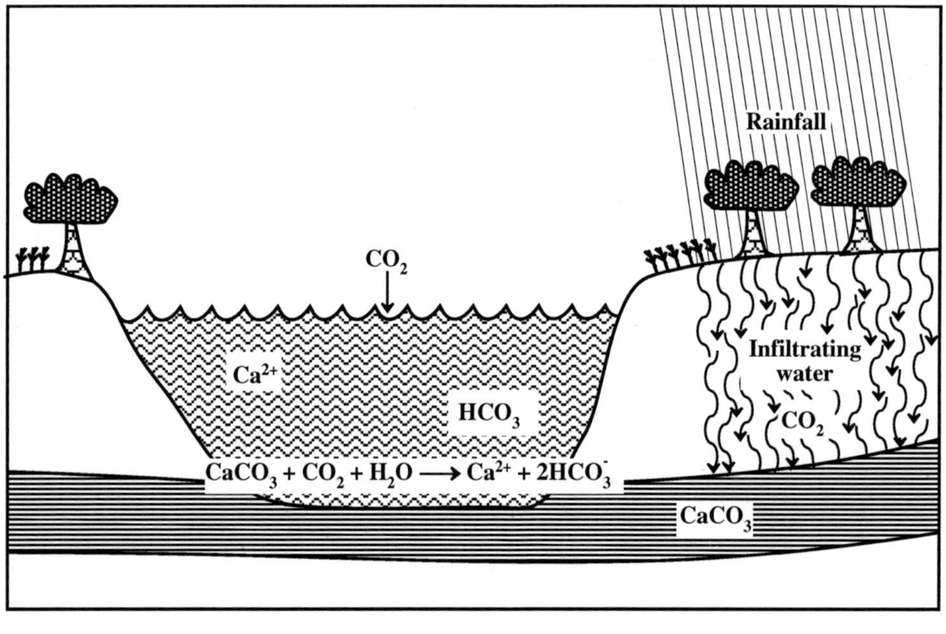

Figure 7.5. Water containing dissolved carbon dioxide from the atmosphere or from bacterially produced sources underground reacts with limestone, $CaCO_2$, to produce dissolved Ca^{2+} (water hardness) and dissolved HCO_3^- (water alkalinity).

$$CO_2(aq) + CaCO_3(s) + H_2O \leftrightarrow Ca^{2+}(aq) + 2HCO_3^- \quad (7.8.2)$$

7.9. METALS IN WATER

Although we commonly write the formulas of metal ions dissolved in water with simply the symbol and charge of the metal, such as Ca^{2+} for calcium discussed above, metal ions are always bound to water molecules in aqueous solution. There are usually 6 water molecules bound to each metal ion, so that dissolved calcium ion can be represented as $Ca(H_2O)_6^{2+}$.

Sometimes the water in the metal ion is displaced by other species. Such species include organic **chelating agents**, which can bond to metal ions in 2 or more places. Chelate formation is reversible, and a metal chelate can dissociate to reform the chelate and the metal ion. Chelates tend to be particularly stable, and they are very important in natural water systems. They are also involved in life systems; for example, blood hemoglobin is a chelate that contains Fe^{2+} ion bonded simultaneously to 4 N atoms on the hemoglobin protein molecule. This particular group binds with molecular O_2 and enables hemoglobin to carry oxygen to all the organs of the body.

Water in nature may contain naturally-occurring chelating agents called **humic substances** that are complex molecules of variable composition left over from the biodegradation of plant material. One of their most important effects in water is their ability to bind with Fe^{2+} ion, which is normally present as insoluble species in water. Soluble humic substance chelates of iron may be encountered in water as colored matter commonly called **gelbstoffe** (German for "yellow stuff"). It is very difficult to remove such chelated iron by water treatment processes.

Humic substances are of special concern because when they are present in water during disinfection by treatment with chlorine, they give rise to the formation of **trihalomethanes**, such as chloroform, $HCCl_3$. These pollutants are suspected of being carcinogens (cancer-causing agents), and permissible levels are extremely low. Therefore, before chlorination water has to be treated carefully to remove humic substances.

7.10. WATER INTERACTIONS WITH OTHER PHASES

So far, the discussion of water chemistry has been centered around species dissolved in water. In fact, most important processes that occur in natural waters and wastewaters and during water purification involve other solid or gaseous phases. This is illustrated in Figure 7.6. Bodies of water rest

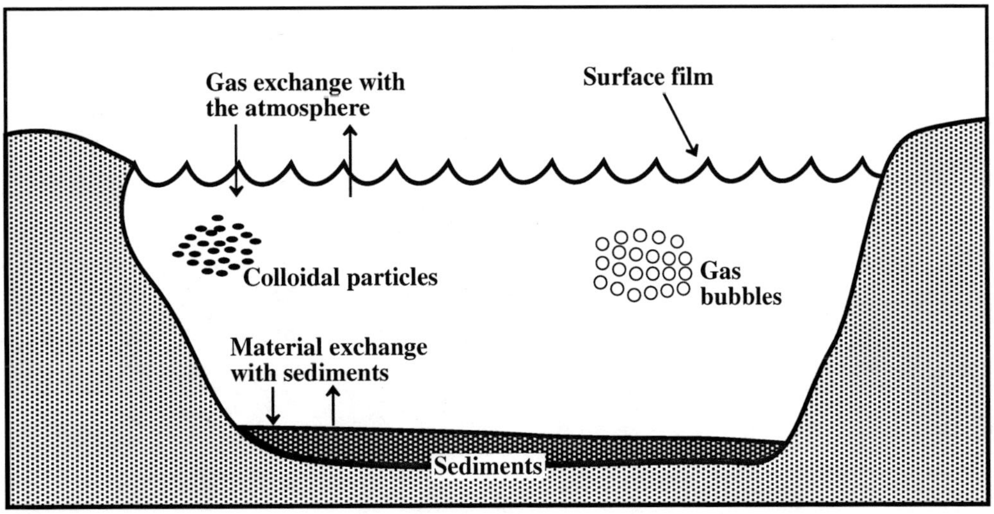

Figure 7.6. Most important chemical processes associated with water involve other phases and commonly occur at the interfaces of these materials with water.

on layers of **sediments** and are in contact with the atmosphere. Materials are exchanged between water and sediments. Sediments are especially susceptible to the effects of insoluble pollutants and tend to collect pollutants such as heavy metals and dense organic liquids. Organisms living in contact with sediments may accumulate high levels of such pollutants, which may be passed on to fish and birds that feed on fish. Anaerobic bacteria that do not require molecular oxygen living in sediments can generate gases such as methane, CH_4, and hydrogen sulfide, H_2S.

There is a constant exchange of carbon dioxide between water and the atmosphere. The rapid transport of oxygen from air to water can be essential to fish in water. The surface of water tends to be covered by a thin film of water-insoluble organic matter in which compounds exposed to the sun may undergo photochemical degradation.

A particularly important phase associated with water consists of **colloidal particles**. These are very small particles of around a micrometer (μ m, one millionth of a meter) in size or less. Because of their very small sizes, colloidal particles stay suspended in water and do not settle out. Such small particles have very large ratios of surface area to mass, which makes them quite reactive. Some organisms, such as single-celled algae and bacteria exist as colloidal particles in water. Colloidal particles may be **hydrophobic colloids** that are not attracted to water and remain as colloidal particles because of their like electrical charges, **hydrophilic colloids**, such as proteins or microbial cells, that remain suspended in water because of their binding to it by hydrogen bonds, and **association colloids** consisting of aggregates of molecules or ions. A commonly encountered association colloid consists of soap anions. Soap is a salt of a fatty acid that has a long hydrocarbon chain at the end of which is a carboxylate anion group as shown below for sodium stearate, the kind of soap that can be made by reacting beef fat with sodium hydroxide:

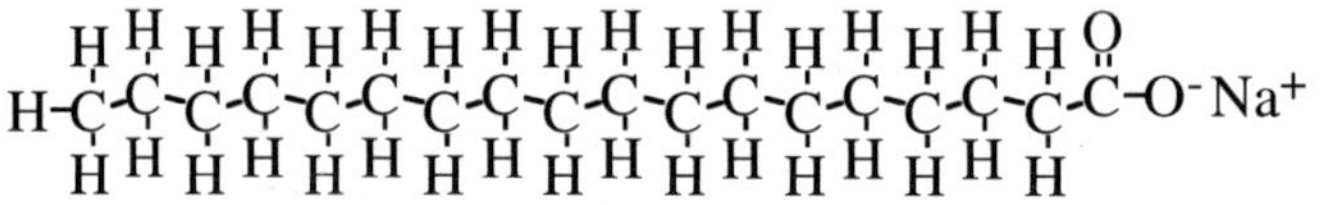

When placed in water, the long organophilic tails of these anions, $CH_3(CH_2)_{16}^-$, behave like water-insoluble hydrocarbons and cluster together in a microscopic ball called a **micelle** composed of around 100 of the anions, on the surface of which are located the negatively charged $-CO_2^-$ "heads" of the anions. One of the reasons that soap is an effective cleaning agent is that the micelles engulf oil and grease droplets attracted by the hydrocarbon portions of the micelles and carry them into suspension in water.

Soap, Candy, and Bathtub Ring

Earlier in this chapter *water hardness* was mentioned in association with dissolved calcium. Water hardness is manifested by the reaction of the dissolved calcium with soap and refers to the formation of insoluble calcium salts of the soap anions. Sodium stearate (see above), is a form of soap, and it reacts with calcium ion,

$$2CH_3(CH_2)_{16}CO_2Na + Ca^{2+} \rightarrow \{2CH_3(CH_2)_{16}CO_2\}_2Ca(s) + 2Na^+ \quad (7.10.1)$$

to form a solid precipitate of calcium stearate, $\{2CH_3(CH_2)_{16}CO_2\}_2Ca$. This material does not work as a cleaning agent and deposits on bathtubs as infamous "bathtub ring." Some sugarless candies have calcium stearate as the main ingredient, mixed with artificial sweeteners, flavors, and food coloring.

7.11. HEAVY METAL WATER POLLUTANTS

The **heavy metals** are those metals of relatively higher atomic numbers. Some heavy metals are considered among the most troublesome and toxic water pollutants. The heavy metals of most concern in water are addressed briefly here.

Cadmium, Cd, is widely used in metal plating and in making small batteries, such as those used in some cameras. Cadmium is very toxic, damaging red blood cells, kidney tissue, and testicular tissue. Cadmium can enter water from industrial pollution sources.

Lead, Pb (from its Latin name of plumbum), is arguably the most common heavy metal pollutant because of its widespread use in industry, in the manufacture of lead storage batteries, formerly as a leaded additive to gasoline, as a pigment in white house paint, and as an anticorrosive primer applied prior to painting steel. Exposure to lead causes a number of adverse health effects and is suspected of causing mental retardation in exposed children. Lead was widely used in plumbing, and its use in solder to join together copper water pipe was banned only recently.

The most tragic modern incident of poisoning from **mercury**, Hg, occurred in the Minamata Bay area of Japan from 1953 through 1960 when people consumed seafood from the bay, which had been polluted by drainage of mercury wastes from a chemical plant. The total number of cases of mercury poisoning reported was 111 and there were 43 deaths and 19 cases of congenital birth defects in babies whose mothers had eaten the contaminated seafood. Hazardous methylated forms of mercury are discussed under organometallic compounds below.

Arsenic, As, is a metalloid (among the elements bordering metals and nonmetals in the periodic table), but its environmental and toxicological effects are much like those of heavy metals. Arsenic has been employed in hundreds of dastardly murder plots over the years, and can cause both chronic poisoning over a long period of time and acute poisoning from the ingestion of as little as 100 mg of the element. Before the advent of more modern pesticides, arsenic compounds were used in huge quantities to kill pests in orchards and on crops. Among the deadly compounds used for this purpose were lead arsenate, $Pb_3(AsO_4)_2$; sodium arsenite, Na_3AsO_3; and Paris Green, $Cu_3(AsO_3)_2$.

In modern times, the most tragic occurrence of water pollution by arsenic has taken place in Bangladesh, an impoverished nation on the Indian subcontinent. Around 1980 a program sponsored by the United Nations to drill wells in this country brought abundant, pathogen-free drinking water to many areas of Bangladesh. However, about 20 years later, symptoms of arsenic poisoning appeared among the people using this water source, leading to debilitating illnesses and death. These otherwise wholesome water sources were contaminated with dangerous levels of arsenic.

Organically Bound Metal Water Pollutants

In a number of instances, simple hydrated metal ions are not soluble enough in water to cause pollution problems. However, organically bound metals may be considerably more mobile and in some cases more toxic. Binding of metals as metal chelates was discussed in Section 7.9. Another form of metal binding occurs when metals are bonded directly to carbon in hydrocarbon groups such as the methyl group ($-CH_3$) to produce **organometallic compounds**.

A nasty surprise related to the formation of organometallic compounds in water was revealed in 1970 when it was found that fish in some areas, such as in Lake Saint Clair located between Michigan and Ontario, Canada, had dangerously high levels of mercury. It was known that the electrically driven chloralkali method of producing sodium hydroxide and elemental chlorine — both important industrial chemicals — used mercury electrodes in electrolyzing sodium chloride solutions, and that each unit in this process was releasing up to 14 kilograms of mercury per day. However, it was known that the inorganic forms of mercury released formed very insoluble precipitates in water and were thought to be safely buried with lake and river sediments. Subsequent investigation showed that anaerobic bacteria growing in the sediments were attaching methyl groups, $-CH_3$, to mercury:

$$HgCl_2 \xrightarrow[\text{bacteria}]{\text{Anaerobic}} CH_3HgCl + Cl^- \qquad (7.11.1)$$

The monomethylmercury ion in this compound, CH_3Hg^+, is soluble and mobile in water and the dimethylmercury, $(CH_3)2Hg$, also produced is volatile as well. These mobile species were released from the sediments and concentrating in fish tissue.

In recent years, the organometallic compounds of most concern in water have been the organotin compounds. Up to 40,000 metric tons per year of organotin compounds, such as tetra-*n*-butyltin,

$$\begin{array}{c} (C_4H_9) \\ | \\ (C_4H_9)-Sn-(C_4H_9) \\ | \\ (C_4H_9) \end{array} \qquad \textbf{Tetra-}\textbf{\textit{n}}\textbf{-butyltin}$$

have been produced each year as industrial biocides to prevent biological growths on surfaces. Boat and ship hulls were painted with organotin-containing paints to prevent the growth of "aufwuchs," organisms that attach themselves to such surfaces and greatly increase the friction, hence the fuel costs, of propelling these vessels through water. Because of water pollution concerns, in 1998 the International Maritime Organization agreed to ban organotin antifouling paints on all ships by 2003.

Heavy Metal Pollutants and Green Chemistry

From the discussion above, it is seen that several heavy metals are among the more troublesome water pollutants, with a number of others in more isolated cases. Obviously, it is important to prevent such elements from getting into water. Here the practice of green chemistry plays an important role. One approach is to strictly forbid the release of heavy metals into water. This has worked reasonably well but, as with all command and control measures, it is subject to human oversight, accident, and even deliberate releases to try to avoid disposal costs.

A much better approach, where possible, is to use the principles of green chemistry to avoid any possibility of pollutant release. For example, a command and control approach to preventing the release of cadmium in electroplating operations would be to strictly control any releases. But, a green chemistry approach is to come up with safer substitutes for cadmium in metal treatment so that there is never any cadmium around to be released. Similar approaches can be tried with any other water pollutant.

2 INORGANIC WATER POLLUTANTS

Cyanide is deadly as volatile hydrogen cyanide, HCN, or as cyanide ion, CN^-; as little as 60 mg of cyanide can be fatal to a human. Cyanide is produced from coke ovens and is widely used in the metals industry for metal extraction from ores and for metal cleaning and electroplating. Cyanide is sometimes released to water, especially from metal extraction operations. One such incident occurred in 1995 when cyanide-containing water mixed with red clay from mine tailings was released from a gold mining operation in the South American country of Guyana. A breached dam allowed release of approximately 2.7 billion liters of cyanide-contaminated wastes. The cyanide present in the water at a level of approximately 25 parts per million killed all the fish in the small Omai Creek leading from the site of release to the Essequibo River, where the dilution from the river flow reduced levels of cyanide to below fatal concentrations. In 1992, cyanide and heavy metals spilling from the Summitville mine in southern Colorado killed all life in a 17-mile section of the Alamosa River. The state of Colorado agreed to settle for damages totaling $30 million in late 2000.

Although no human fatalities resulted from these incidents, the scope of the spill and the extreme toxicity of cyanide point to the dangers of using large quantities of such a toxic reagent. Regulations forbidding such a release were not helpful in this case — it happened. A green chemistry approach to this problem would be to find safer alternatives to cyanide so that there is no possibility of its release.

Excessive levels of ammoniacal nitrogen in the form of ammonium ion, NH_4^+, or molecular ammonia, NH_3, cause water-quality problems and may be harmful to aquatic life. However, ammoniacal nitrogen at lower levels is a normal constituent of water and is even added deliberately to drinking water so it can react with chlorine used for disinfection to provide for residual disinfection in water distribution systems.

Hydrogen sulfide, H_2S, is a toxic gas with a foul odor that is produced by anaerobic bacteria acting upon inorganic sulfate (see Section 7.5 and Reaction 7.5.4), from geothermal sources (hot springs) and as a pollutant from chemical plants, paper mills, textile mills, and tanneries. Because of its bad odor and toxicity, it is an undesirable pollutant in water.

Microbial degradation under ground may generate **carbon dioxide**, CO_2, that exists as free carbon dioxide in water (see Section 7.6). Excessive levels can be toxic to aquatic organisms and can make water corrosive because of its acidity and tendency to dissolve protective $CaCO_3$ coatings on pipe.

Nitrite ion, NO_2^-, can be generated by bacteria from inorganic nitrogen species and is added to some industrial process water as a corrosion inhibitor. Normally, levels are low because of the narrow range of conditions under which nitrite is stable. Nitrite is quite toxic, causing methemoglobinemia by converting the hemoglobin in blood to methemoglobin, a form useless for transporting oxygen. **Nitrate ion**, NO_3^-, is a more common water contaminant, but is of less concern because of the high tolerance of adult humans for it. However, infants and multi-stomached ruminant animals (cattle, sheep, goats, deer) have conditions in their stomachs that can result in the reduction of nitrate to toxic nitrite. This has killed a large number of animals and has resulted in fatal cases of methemoglobinemia in infants.

Acidity

Strong acid pollutants that cause water to have a low pH are very damaging to organisms living in water. Although spills of acids can pollute water, the most common acid pollutant comes from the bacterial mediated oxidation of iron pyrite, FeS_2, to produce sulfuric acid. The overall process is represented by the reaction,

$$4FeS_2(s) + 2H_2O + 15O_2 \rightarrow 2H_2SO_4 + 2Fe_2(SO_4)_3 \qquad (7.12.2)$$

which produces sulfuric acid, H_2SO_4, and $Fe_2(SO_4)_3$, which also acts as an acid. The bacteria that carry out these reactions include *Thiobacillus ferrooxidans*, *Thiobacillus thiooxidans* and *Ferrobacillus ferrooxidans*.

Another source of water pollutant acid is from acid rain. Hydrogen chloride, HCl, emitted to the atmosphere forms hydrochloric acid, whereas nitric oxide, NO, and sulfur dioxide, SO_2, emitted to the atmosphere can be oxidized in the presence of atmospheric water vapor to produce nitric acid, HNO_3, and sulfuric acid, respectively. Falling from the atmosphere as acid rain, these acids are especially damaging to life in lake water that does not have contact with the kinds of minerals that can neutralize the acid and keep the pH from falling low enough to damage aquatic life. Lakes in some parts of Canada and in New England in the U.S. are especially susceptible to this kind of damage.

There are not many direct pollutant sources of excessively high **alkalinity**, which is due to salts such as sodium carbonate, Na_2CO_3, that tend to raise the pH to levels high enough to be harmful to aquatic life. Some soils and rocks associated with mining have high alkalinity, and human activities can cause this alkalinity to be leached into water.

Water **salinity** is due to dissolved salts, such as sodium chloride and calcium chloride. Each pass of water through a municipal water system adds salinity, especially from NaCl flushed into water from recharging water softeners. Irrigation also adds salinity and is responsible for the high levels of salt in California's Salton Sea, an artificial body of water with no outlet to the ocean created artificially by runoff from irrigated lands. Fertilizers are salts, and they get into runoff water during irrigation.

Water Pollutants That Are Just Too Nutritious

Some inorganic species are pollutants, not because they are toxic, but because they are very nutritious for algae in water. Algae and other plants require a number of different inorganic nutrients. Those required in the greatest quantity are inorganic phosphorus ($H_2PO_4^-$, HPO_4^{2-}), nitrogen (NH_4^+, NO_3^-), and potassium (K^+).

The plant fertilizers described above can get into water from a number of sources, including fertilizers put on soil to enhance crop growth, from some industrial pollutants, and — especially in the cases of phosphates and nitrates — from the degradation of sewage in wastewater. Until they were phased out during the 1970s, phosphate builders in phosphate detergents were a major source of phosphate in water. So what is wrong with having nutrient-rich water? If the levels of nutrients are too high, algae grow too well and generate too much biomass. This material eventually dies and decays, which uses up all the oxygen in the water and clogs a body of water with dead plant matter. This unhealthy condition of excessive plant growth is called **eutrophication**, derived from the Greek word meaning "well-nourished."

Eutrophication is usually curtailed by limiting phosphate input into bodies of water and streams. The reason for doing this is that phosphorus is usually the limiting nutrient, much like the limiting reactant in a chemical reaction (see Section 4.10). So if phosphate levels are cut down, algal growth and resulting eutrophication are curtailed. Around 1970 the most common source of pollutant phosphorus was from phosphates added as builders to household detergents and discharged with sewage. Treated wastewater discharged to streams and other natural waters added phosphate that caused eutrophication. By the application of green chemistry (though not known as such then) other chemicals were found that could substitute for phosphate in detergents without causing eutrophication.

7.13. ORGANIC WATER POLLUTANTS

A whole host of water pollutants are organic compounds, which include virtually all carbon-containing compounds. These may be nontoxic, highly biodegradable materials, such as waste food in sewage, that are nevertheless bad for water because of the dissolved oxygen consumed when they degrade (see below). At the other extreme are very poorly degradable substances, such as polychlorinated biphenyls (PCBs), that tend to accumulate in sediments and in the lipid (fat) tissues of fish and birds that eat fish. Organic water pollutants are addressed briefly here.

Oxygen-Demanding Substances

One of the most common water pollutants consists of substances that are not toxic, but serve as excellent food sources for bacteria in water. When such substances, represented here as $\{CH_2O\}$, undergo biodegradation,

$$\{CH_2O\} + O_2 \xrightarrow{\text{Microorganisms}} CO_2 + H_2O \qquad (7.13.1)$$

oxygen is consumed. The resulting low oxygen levels are detrimental to fish and some other forms of aquatic life and also include the likelihood that odorous reduced species, particularly H_2S, will be evolved. High levels of oxygen-demanding substances are associated with the water pollution phenomenon of eutrophication discussed above.

The oxygen-consuming potential of biodegradable materials in water is called **biochemical oxygen demand, BOD**. BOD is commonly expressed as the amount of oxygen consumed in biodegrading the organics in a liter of water.

Sewage

One of the most common sources of BOD is sewage from domestic, commercial, and industrial sources. In addition to BOD from fecal matter and food wastes, sewage contains oil, grease, grit, sand, salt, soap, detergents, degradation-resistant organic compounds, and an incredible variety of objects that get flushed down the drain. Sewage used to be simply discarded to the nearest handy stream or body of water, an unfortunate practice that still continues in many parts of the world. There are several reasons that this should not be done, the most obvious of which is that sewage stinks. But it also carries disease-causing (pathogenic) microorganisms (bacteria, virus, protozoa) and it exerts a high BOD in the water to which it is discarded. Sewage treatment to reduce BOD is addressed in Section 7.17.

Detergents in Sewage

A major problem with sewage that was solved by the application of green chemistry before it was known as such was the presence of detergents in the sewage. After detergents came into widespread use, in the 1940s, massive layers of foam that sometimes covered most of the treatment plant developed at sewage treatment plants in which air is blown through sewage to promote biodegradation and at the outfalls where treated sewage was discharged into streams. There were even cases of treatment plant workers who walked into the foam, fell, and were asphyxiated by the gases entrained by the foam. The phenomenon even extended to surface water and groundwater contaminated with treated sewage, which developed troublesome heads of foam upon being drained from a water faucet. Furthermore, fish fingerlings were killed by whatever was getting through the sewage treatment plant.

The culprit in the troublesome incidents of foam associated with sewage was found to be the surfactant (the active ingredient that makes water "wetter") in waste detergents, a material called alkyl benzene sulfonate, ABS, for which the structure is shown below:

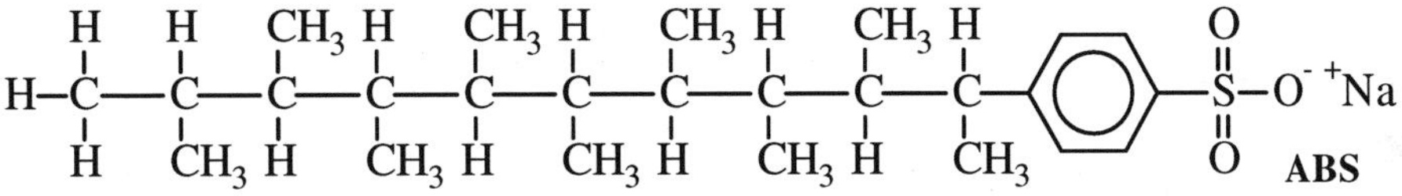

ABS

(Recall from Section 5.2 that the hexagon with a circle in it stands for an aromatic benzene ring.) It is seen in this structure that there is a chain of 10 C atoms (counting the one that is part of the end -CH_3 group), and attached to alternate C atoms on the chain are -CH_3 groups. This molecule and similar ones with different C-atom chain lengths were found to be very effective stable surfactants. There was only one problem, which is that bacteria hate branched chains and would degrade this compound only slowly. Hence it stayed around forming foam in sewage treatment plants and significant amounts remained even after the treated water was discharged.

The solution to the ABS detergent problem was found in the 1960s. Although it predates the concepts of green chemistry and industrial ecology, it is still cited as one of the best applications of green chemistry. The solution to the problem created by chemistry was to use chemistry to devise another kind of molecule, equally effective as a surfactant, but much more to the liking of bacteria. The substance developed is still used today and is known as **linear alkyl sulfonate, LAS** shown below:

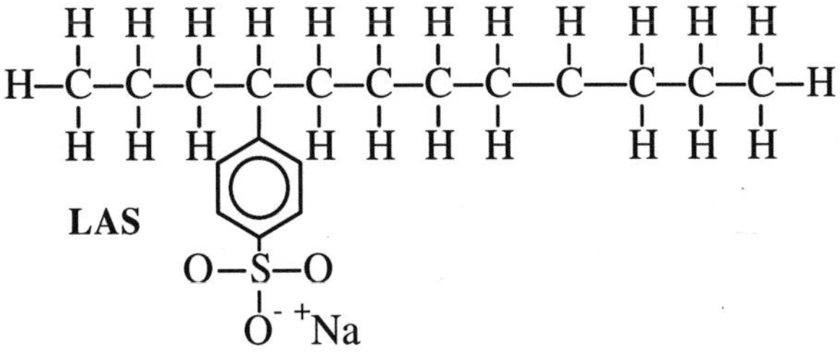

LAS

Like ABS, LAS has a chain of carbon atoms. But it is a straight chain, not a branched one. Bacteria like straight chains, the LAS degraded readily in the sewage treatment plant, and the problem with surfactants in sewage went away.

7.14. PESTICIDES IN WATER

Food producers and gardeners constantly struggle with pests of various kinds that consume food, ruin crops, compete for space, water, and nutrients, and otherwise make life difficult and expensive. In addition to **insecticides** used to kill insects and **herbicides** to control weeds, there are many other kinds of pesticides, including **bactericides** to control bacteria, **slimicides** to control slime-causing organisms in water, and **algicides** used against algae, all potential water pollutants. When the major water pollution control laws were enacted around 1970, a great concern was DDT and other very persistent insecticides that polluted water and sediments. In rural areas the greater problem is from herbicides which, because they must be applied directly to cropland, have an inherent tendency to get into runoff and into water sources. Fortunately, pesticides have become generally safer and less persistent in the environment. Although pesticide use has leveled off, it is still enormous, around 350 million kilograms per year for U.S. agriculture as well as large amounts for nonagricultural uses including forestry, landscaping, gardening, food distribution, and home pest control.

Insecticides

Figure 7.11 shows some insecticides that either are potential water pollutants or have been water pollutants in the past. The insecticides that caused the greatest water pollution problems, and that still do so in developing countries where their use is still allowed, are the **organochlorine** insecticides. There were many of these, probably best exemplified by DDT and including others, such as chlordane. DDT was widely sprayed in the environment and was instrumental in killing mosquitos that carry malaria. Highly persistent chlordane was the most effective insecticide against termites and was buried around buildings to prevent termite infestation.

The organochlorine insecticides were dominant from the 1940s until the 1960s. In general, they are not particularly toxic to humans and other animals. However, DDT and related compounds have an even more detrimental characteristic because of their tendency to undergo **bioaccumulation** in fish and other organisms, concentrating in fat tissue. Furthermore, as organisms that have accumulated these compounds are eaten, and these organisms, in turn, are eaten by larger animals, the organochlorine compounds become progressively more concentrated in fat tissue, a process called **biomagnification**. It was this phenomenon that caused the birds of prey — eagles, falcons, hawks — to become so contaminated with DDT that they produced soft egg shells that broke before young could hatch. This threatened a number of species of birds, including the Bald Eagle that is the U.S. symbol, leading to the banning of DDT and most other organochlorine insecticides.

For a time, as DDT and other organochlorine insecticides lost favor, **organophosphate insecticides**, which are organic derivatives of phosphoric acid, H_3PO_4, came into common use. These insecticides had a big advantage in being biodegradable with no tendency to undergo bioaccumulation. Two examples are methyl parathion and malathion. Parathion, once the most widely used organophosphate insecticide, is a very effective insecticide and, because of its rather rapid biodegradability, it was not usually a significant water pollutant. It acts by inhibiting the action of acetylcholinesterase, an enzyme essential to nerve function. This is the same mode of action of military poison "nerve gases," such as sarin, and a significant number of fatal poisonings occurred due to parathion exposure. Although parathion is now banned, malathion remains on the market and is only about 1/100 as toxic to mammals as is parathion. This is because, as shown by the arrows in its structural formula in Figure 7.7, malathion can be cleaved with addition of water by enzymes possessed by humans and other mammals, but not by insects.

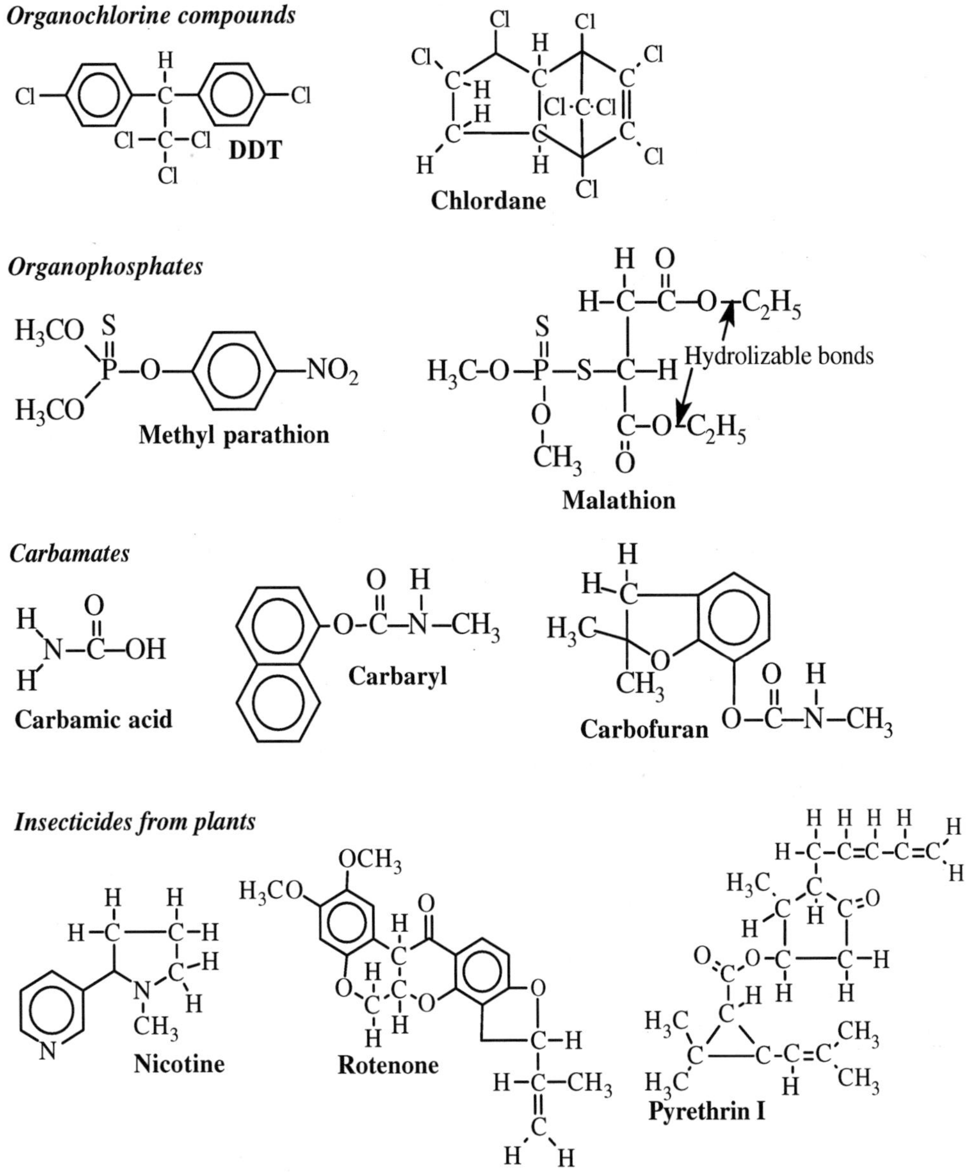

Figure 7.7. Some examples of insecticides that are, or have been, potential water pollutants.

Carbamates, in which various hydrocarbon groups are substituted for H on carbamic acid, largely replaced phosphates, and several carbamates are still widely used. **Carbaryl** (Sevin) is used to kill insects on lawns or gardens and, because of its low toxicity to mammals, can be sprinkled on pets in (often futile) attempts to rid them of flea infestations. Highly water-soluble **carbofuran** is a plant **systemic insecticide** that is taken up by roots and leaves and distributed through the plants, killing insects that feed on the plants. Like organophosphates, carbamates are acetylcholinesterase inhibitors, but are not highly toxic to animals other than insects. They are very biodegradable and are not generally serious water pollutants.

Some important insecticides have been derived from plants. These include **nicotine** from tobacco and **rotenone** extracted from the roots of some kinds of plants. There has been some concern about rotenone because of a year 2000 report that it causes a syndrome resembling Parkinson's disease in mice. **Pyrethrins** are excellent, biodegradable insecticides extracted from dried chrysanthemum or pyrethrum flowers. Noted for their low toxicities to mammals and excellent abilities to "knock down" flying insects, pyrethrins were probably used to kill insects in China 2000 years ago. The current major source of these environmentally-friendly insecticides are chrysanthemum varieties grown in Kenya.

Because of the excellent insecticidal properties of pyrethrins, their synthetic analogs, **pyrethroids**, have been synthesized and widely produced. These substances are now insecticides of choice for household applications. Because of their biodegradability, pyrethrins and pyrethroids are not serious water pollutants.

Herbicides

Herbicides are applied to vast areas of farmland to control weeds that crowd out corn, soybeans, cotton, and other economically important crops. The manner in which herbicides must be applied ensures that they are susceptible to being washed off fields by rainfall with a high potential to become water contaminants. Herbicides are commonly found in drinking water supplies and some municipalities are required to use activated carbon filtration to remove herbicides from municipal drinking water.

Herbicides come in an enormous variety of chemical compounds. One of the most widely produced types consists of the **triazines**, which have 6-membered rings in which C atoms alternate with N atoms ("triazine" denotes 3 carbon atoms). Three of these widely used to control weeds on corn and soybeans are atrazine, simazine, and glyphosate:

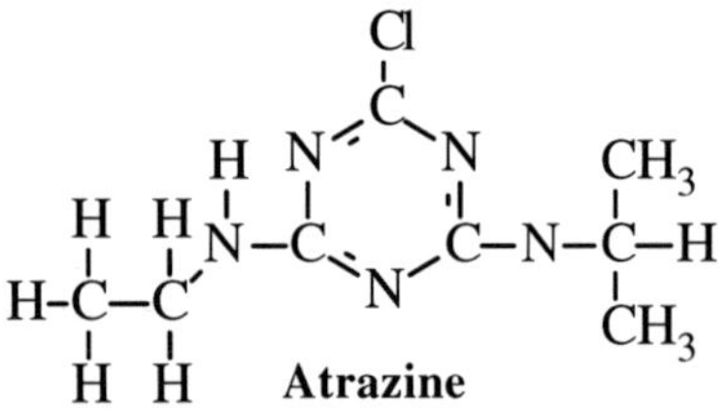
Atrazine

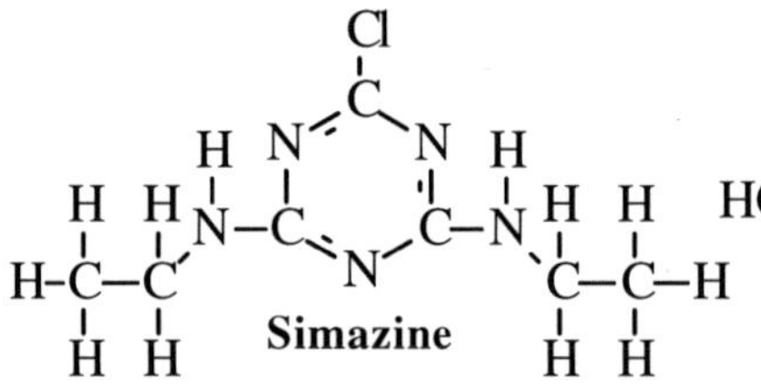
Simazine

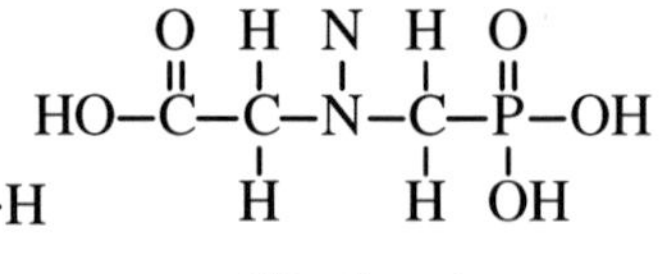
Glyphosate

Glyphosate kills weeds by interfering with the synthesis of some kinds of amino acids essential for plant proteins. It is a postemergence herbicide, which means that it is applied directly to weeds after they have started to grow. It is effective against broadleaf weeds, grasses, and perennial plants. Glyphosate is the active ingredient in Monsanto's Roundup herbicide. Monsanto has developed genetically modified soybeans and other crops that are "Roundup ready," meaning that they are not harmed by the direct application of this herbicide, which kills competing weeds.

The Infamous Dioxin

Some of the more severe pollution problems associated with pesticides have come from their manufacture. One of the more notorious of these was **dioxin**, known chemically as 2,3,7,8-tetrachlorodibenzo-*p*-dioxin (TCDD), the structure of which is shown below:

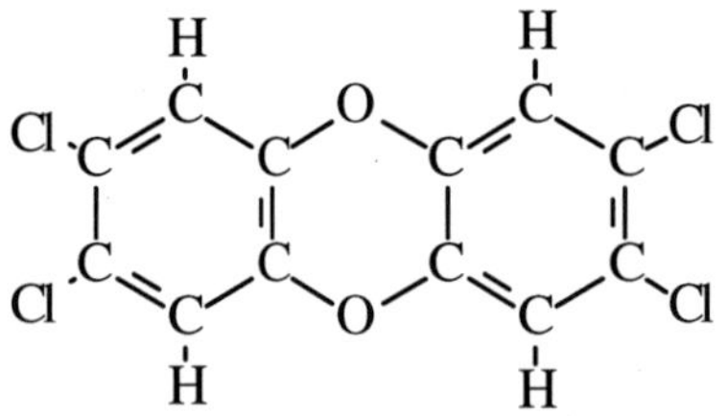
2,3,7,8-Tetrachlorodibenzo-*p*-dioxin (commonly called dioxin)

Dioxin was produced as a byproduct of the manufacture of 2,4,5-trichlorophenoxyacetic acid, 2,4,5-T, the infamous "Agent Orange" used to defoliate jungles in the Vietnam war. Dioxin is essentially insoluble in water, melts at 305°C (very high for an organic compound), is chemically stable up to 700°C, and does not undergo biodegradation well. Although it is remarkably toxic to some animals (especially guinea pigs) it is not extremely toxic to humans. It has no uses, but is generated as a byproduct of the manufacture of some chemicals and during the incineration of chlorine-containing plastics, such as polyvinyl chloride. It was a highly undesirable impurity from the synthesis of 2,4,5-T herbicide mentioned above.

Because of its extremely low water solubility, dioxin is not a common water pollutant. However, it can accumulate in sediments. It gained notoriety as a hazardous waste in the early 1970s when dioxin-contaminated wastes were mixed with waste oil and sprayed for dust abatement on roads, horse arenas, and other areas in the state of Missouri. Horses and birds were killed, and the entire town of Times Beach, Missouri, was contaminated with the contaminated oil. The U.S. Environmental Protection Agency bought out the whole town in March 1983, at a cost of $33 million. Soil from the town along with other dioxin-contaminated soil was subsequently incinerated at a total estimated cost of about $80 million.

7.14. POLYCHLORINATED BIPHENYLS (PCBs)

Polychlorinated biphenyls (**PCB**s), discussed in Section 5.4 and shown in Figure 5.5, consist of a class of 209 compounds made by substituting from 1 to 10 chlorine atoms for H atoms on biphenyl. PCBs are notable for their extreme chemical and thermal stability, resistance to biodegradation, low vapor pressure, and high dielectric constants. They even survive ordinary combustion processes and are dispersed to the atmosphere with stack gases. Until their manufacture and use were banned by the Toxic Substances Control Act of 1976, they were widely used as coolant-insulation fluids in transformers and capacitors, as plasticizers to make plastics more flexible, to impregnate cotton and asbestos, and as additives to some epoxy paints. The extreme stability that led to these uses also contributed to their widespread dispersion and persistence in the environment.

PCBs are denser than water and tend to accumulate in sediments. One of the worst incidents of sediment pollution by PCBs occurred during 1950 to 1976 as thousands of kilograms of PCBs were dumped — legally at the time — into New York's Hudson River from electrical equipment manufacturing operations. The U.S. Environmental Protection Agency and New York state agencies came to an agreement in 2002 with General Electric that could lead to the company having to spend up to several hundred $ million to dredge and dispose of contaminated Hudson River sediment.

7.16. RADIOACTIVE SUBSTANCES IN WATER

Radioactive isotopes, or **radionuclides** can get into water from either natural sources or from the fission of uranium or plutonium in nuclear power reactors or (formerly) above-ground weapons testing. Radionuclides have unstable nuclei that change spontaneously to nuclei of atoms of different elements by the emission of **ionizing radiation** in the form of alpha particles, beta particles, and gamma rays, called ionizing radiation because they produce reactive ions in materials including flesh. These ions and other reactive species can damage DNA, impede the body's ability to make hemoglobin, and cause a number of biological effects, including severe anemia, mutations, cancer, and death.

There are 3 major kinds of ionizing radiation that are most commonly given off by radionuclides. Some of the heavier elements, such as uranium and radium emit **alpha particles**, a helium atom nuclei composed of two neutrons and two protons, denoted $^{4}_{2}\alpha$. The penetrating ability of these very heavy particles is low, so that they do not pose a hazard outside the body. However, if they enter the body, such as by ingestion of contaminated drinking water, they cause enormous damage to exposed tissue. **Beta radiation** is in the form of high-energy electrons designated $^{0}_{-1}\beta$, or, less commonly, positive electrons, called positrons. These particles are more penetrating than alpha particles. **Gamma rays** are not particles, but are electromagnetic radiation that behave like very short wavelength, high-energy X-rays. The energies of gamma rays are highly specific and can be measured by sophisticated gamma ray spectrometers as a means of identifying the kinds and quantities of radionuclides emitting the radiation. Radionuclides decay with specific half-lives that can range from fractions of a second to millions of years. After the passage of each half-life, the radioactivity of a specific radionuclide is half of what it was at the beginning of the half-life. This means that all radionuclides are eventually converted to nonradioactive forms, although this may take a very long time.

Aside from some special circumstances involving production of radionuclides in military production facilities, the greatest concern with respect to these materials in water arises from natural geological sources. Specifically, the alpha particle emitter radium-226, ^{226}Ra, half-life 1620 years, is a particular concern in drinking water. A number of municipal water supplies contaminated with this radionuclide by leaching into groundwater from rock formations underground have been shut down because of the hazard presented.

Immediately following World War II, aboveground testing of nuclear weapons was a significant source of environmental pollution of radionuclides. Nuclear bombs typically detonated on towers in remote regions of New Mexico, Nevada, and Russia would generate large quantities of uranium fission products and would produce more radionuclides by the absorption of neutrons from the bomb by dirt and sand. A large mass of radioactive dust and debris would be entrained in a rapidly ascending column of hot gases and dispersed throughout the world. "Fallout" from these tests got onto land, causing particular concern about strontium-90 falling on pastureland and getting into cow's milk. Radionuclides were also scavenged from the atmosphere by rainfall, which could get into water supplies. The radioactive products of most concern from these tests were of elements that the body recognizes as material to be incorporated into tissue. These were strontium-90 (^{90}Sr), half-life 28 years, which is in the same chemical group as calcium and is incorporated into bone; cesium-137, (^{137}Cs), an alkali metal that the body handles much like sodium and potassium ions; and iodine-131 (^{131}I), half-life 8 days, that is attracted to the thyroid and can impair its function and even cause thyroid cancer. The banning of above-ground nuclear weapons testing has largely stopped release of these elements to the environment. The last such release, and perhaps the largest, was the catastrophic explosion and fire at the Soviet power reactor at Chernobyl in 1986 (see Section 6.9).

Fortunately, radionuclides are easily and sensitively detected in water. The most common such contaminant, radium-226, is readily removed by water softening processes that involve treatment of water with lime.

All nations seem to be abiding well with international treaties forbidding aboveground testing of nuclear weapons, so that should not be a source of radionuclide contamination. We can hope that there will not be any hostile exchanges of nuclear weapons or nuclear terrorist attacks in the future. Improperly disposed radionuclides remain a threat to some water supplies. One example is a reported plume of radioactive groundwater flowing from the Hanford atomic energy installation to the Columbia River in Washington State.

7.17. WATER TREATMENT

Municipal Water

Water requires varying degrees of treatment before it is used for households or for industrial applications. Furthermore, wastewater from sewage or from industrial applications normally must be treated before it is released. Some of the basic treatment processes are similar, regardless of the intended use or mode of disposal of the water. They are discussed briefly here for municipal water, and some of the specialized treatment processes applied to wastewater are summarized at the end of the chapter.

Water being purified for municipal use is usually **aerated** by contacting it with air. This treatment blows out odor-causing impurities. When, as is often the case, dissolved iron is present as soluble iron(II) ion, Fe^{2+}, oxygen from the air oxidizes it,

$$4Fe^{2+} + 4O_2 + 4H_2O \rightarrow 4Fe(OH)_3(s) + 8H^+ \quad (7.17.1)$$

forming a gelatinous solid of $Fe(OH)_3$. As it settles, this solid entrains and carries with it very small colloidal solid impurities suspended in the water, such as mud, a process called **coagulation**.

Water often contains excessive levels of dissolved calcium along with bicarbonate ion, HCO_3^-. This *temporary* hardness can be removed from water by adding **lime**, $Ca(OH)_2$:

$$Ca^{2+} + 2HCO_3^- + Ca(OH)_2 \rightarrow 2CaCO_3(s) + 2H_2O \quad (7.17.2)$$

The $CaCO_3$ solid formed settles in a settling basin. The process can be somewhat slow and incomplete. So to avoid deposition of solid $CaCO_3$ precipitation in the water distribution system, which can clog pipes, CO_2 gas is added to the water.

If residual organic matter, such as humic substances (Section 7.9) is present in the water, it may be treated by running it over granular activated carbon. This is a form of carbon that has been reacted with steam at a high temperature to cause the formation of an enormous number of microscopic pores in the carbon, giving it a huge internal surface area of up to 2000 square meters per gram! Carbon so treated is very effective in removing dissolved organic matter from water.

As a final step in water purification, water is disinfected. This may be done by adding elemental chlorine to the water, which reacts with water as follows:

$$Cl_2 + H_2O \rightarrow H^+ + Cl^- + HOCl \quad (7.17.3)$$

Hypochlorous acid, HOCl, is a good disinfecting agent that kills virus and bacteria in the water. Salts of HOCl including NaOCl and $Ca(OCl)_2$ can also be used for water disinfection.

When used as a disinfectant, chlorine may form undesirable organic compounds called trihalomethanes. To avoid this problem **chlorine dioxide**, ClO_2, an effective water disinfectant that does not produce trihalomethanes, is now commonly used. This compound is a dangerously reactive gas that is unsafe to move from a manufacturing site to where it is needed, so it is made by reacting sodium chlorite with elemental chlorine:

$$2NaClO_2(s) + Cl_2(g) \longleftrightarrow 2ClO_2(g) + 2NaCl(s) \quad (7.17.4)$$

This process of making potentially dangerous chlorine dioxide only in the quantities needed, where needed, when needed is in keeping with the best practice of green chemistry.

Green Ozone for Water Disinfection

Another disinfection agent that is superior to chlorine in some respects is **ozone**, O_3. An ozone-based water disinfection system is outlined in Figure 7.8. This reactive form of elemental oxygen is made by passing an electrical discharge at approximately 20,000 volts through air that has been filtered, cooled, dried, and pressurized:

$$3O_2(g) \xrightarrow[\text{discharge}]{\text{Electrical}} 2O_3(g) \quad (7.17.5)$$

The ozone contacts water in a contact chamber for 10–15 minutes. Although the low water solubility of ozone limits its disinfective ability, there is no formation of organochlorine disinfection byproducts. And ozone is more effective than chlorine against viruses. Because ozone lasts for only a short period of time in water, some chlorine must be added to maintain residual disinfection in the water distribution system.

The use of ozone for water disinfection is a virtually ideal example of the practice of green chemistry. The only raw material required is air, which is available without cost everywhere. The ozone is produced only as needed, where it is needed, and it is not stored. The ozone does not persist, and it decomposes to benign molecular oxygen. The likelihood of forming harmful byproducts is very low.

Water for Industrial Use

Water destined for industrial use may range in quality all the way from cooling water, for which the major requirement is that it be wet, to hyperpure water used in the semiconductor industry. For economic reasons, water destined for industrial applications is usually treated only to meet the requirements of the intended use. As examples, water used in food processing must be treated to destroy pathogens and boiler feedwater must be treated to remove corrosive and scale-forming solutes.

There are a number of specific treatment operations to which industrial feedwater may be subjected. Dissolved oxygen that may be corrosive can be removed with hydrazine or sulfite. Precipitants may be added to remove specific contaminants, the most common of which is calcium ion, Ca^{2+}. Calcium can form harmful deposits, especially when the water is heated. The addition of phosphate precipitates calcium, reducing the levels of soluble calcium to very low values. In some cases, the calcium can be tolerated in the water if chelating agents are added to prevent the precipitation of calcium solids. Scale can be inhibited by the addition of dispersants. It may be necessary to adjust pH by adding acid or base. Even water not destined for food use may have to be disinfected to prevent growth of harmful bacteria, such as in cooling systems.

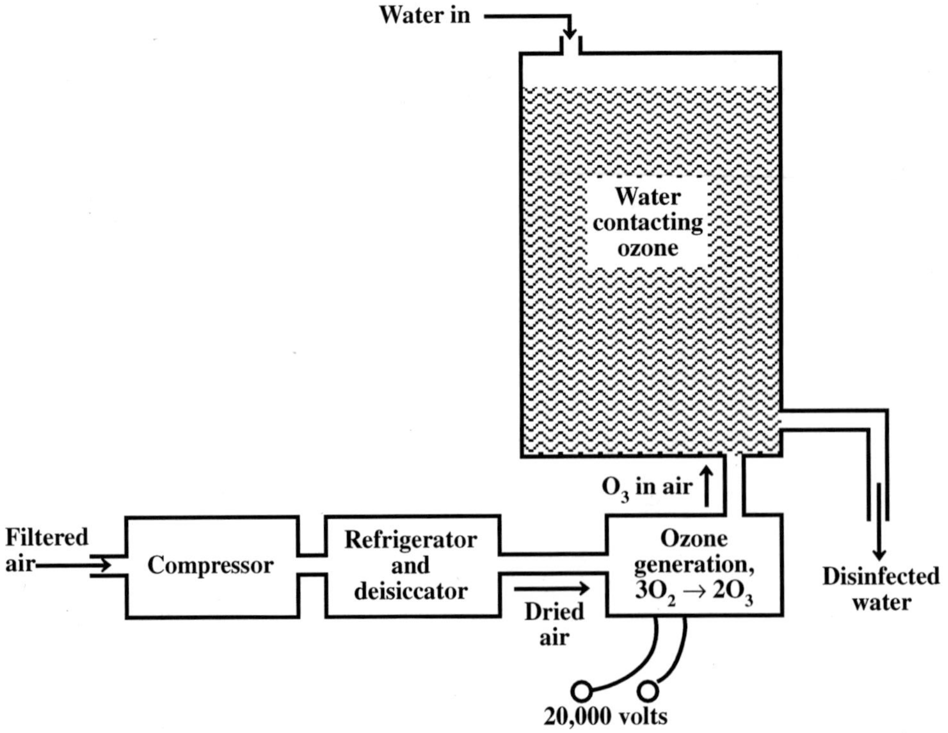

Figure 7.8. A schematic diagram of a system to generate ozone by electrical discharge through refrigerated, dried air followed by disinfection of water by the ozone product.

An important consideration with industrial water is the possibility for **recycling** and **sequential use**. As the name implies, recycling refers to running water back through a system for essentially the same use. Sequential use recognizes that several applications may require water of successively lower quality. The water is first used for the application requiring the best quality of water. The next use of the water requires a somewhat lower quality, and the last use before discharge requires the least quality. In some favorable cases, the water leaving the system can be applied to grass or fields for irrigation.

Both organic and inorganic (salt) solutes can make water unsuitable for recycling. Biodegradable organic materials can be degraded by biological waste treatment measures, which are discussed under the category of sewage treatment below. Filtration over activated carbon mentioned earlier in this section can remove harmful organic solutes. The most straightforward way to remove dissolved salts is distillation in which water is evaporated leaving the salts behind, then condensed to a pure H_2O product. Although used to get fresh water from seawater in some energy-rich arid Middle Eastern countries, this approach is too costly for water recycling in most parts of the world. A more cost-effective method of water purification is **reverse osmosis**, the basic principle of which is shown in Figure 7.9. With this method, water is forced under pressure through a semipermeable membrane that has an attraction for pure water, but a tendency to reject dissolved salts. The purified water is readily recycled for a variety of uses, but disposal is required of the concentrated salt brine left behind.

Sewage

The primary objective of sewage treatment is to remove oxygen-demanding substances from wastewater. These are substances of mostly biological origin, abbreviated $\{CH_2O\}$, that undergo biodegradation and consumption of dissolved oxygen, thus exerting a biochemical oxygen demand (BOD). Sewage treatment can be divided into the three main categories of (1) primary treatment to remove grit, grease, and solid objects, (2) secondary treatment to reduce BOD, and (3) tertiary treatment to further refine the quality of the effluent water. This section addresses primarily secondary treatment.

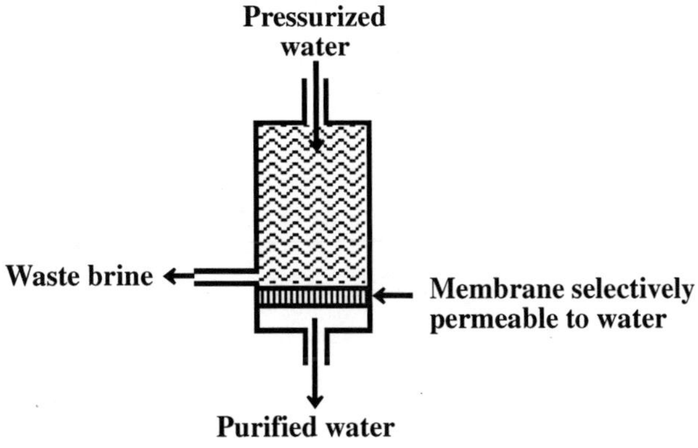

Figure 7.9. Dissolved impurities can be removed from water by forcing pure water for a membrane selectively permeable to water in a process called reverse osmosis.

Secondary wastewater treatment uses a mass of microorganisms in contact with sewage and atmospheric oxygen to eliminate BOD by the reaction,

$$\{CH_2O\} + O_2 \rightarrow CO_2 + H_2O \quad (7.17.6)$$

a process that also generates additional biomass capable of degrading more BOD. The microorganisms can be on a support that alternately contacts liquid sewage and air. One system that uses that approach is the **trickling filter** in which sewage is sprayed over rocks coated with microorganisms. However, the most commonly used process at present is the **activated sludge process** shown in Figure 7.10. In this process wastewater is pumped into one end of a large aeration tank through which air is bubbled. Viable microorganisms suspended in the tank metabolize the degradable materials composing the BOD in the sewage, converting some of the carbon to carbon dioxide and synthesizing additional microorganisms by an overall reaction that can be represented by the following equation:

$$\{CH_2O\} + O_2 \rightarrow CO_2 + H_2O + \text{Biomass} \qquad (7.17.7)$$

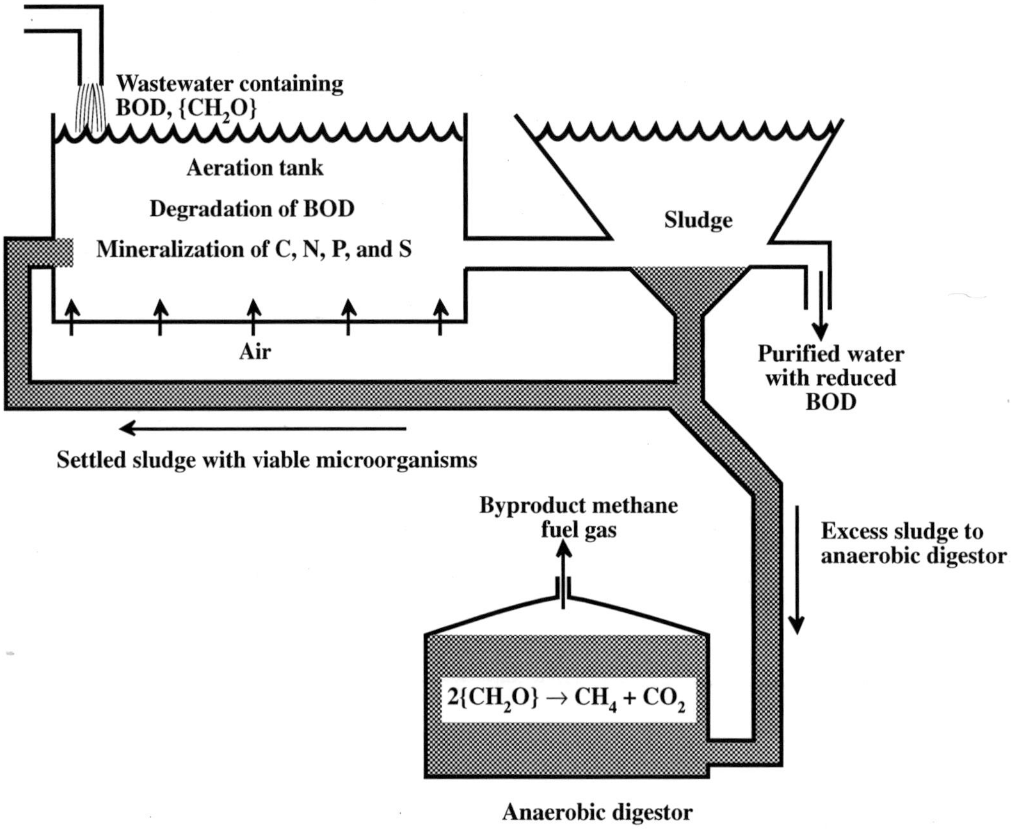

Figure 7.10. Activated sludge process.

In addition to converting organic C to CO_2, the bacteria in the aeration tank convert organic N, P, and S to simple inorganic ions, such as NH_4^+, $H_2PO_4^-$, SO_4^{2-}, a process called **mineralization**. After an appropriate residence time in the tank, the treated water is taken to a settling tank where the microorganisms settle out as sludge. Most of this sludge is pumped back to the front of the aeration tank to contact additional BOD. Some excess sludge is taken to an anaerobic digester, where the sludge is acted upon by methane-forming bacteria in the absence of oxygen to generate methane gas, CH_4. The purified water from the treatment plant can be discharged directly or subjected to additional treatment. One option is to allow the water to flow slowly through a constructed wetland system where it is purified by natural processes.

Sewage treatment has a significant potential for the practice of green chemistry. One of the most obvious ways in which this can occur is by efficient utilization of the methane produced in the anaerobic digestion of the spent sludge from the activated sludge process. Processes that rely primarily on anaerobic digestion have the potential to produce even more methane. Such anaerobic processes are used in some large livestock feeding operations, providing significant amounts of useful methane. Even after anaerobic digestion, the leftover sludge has value as fertilizer to provide nitrogen and phosphorus for plant growth and is commonly spread on soil in locations where sufficient land is available.

Another alternative is to chemically gasify the leftover sludge with oxygen by high-temperature processes similar to those used to gasify coal (see Section 6.7). Properly designed and operated wetlands for advanced treatment of secondary sewage effluent are a means of purifying the effluent without consuming additional chemicals and energy while providing desirable areas for wildlife. It should be kept in mind that all of these applications require sewage that is relatively free of persistent pollutants, particularly poorly degradable organic compounds and heavy metals, so control of materials discharged with sewage is important.

QUESTIONS AND PROBLEMS

1. Why does the water molecule have a bent structure?
2. What does it mean that the water molecule is a dipole? Why is it a dipole?
3. In addition to its polar nature, what is the other major characteristic of the water molecule and its interaction with other water molecules that largely determines water's behavior?
4. How does water behave in respect heat? What are the practical implications of this behavior?
5. What is shown by the hydrologic cycle?
6. What happens to a body of water that is not stirred significantly by wind or currents as it is heated by sunlight during the summer?
7. What forms the basis of the food chain in an aquatic ecosystem?
8. What is meant by biochemical oxygen demand?
9. How is photosynthesis involved in the production of mineral calcium carbonate in water?
10. What happens to sulfur during the reaction $H_2S + 2O_2 \rightarrow SO_4^{2-} + 2H^+$?
11. How is the solubility of a gas in water related to the partial pressure of the gas in contact with the water?
12. Why is rainwater a weak acid?
13. Which two important properties of water are due to the presence of HCO_3^- ion and Ca^{2+} ion?
14. Why is it not completely accurate to represent dissolved calcium ion in water simply as Ca^{2+}?
15. What is a metal chelate in water? What is a common kind of naturally occurring chelating agent?
16. What are colloidal particles?
17. Name 3 important heavy metal water pollutants and one metalloid.
18. What is an organometallic compound?
19. Cite an example in which bacteria act to produce toxic water pollutant species containing a heavy metal.
20. What is an inorganic water pollutant that is involved with metal extraction?
21. What is a health condition caused by nitrite ion? Under what circumstances may nitrate ion cause the same problem?

22. Name a mineral source and an atmospheric source of strong acid in water.
23. Are inorganic nitrogen, phosphorus, and potassium toxic to organisms in water? If not, why are they regarded as water pollutants?
24. Explain why both highly degradable organic compounds and ones that are not degraded well at all are undesirable water pollutants.
25. Aside from obvious esthetic problems, why is sewage a pollutant that requires treatment?
26. What change in the chemical structure of the surfactants in sewage made them much more environmentally friendly?
27. What are 5 different classes of pesticides?
29. What is the least biodegradable class of insecticides? What do they do in the environment?
30. What was a major advantage of organophosphate insecticides? What was a disadvantage of some of them, such as parathion?
31. What is an important characteristic of the mode of action of carbofuran?
32. Name three insecticides derived from plants. Why would plants make insecticides?
33. What is the connection between herbicidal glyphosate and modern genetically modified plants?
34. What purpose is served by dioxin? How has it become an environmental problem?
35. What are PCBs? What is a major case of water pollution by PCBs in the United States?
36. What do radioactive materials produce and how is it harmful?
37. Why are ^{90}Sr, ^{137}Cs, ^{131}I particularly dangerous to animals?
38. What was the last big uncontrolled release of radioactivity to the environment?
39. Why is water aerated? How may aeration indirectly lead to the removal of colloidal solids from water? What is the latter process called?
40. Explain the seemingly contradictory statement that "calcium is used to remove calcium from water."
41. Why is chlorine added to drinking water? Name a "somewhat greener" alternative to chlorine in this application. Name a "much greener" alternative.
42. What is meant by sequential use of water?
43. Describe reverse osmosis? What kinds of impurities does it remove from water?
44. For what purpose is the activated sludge process used? Describe how the process works.

8. AIR AND THE ATMOSPHERE

8.1. MORE THAN JUST AIR TO BREATHE

A Sea of Gas

We live and breathe in the **atmosphere**, a sea of gas consisting primarily of elemental O_2 and N_2. So it is appropriate to first consider some of the fundamental properties of gases that determine the properties of the atmosphere. Recall that gases consist of molecules and (in the case of noble gases) atoms with large amounts of space between them. The gas molecules are in constant, rapid motion, which cause gases to exert pressure. The motion of gas molecules becomes more rapid with increasing temperature. This motion also means that gas molecules move by a process called **diffusion**.

Whereas seawater in the ocean has a well-defined volume and a distinct surface, the same cannot be said for the mass of gases comprising the atmosphere. Although most of the atmosphere is within a few kilometers of Earth's surface, there is no distinct point at higher altitude where the atmosphere ends. Instead, air becomes progressively thinner with increasing altitude. This is noticeable to humans who have traveled to higher altitudes on mountains where the thinner air makes breathing more difficult. Indeed, climbers who scale the highest mountain peaks commonly carry oxygen to aid breathing.

The Gas Laws

The behavior of gases in the atmosphere is governed by several fundamental **gas laws** which are covered briefly here. In using these laws, it should be kept in mind that the quantity of gas is most usefully expressed in numbers of moles. There are many units of pressure, but the most meaningful conceptually is the **atmosphere** (atm) where 1 atmosphere is the average pressure of air in the atmosphere at sea level. (Air has pressure because of the mass of all the molecules of air pressing down from the atmosphere above; as altitude increases, this pressure becomes less.) For calculations involving temperature, the **absolute** temperature scale is used in which each degree is the same size as a degree Celsius (or Centigrade, the temperature scale used for scientific measurements and for temperature readings in most of the world), but zero is 273 degrees below the freezing point of water, which is taken as zero on the Celsius scale. Three important gas laws are the following:

Avogadro's law: At constant temperature and pressure the volume of a gas is directly proportional to the number of moles; doubling the number of moles doubles the volume.

Charles' law: At constant pressure the volume of a fixed number of moles of gas is directly proportional to the absolute temperature (degrees Celsius +273) of the gas; doubling the absolute temperature at constant pressure doubles the volume.

Boyle's law: At constant temperature the volume of a fixed number of moles of gas is inversely proportional to the pressure; doubling the pressure halves the volume.

These three laws are summarized in the **general gas law** relating volume (V), pressure (P), number of moles (n), and absolute temperature (T) expressed as

$$PV = nRT \tag{8.1.1}$$

where R is a constant.

Mathematical calculations involving the gas laws are simple. One of the most common such calculations is that of changes in volume resulting from changes in pressure, temperature, or moles of gas. The parameter that does not change is the constant R. Using subscripts to represent conditions before and after a change yields the following relationship:

$$R = \frac{P_1V_1}{n_1T_1} = \frac{P_2V_2}{n_2T_2} \tag{8.1.2}$$

This equation can be arranged in a form that can be solved for a new volume resulting from changes in P, n, or T:

$$V_2 = V_1 \times \frac{n_2T_2P_1}{n_1T_1P_2} \tag{8.1.3}$$

As an example, calculate the volume of a fixed number of moles of gas initially occupying 12.0 liters when the temperature is changed from 10°C to 90°C at constant pressure. In order to use these temperatures, they must be changed to absolute temperature by adding 273° . Therefore, T_1 = 10° + 273° = 283°, and T_2 = 90° + 273° = 363° . Since n and P remain constant, they cancel out of the equation yielding

$$V_2 = V_1 \times \frac{T_2}{T_1} = 12.0\ \text{L} \times \frac{363^\circ}{283^\circ} = 15.4\ \text{L} \tag{8.1.4}$$

As another example consider the effects of a change of pressure, holding both the temperature and number of moles constant. Calculate the new volume of a quantity of gas occupying initially 16.0 L at a pressure of 0.900 atm when the pressure is changed to 1.20 atm. In this case, both n and P remain the same and cancel out of the equation giving the following relationship:

$$V_2 = V_1 \times \frac{P_1}{P_2} = 16.0\ \text{L} \times \frac{0.900\ \text{atm}}{1.20\ \text{atm}} = 12.0\ \text{L} \tag{8.1.5}$$

Note that an increase in temperature *increases* the volume and an increase in pressure *decreases* the volume.

The Protective Atmosphere

The atmosphere is the air around and above us. We know we must have air to breathe. A human deprived of air's life-giving oxygen for just a brief time will lose consciousness, and within a few minutes will die. But air is far more than just a source of oxygen. That is because it protects Earth's organisms in ways that are absolutely essential for their existence. One major protective function is to act as a blanket to keep us warm. It does that by reabsorbing the infrared radiation by which Earth radiates the energy that it receives from the sun. By delaying the exit of this energy into outer space, the average temperature of Earth's surface remains at about 15°C at sea level, though much colder at certain times and places and significantly warmer at others. Without this warming effect, plants could not grow and most other known organisms could not exist. The second protective function of the atmosphere is absorption of very short wavelength ultraviolet solar radiation. Were this radiation to reach our level, it would tear apart biomolecules, making it impossible for most life forms to exist.

What is air? At our level, it is a mixture of gases of uniform composition, except for water vapor, which composes 1-3% of the atmosphere by volume, and some of the trace gases, such as pollutant sulfur dioxide. On a dry basis, air is 78.1% (by volume) nitrogen, 21.0% oxygen, 0.9% argon, and 0.04% carbon dioxide. Normally, air is 1–3% water vapor by volume. Trace gases at levels below 0.002% in air include ammonia, carbon monoxide, helium, hydrogen, krypton, methane, neon, nitrogen dioxide, nitrous oxide, ozone, sulfur dioxide, and xenon.

Although one might get the impression that the atmosphere is very thick, it is "tissue thin" compared to Earth's diameter. Consider a corporate jet aircraft cruising at 35,000 feet (about 6.6 miles or 10.7 kilometers). In the unlikely event of sudden, catastrophic loss of pressure in the pressurized cockpit, the pilot has only about *15 seconds* to grab an oxygen mask before losing consciousness (the passengers in the cabin have an equally short time, but it is more important for the pilot to stay conscious and dive to a lower altitude). The reason for this is that virtually all the air in the atmosphere is below the altitude of around 11 km. By way of comparison, Earth's diameter is almost 13,000 km.

The altitude at which high-flying jet aircraft cruise marks the upper limit of the lowest of several layers of the atmosphere, the **troposphere**, which extends from sea level to about 11 km (Figure 8.1). As anyone who has driven to the top of Pike's Peak or some other mountain knows, the troposphere gets cooler with increasing altitude, from an average temperature of 15°C at sea level to an average at 11 km of -56°C. Above the layer of the troposphere, however, atmospheric temperature increases to an

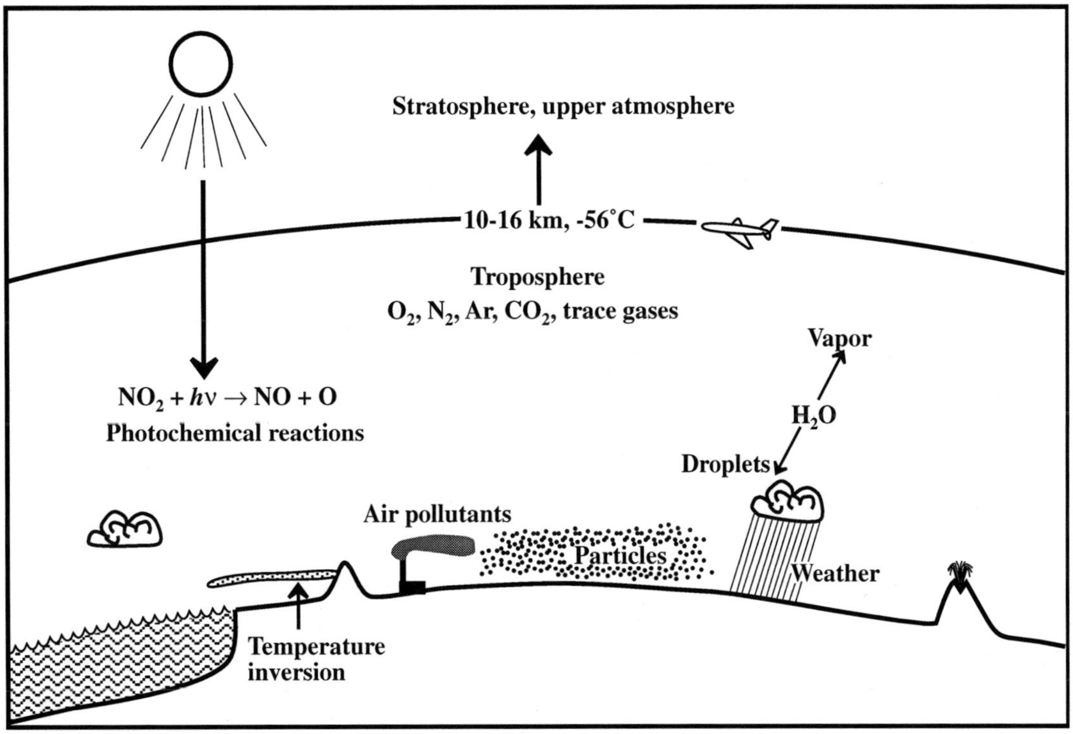

Figure 8.1. The troposphere is the very thin layer of the atmosphere closest to Earth, containing most of the atmosphere's air and water vapor. It is the source of oxygen, carbon dioxide, nitrogen, and water used by living organisms and as raw materials for manufacturing. With the important exception of stratospheric ozone destruction, it is where most air pollution phenomena occur.

average of -2°C at 50 km altitude. The layer above the troposphere is the **stratosphere**, which is heated by the absorption of intense ultraviolet radiation from the sun (Figure 8.2). There is virtually no water vapor in the stratosphere, and it contains ozone, O_3, and O atoms as the result of ultraviolet radiation acting upon stratospheric O_2. Beyond the stratosphere are layers called the mesosphere and thermosphere, but they are relatively less important in the discussion of the atmosphere.

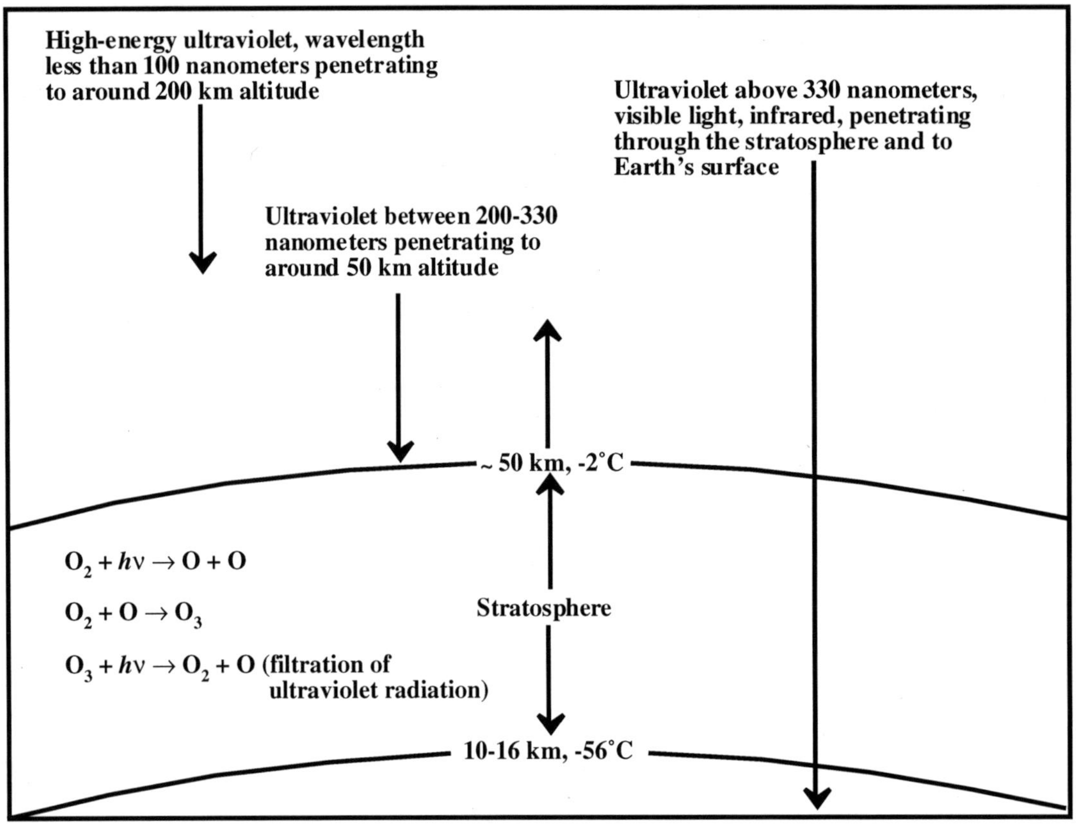

Figure 8.2. The upper atmosphere including the stratosphere and regions beyond is particularly important in the absorption of radiation that would make life impossible if it reached Earth's surface. The layer of ozone, O_3, in the stratosphere is of the utmost importance and one that is subject to damage from anthropogenic species released into the atmosphere.

8.2. ATMOSPHERIC CHEMISTRY AND PHOTOCHEMICAL REACTIONS

Atmospheric chemistry describes chemical processes that occur in the atmosphere. One notable aspect of atmospheric chemistry is that it occurs in the gas phase where molecules are relatively far apart, so a molecule or a fragment of a molecule may travel some distance before bumping into another species with which it reacts. This is especially true in the highly rarefied regions of the stratosphere and above.

A second major aspect of atmospheric chemistry is the occurrence of **photochemical reactions** that are initiated when a photon (see Section 6.2) of ultraviolet radiation is absorbed by a molecule. The energy of a photon, E, is given by $E = h\nu$ where h is Planck's constant and ν is the frequency of the radiation. Electromagnetic radiation of a sufficiently short wavelength breaks chemical bonds in molecules, leading to the formation of reactive species that can participate in reaction sequences called **chain reactions**.

An example of an important chain reaction sequence that begins with photochemical dissociation of a molecule is the one that occurs when chlorofluorocarbons get into the stratosphere. Chlorofluorocarbons are given the trade name of Freons and consist of carbon atoms to which are bonded fluorine and chlorine atoms. Noted for their extreme chemical stability, they were once widely used as refrigerant fluids in air conditioners, as aerosol propellants for products such as hair spray, and for foam blowing to make very porous plastic or rubber foams. Dichlorodifluoromethane, CCl_2F_2, was used in automobile air conditioners. Released to the atmosphere, this compound remained as a stable atmospheric gas until it got to very high altitudes in the stratosphere. In this region, ultraviolet radiation of sufficient energy ($h\nu$) is available to break the very strong C-Cl bonds,

$$CCl_2F_2 + h\nu \rightarrow \cdot CCl_2F_2 + Cl\cdot \qquad (8.2.1)$$

releasing Cl atoms. The dot represents a single unpaired electron remaining with the Cl atom when the bond in the molecule breaks. Species with such unpaired electrons are very reactive and are called **free radicals**. As discussed in Section 1.13 and shown by reactions 1.13.1 and 1.13.2, there are oxygen atoms and molecules of ozone, O_3, also formed by photochemical processes in the stratosphere. A chlorine atom produced by the photochemical dissociation of CCl_2F_2 as shown in Reaction 8.2.1 can react with a molecule of O_3 to produce O_2 and another reactive free radical species, $ClO\cdot$. This species can react with free O atoms which are present along with the ozone to regenerate Cl atoms, which in turn can react with more O_3 molecules. These reactions are shown below:

$$Cl\cdot + O_3 \rightarrow O_2 + ClO\cdot \qquad (8.2.2)$$

$$ClO\cdot + O \rightarrow O_2 + Cl\cdot \qquad (8.2.3)$$

These are chain reactions in which $ClO\cdot$ and $Cl\cdot$ are continually reacting and being regenerated, the net result of which is the conversion of O_3 and O in the atmosphere to O_2. One Cl atom can bring about the destruction of as many as 10,000 ozone molecules! Ozone serves a vital protective function in the atmosphere as a filter for damaging ultraviolet radiation, so its destruction is a very serious problem that has resulted in the banning of chlorofluorocarbon manufacture.

Very small particles of the size of a micrometer or less called **aerosols** are important in atmospheric chemical processes. Photochemical reactions often result in the production of particles. Particle surfaces can act to catalyze (bring about) atmospheric chemical reactions. Some particles in the atmosphere consist of water droplets with various solutes dissolved in them. Solution chemical reactions can occur in these droplets. One such process is believed to be the conversion of gaseous atmospheric sulfur dioxide to droplets of dilute sulfuric acid, which contribute to acid rain. Some very small particles, such as sea salt crystals entrained into the atmosphere and formed by evaporation of water from ocean spray, act as **condensation nuclei** around which raindrops form.

8.3. ENERGY AND MASS TRANSFER IN THE ATMOSPHERE

As discussed in Chapter 6, Section 6.2, and illustrated in Figure 6.2, the flux of energy reaching Earth's atmosphere is 1,340 watts/m^2. This is an enormous amount of energy. As shown in Figure 8.3, some of the incoming energy reaches Earth's surface, some is absorbed in the atmosphere, warming it, and some is scattered back to space. The energy that comes in as visible and short wavelength electromagnetic radiation (at a maximum intensity of 500 nanometers in the visible region), must go out, which it does as infrared radiation (with maximum intensity at about 10 micrometers (μm), primarily between 2 μm and 40 μm). Water molecules, carbon dioxide, methane, and other minor species in the atmosphere absorb some of the outbound infrared, which eventually is all radiated to space. This temporary absorption of infrared radiation warms the atmosphere — a greenhouse effect.

The fraction of electromagnetic radiation from the sun that is reflected by Earth's surface varies with the nature of the surface. The percentage reflected is very important because it determines how

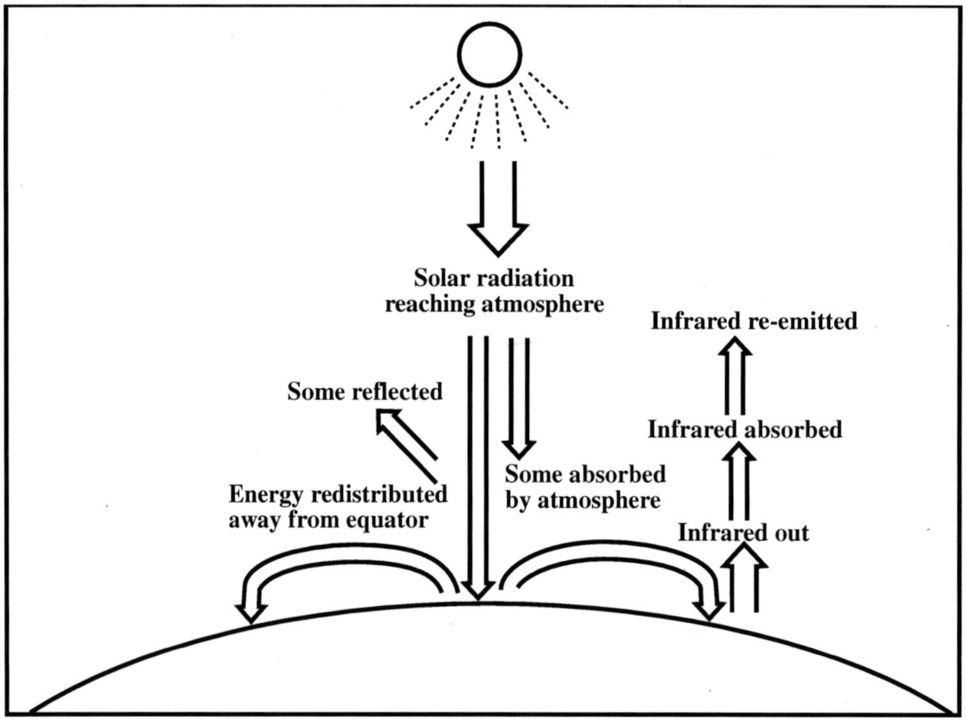

Figure 8.3. Some of the solar energy reaching the top of the atmosphere penetrates to Earth's surface, some is absorbed in the atmosphere, which warms it, and some is scattered by the atmosphere and from Earth's surface. Outgoing energy is in the infrared region. Some of it is temporarily absorbed by the atmosphere before being radiated to space, causing a warming (greenhouse effect). The equatorial regions receive the most energy, part of which is redistributed by warm air masses and latent heat in water vapor away from the equator.

effective incoming radiation is in warming the surface and it is expressed as **albedo**. Freshly plowed black topsoil has a very low albedo of only about 2.5%. In contrast, the albedo of a covering of fresh snow is about 90%. The anthrosphere affects albedo. One of the ways that this is done is in cultivating land, turning over relatively high albedo grass and covering it with exposed black soil. Another way is covering of large areas with asphalt paving, which reflects sunlight poorly.

The maintenance of Earth's heat balance to keep temperatures within limits conducive to life is very complex and not well understood. Geological records show that in times past, Earth was sometimes relatively warm and that at other times there were ice ages in which much of Earth's surface was covered by ice a kilometer or two thick. The differences in average Earth temperature between these extremes and the relatively temperate climate conditions that we now enjoy were only a matter of a few degrees. It is also known that massive volcanic eruptions and almost certainly hits by large asteroids have caused cooling of the atmosphere that has lasted for a year or more. As addressed later in this chapter, there is now concern that anthropogenic gas emissions, particularly of carbon dioxide from fossil fuel combustion, may be having a warming effect upon the atmosphere.

Earth receives solar energy most directly at the equator, so equatorial regions are warmer than regions farther north and south. A significant fraction of this energy moves away from the equator. This is largely done by **convection** in which heat is carried by masses of air. Such heat can be in the form of **sensible heat** from the kinetic energy of rapidly moving air molecules (the faster their average velocities, the higher the temperature). Heat can also be carried as **latent heat** in the form of water vapor. Recall from Section 7.2, that the **heat of vaporization** of water is 2,259 joules per gram (J/g). This is a very high value, meaning that the evaporation of ocean water by solar energy falling on it in warmer regions absorbs an enormous amount of heat to form water vapor. This vapor may be carried elsewhere and condense to form rainfall. As the heat is released, the temperature increases.

Meteorology

The movement of air masses, cloud formation, and precipitation in the atmosphere are covered by the science of **meteorology**. Meteorologic phenomena have a strong effect upon atmospheric chemistry by processes such as the following:

- Movement of air pollutants from one place to another, such as the movement of air pollutant sulfur dioxide from the U.S. Ohio River Valley to New England and southern Canada, where it forms acid rain.
- Conditions under which stagnant pollutant air masses remain in place so that secondary pollutants, such as photochemical smog, can form.
- Precipitation, which can carry acidic compounds from the atmosphere to Earth's surface in the form of acid rain.

Atmospheric chemical processes can influence meterological phenomena. The most obvious example of this is the formation of rain droplets around pollutant particles in the atmosphere.

Weather refers to relatively short term variations in the state of the atmosphere as expressed by temperature, cloud cover, precipitation, relative humidity, atmospheric pressure and wind. Long-term trends in weather are expressed by **climate**. For example, the climate in desert regions of the world may be relatively hot and dry, but the weather in such regions may at times produce torrential rainfall or frigid temperatures.

Weather is driven by redistribution of energy in the atmosphere. A particularly important aspect of this redistribution is the energy released when precipitation forms, which can be enormous because of the high heat of vaporization of water. As an example, heat energy from sunlight and from hot masses of air is converted to latent heat by the evaporation of ocean water off the west coast of Africa. Wind-driven moist air masses travel westward across the ocean. Rainfall forms later, releasing the energy from the latent heat of water and warming the air mass. The resulting hot air mass rises, creating a region of low pressure into which air flows in a circular manner. This can result in the formation of a whirling mass of air in the form of a hurricane that may strike Puerto Rico, Cuba, Florida, or other areas thousands of miles from the area where the water was originally evaporated from the ocean.

A very obvious manifestation of weather consists of very small droplets of liquid water composing **clouds**. These very small droplets may coalesce under the appropriate conditions to form raindrops large enough to fall from the atmosphere. Clouds may absorb infrared radiation from Earth's surface, warming the atmosphere, but they also reflect visible light, which has a cooling effect. Pollutant particles are instrumental in forming clouds. One of the more active kinds of cloud-forming pollutants are atmospheric strong acids, particularly H_2SO_4,

Air masses characterized by pressure, temperature, and moisture contents flow from regions of high atmospheric pressure to regions of low pressure. The boundaries between air masses are called **fronts**. The movement of air associated with moving air masses is **wind** and vertically moving air is an **air current**. Wind is involved in the movement of pollutants from one place to another and is also responsible for dispersing pollutants to harmless levels. An enormous amount of energy is contained in wind, and it can be harnessed to generate electricity (see Figure 6.11 and the discussion of renewable energy resources in Section 6.8).

A lack of wind and air currents often occurs under conditions of **temperature inversion** in which warmer air masses overlay cooler ones (see Figure 8.1). As shown in this figure, topographical features, such as a mountain range that limits horizontal air movement, may make temperature inversion much more effective in trapping polluted masses of air. These conditions occur in the Los Angeles basin noted for photochemical smog formation.

8.4. ATMOSPHERIC OXYGEN AND NITROGEN

By a wide margin, oxygen and nitrogen are the most abundant gases in the atmosphere. Because of the extremely high stability and low reactivity of the N_2 molecule, the chemistry of atmospheric elemental nitrogen is singularly unexciting, although nitrogen molecules are the most common "third bodies" that absorb excess energy from atmospheric chemical reactions, preventing the products of addition reactions in the atmosphere from falling apart. Oxides of nitrogen actively participate in atmospheric chemical reactions. Elemental nitrogen is an important commercial gas extracted from the atmosphere by nitrogen-fixing bacteria and in the industrial synthesis of ammonia.

Oxygen is a reactive species in the atmosphere. It provides the oxygen to produce oxidation products from oxidizable gases in the atmosphere. Two such species that are particularly important are sulfur dioxide gas, SO_2, and pollutant hydrocarbons. Oxygen reacts with these substances only indirectly through the action of reactive intermediates, especially **hydroxyl radical**, HO•.

A crucially important atmospheric chemical phenomenon involving oxygen is the formation of stratospheric ozone, O_3. The formation of this gas in the stratosphere is discussed in Section 1.13 and shown by Reactions 1.13.1 and 1.13.2.

Oxygen in the atmosphere is consumed in the burning of hydrocarbons and other carbon-containing fuels. It is also consumed when oxidizable minerals undergo chemical weathering, such as

$$4FeO + O_2 \rightarrow 2Fe_2O_3 \qquad (8.4.1)$$

All of the oxygen in the atmosphere was originally placed there by photosynthesis shown by

$$CO_2 + H_2O \rightarrow h\nu \rightarrow \{CH_2O\} + O_2 \qquad (8.4.2)$$

where $\{CH_2O\}$ is a generic formula representing biomass.

Unlike molecular oxygen, which can undergo direct photodissociation in the stratosphere, the very stable N_2 molecule does not encounter ultraviolet radiation sufficiently energetic to cause its photodissociation at altitudes below 100 km. However, nitrogen dioxide, NO_2, readily undergoes photodissociation in the troposphere,

$$NO_2 + h\nu \rightarrow NO + O \qquad (8.4.3)$$

to generate highly reactive O atoms. These in turn can attack hydrocarbons in the atmosphere, leading to the formation of photochemical smog discussed later in this chapter.

Green Oxygen and Nitrogen from the Air

Elemental oxygen and nitrogen are important commercial products extracted from air. Most commonly this is done by distilling cold liquid air, a process that can also produce noble gas neon, krypton, and xenon, if desired. The initial step in air distillation is to compress air to about 7 times atmospheric pressure and cool it to remove water vapor and carbon dioxide. Further compression and cooling yields a liquid air product that can then be fractionally distilled to give relatively pure oxygen, nitrogen, and other gases. These can be stored as cold liquids or as the compressed gases.

Essentially pure oxygen has a number of applications, such as for breathing by people with pulmonary insufficiencies. Huge amounts are consumed in steel making. Pure nitrogen is used to provide inert atmospheres free of oxygen. Large quantities of liquid nitrogen are used in the science of **cryogenics** involving very low temperatures.

Emergency Oxygen

Emergency oxygen is required on aircraft that fly at high altitudes. The containers required to transport pure oxygen are too heavy to put on aircraft, so emergency oxygen is generated by a chemical process using a **chlorate candle**. This device contains sodium chlorate, $NaClO_3$, which decomposes when heated to generate oxygen gas:

$$2NaClO_3 \rightarrow 2NaCl + 3O_2 \qquad (8.4.4)$$

Some of the oxygen generated reacts with a fuel, commonly elemental iron, mixed with the sodium chlorate,

$$4Fe + 3O_2 \rightarrow 2Fe_2O_3 \qquad (8.4.4)$$

a heat-generating reaction that provides heat for the decomposition of the sodium chlorate. Chlorate candles can be stored for many years before being activated and still perform well. They are generally safe. However, chlorate candles improperly shipped in the baggage compartment of a ValuJet DC-9 aircraft caused an uncontrollable fire that brought the aircraft down in the Florida Everglades with the loss of all aboard in 1997.

8.5. ATMOSPHERIC POLLUTANT PARTICLES

Because of their ability to reduce visibility and light, atmospheric particles are the most visible form of air pollution. Commonly called **particulates** in air pollution parlance, atmospheric *aerosols* are solids or liquids less than 100 micrometers (millionths of a meter, μm) in diameter, and commonly in a size range of 0.001 to 20 μm. They may be inorganic or organic materials and may belong to the two general classes of **dispersion aerosols** formed by grinding solids, dispersing dusts, or atomizing liquids, and **condensation aerosols** produced by the condensation of gases or vapors, often formed as the result of atmospheric chemical processes. Common dispersion aerosols include water droplets from sea spray, solid particles of NaCl left over when water evaporates from sea spray droplets, cement dust, soil dust dispersed by wind, foundry dust, and pulverized coal. Carbon black, metal fumes, and combustion nuclei form as condensation aerosols from combustion or partial combustion. Liquid particle **mists** include raindrops, fog, cloud droplets, and droplets of sulfuric acid produced when atmospheric SO_2 is oxidized. Organisms produce an abundance of particles. For those afflicted with allergies, the most annoying such particles are plant pollen. Other particles of biological origin include viruses, bacteria, and spores of bacteria and fungi.

In the past, one of the more troublesome sources of atmospheric particles was **fly ash**, a residue left over from the combustion of liquids or very finely divided coal. Often the most abundant component of fly ash is elemental carbon left over from incompletely burned fuel. Fly ash commonly includes oxides of aluminum, calcium, iron, and silicon, as well as some magnesium, sulfur, titanium, phosphorus, potassium, and sodium. With properly operating emission control devices, fly ash emissions are now well controlled.

A health concern with particles, especially from combustion sources, is their ability to carry toxic metals. Of these, lead is of the greatest concern because it usually comes closest to being at a toxic level. Problems with particulate lead in the atmosphere have been greatly reduced by the elimination of tetraethyllead as a gasoline additive, an application that used to spew tons of lead into the atmosphere every day. Another heavy metal that causes considerable concern is mercury, which can enter the atmosphere bound to particles or as vapor-phase atomic mercury. Other metals that can cause problems in particulate matter are beryllium, cadmium, chromium, vanadium, nickel, and arsenic (a metalloid).

Some areas of the world, including parts of the United States, have problems with radioactive particles resulting from underground sources of radioactive **radon**, a noble gas product of radium decay. The two major radium isotopes, ^{222}Rn (half-life 3.8 days) and ^{220}Rn (half-life 54.5 seconds) are alpha particle emitters that decay to radioactive ^{218}Po and ^{216}Po, respectively. These radionuclides are nongaseous and adhere readily to atmospheric particulate matter, which, along with gaseous radon, can cause significant indoor air pollution problems.

Pollutant particles in the atmosphere have both direct and indirect effects. The most obvious direct effects are reduction and distortion of visibility. The light scattering effects of particles in a size range of 0.1 μm–1 μm are especially pronounced due to interference phenomena resulting from the particles being about the same size as the wavelengths of visible light. Particles also have direct health effects when inhaled. This is especially true of very small particles that can be carried into the innermost parts (alveoli) of lungs. An indirect effect of particles is their ability to serve as reaction sites for atmospheric chemical reactions. They also act as nucleation bodies upon which water vapor condenses.

Limiting Particulate Emissions

The first widespread measures to limit air pollution were directed at control of particle emissions. These measures have become very effective so that the "smoke" that one sees emanating from smokestacks usually consists of droplets of water formed by condensation of steam.

The simplest method of particle control from stack gas and other gases released to the atmosphere consists of **sedimentation** in which particles entrained in stack gas are allowed to settle by gravity in relatively large chambers. Sedimentation is most effective for relatively large particles. **Inertial mechanisms** operate by spinning a gas in a round chamber such that particles impinge upon the container walls by centrifugal force. **Fabric filters** contained in **baghouses** act to filter particles from air or stack gas. The mechanism employed provides for periodic shaking of the fabric filters to collect particles held on their walls, thus restoring gas flow through the fabric. **Scrubbers** that spray water or solutions into stack gas are employed to literally wash particles out of gas. In some cases these are operated with a minimal amount of water, which evaporates, so that a solid material is collected. One of the most effective means of particle control consists of **electrostatic precipitators**. These devices use a very high voltage to impart a negative charge onto particles from a central electrode, and the particles are attracted to, and collect on the positively charged walls of the precipitator.

8.6. POLLUTANT GASEOUS OXIDES

Carbon Monoxide

Carbon monoxide, CO, is an air pollutant of some concern because of its direct toxicity to humans. As discussed in Chapter 13, carbon monoxide binds to blood hemoglobin and prevents the hemoglobin from transporting oxygen from the lungs to other tissues. Global and regional levels of atmospheric carbon monoxide are too low to be of concern. However, local levels in areas with heavy automobile traffic can become high enough to pose a health hazard and on some congested urban streets have reached levels of 50–100 ppm. The use of exhaust pollution control devices on automobiles have lowered these levels significantly during the last 30 years. The numerous fatal cases of carbon monoxide poisoning that occur each year are almost always the result of improperly vented heating devices in indoor areas.

Carbon monoxide is produced by partial combustion of fuels, largely in the internal combustion engine. Carbon monoxide emissions can be reduced by careful control of combustion conditions. Running the exhaust along with pumped air over an exhaust catalyst can oxidize carbon monoxide to nontoxic carbon dioxide:

$$2CO + O_2 \rightarrow 2CO_2 \tag{8.6.1}$$

Modern automobile engines use precise computerized control of engine operating parameters along with exhaust catalysts to control carbon monoxide emissions.

Sulfur Dioxide

Sulfur dioxide enters the atmosphere as the result of the following:

- Direct emissions from volcanoes
- Atmospheric oxidation of H_2S emitted to the atmosphere by bacteria and from geothermal sources (volcanoes, hot springs, geysers)
- Atmospheric oxidation of dimethyl sulfide, $(CH_3)_2S$, emitted to the atmosphere from marine organisms
- Pollutant sources from the combustion of organic sulfur and iron pyrite, FeS_2, in fossil fuels

The pollutant sources are of most concern because of their contribution to local and regional air pollution problems and because they are sources that humans can do something about.

The fate of sulfur dioxide in the atmosphere is oxidation and reaction with water to produce sulfuric acid. The overall process is complex and not completely understood, but it can be described by the reaction

$$2SO_2 + O_2 + 2H_2O \rightarrow 2H_2SO_4 \tag{8.6.2}$$

This process is generally rather slow in the atmosphere, but it can be quite rapid under conditions of photochemical smog formation (see Section 8.10) in which highly reactive oxidizing species are present. It is very important because it is the main mechanism for forming acid rain, which can be directly harmful to vegetation, fish (especially fingerlings), and materials, such as building stone that can be attacked by acid. Sulfur dioxide forms aerosol droplets of sulfuric acid in the atmosphere. As a result, much of the Eastern United States is covered by a slight haze of sulfuric acid droplets during much of the year. In recent years, some volcanic eruptions have blasted enough sulfur dioxide into the atmosphere to produce a sufficient amount of sunlight-reflecting sulfuric acid aerosol to cause a noticeable cooling of the atmosphere.

In addition to indirect effects from the formation of acid rain, sulfur dioxide affects some plants directly, causing leaf necrosis (death of leaf tissue). Another symptom of sulfur dioxide phytotoxicity (toxicity to plants) is chlorosis, a bleaching or yellowing of green leaves.

The most straightforward means of reducing sulfur dioxide emissions is to avoid having sulfur in fuels. As discussed below, sulfur compounds are removed from natural gas and petroleum. Coal often has high levels of sulfur, and during recent years there has been a major shift to low sulfur coal in power plants. Much of the pyritic sulfur, FeS_2, can be washed from coal because it exists in a separate mineral phase that can be separated from the combustible organic matter in coal. However, about half of the sulfur in coal typically is bound to the coal as organic sulfur, and cannot be removed.

A number of coal-fired power plants have installed systems for removing sulfur dioxide resulting from the combustion of coal. One such approach uses **fluidized bed combustion** in which pulverized coal is blasted into a hot bed of calcium oxide, where the coal is burned, and sulfur dioxide is bound by the following reaction:

$$CaO + SO_2 \rightarrow CaSO_3 \tag{8.6.3}$$

Some of the $CaSO_3$ product is oxidized to $CaSO_4$. Another approach uses a slurry of calcium hydroxide (lime, $Ca(OH)_2$) to react with sulfur dioxide in stack gas:

$$Ca(OH)_2 + SO_2 \rightarrow CaSO_3 + H_2O \tag{8.6.4}$$

Although effective in removing sulfur dioxide, this process uses enormous quantities of limestone ($CaCO_3$) as a source of lime and produces huge quantities of byproduct.

Green Chemistry and Sulfur Dioxide

The problem of sulfur in fuel provides an excellent illustration of the potential for the application of green chemistry to the elimination of a pollution problem. Sulfur is a valuable raw material required in the manufacture of sulfuric acid, one of the largest volume chemicals made. As discussed in Section 2.7, hydrogen sulfide, H_2S, occurs in large quantities in natural gas, such as that produced in the Canadian province of Alberta. This hydrogen sulfide must be removed from the natural gas. Rather than presenting a pollution problem, it is converted to elemental sulfur, then used to make sulfuric acid.

Another green chemistry approach to the reclamation of waste sulfur is practiced in Kalundborg, Denmark, the site of the world's most clearly recognizable system of industrial ecology (see Chapter 11). The huge coal-fired power plant in Kalundborg uses lime scrubbing to remove sulfur dioxide from stack gas. The calcium sulfite product of this process is oxidized,

$$CaSO_3 + {}^1/_2O_2 + 2H_2O \rightarrow CaSO_4{\cdot}2H_2O \tag{8.6.5}$$

to generate gypsum, $CaSO_4{\cdot}2H_2O$. This mineral is then used to make wallboard, thus solving a pollution problem from the production of spent lime and a raw materials problem arising from the need for gypsum to make wallboard needed for building construction.

Nitrogen Oxides

Nitrous oxide (N_2O), colorless, odorless, nitric oxide (NO), and pungent-smelling, red-brown nitrogen dioxide (NO_2) occur in the atmosphere. Of these, nitrous oxide is generated by bacteria and its release is one of the ways in which chemically fixed nitrogen in the biosphere is returned to the atmosphere. It is not involved much with chemical processes in the troposphere, but undergoes photochemical dissociation in the stratosphere:

$$N_2O + h\nu \rightarrow N_2 + O \tag{8.6.6}$$

Both NO and NO_2, collectively designated as NO_x, are produced from natural sources, such as lightning and biological processes, and from pollutant sources. Pollutant concentrations of these gases can become too high locally and regionally, causing air pollution problems. A major pollutant source of these gases is the internal combustion engine in which conditions are such that molecular elemental nitrogen and oxygen react,

$$N_2 + O_2 \rightarrow 2NO \tag{8.6.7}$$

to produce NO. Combustion of fuels that contain organically bound nitrogen, such as coal, also produces NO. Atmospheric chemical reactions convert some of the NO emitted to NO_2.

Exposed to electromagnetic radiation of wavelengths below 398 nm, nitrogen dioxide undergoes photodissociation,

$$NO_2 + h\nu \rightarrow NO + O \tag{8.6.8}$$

to produce highly reactive O atoms. The O atoms can participate in a series of chain reactions through which NO is converted back to NO_2, which can undergo photodissociation again to start the cycle over. Nitrogen dioxide is very reactive, and photodissociates within a minute or two in direct sunlight.

Nitrogen dioxide, NO_2, is significantly more toxic than NO, although concentrations of NO_2 in the outdoor atmosphere rarely reach toxic levels. Accidental releases of NO_2 can be sufficient to cause toxic effects or even death. Brief exposures to 50–100 parts per million (ppm) of NO_2 in air inflames lung tissue for 6-8 weeks followed by recovery. Exposure to 500 ppm or more of NO_2 causes death within 2–10 days. Exposure to 100-500 ppm of NO_2 causes a lung condition with the ominous name of *bronchiolitis fibrosa* obliterans that is fatal within 3–5 weeks after exposure. Fatal incidents of NO_2 poisoning have resulted from accidental release of the gas used as an oxidant in rocket fuels and from burning of nitrogen-containing celluloid and nitrocellulose moving picture film (long banned because of catastrophic fires that killed numerous people). Plants exposed to nitrogen dioxide may suffer decreased photosynthesis, leaf spotting, and breakdown of plant tissue.

The release of NO from combustion sources can be reduced by limiting excess air so that there is not enough excess oxygen to produce NO according to Reaction 8.6.7. Exhaust catalytic converters reduce NO_x emissions from automobile exhausts. Scrubbing NO_x from furnace and power plant stack gases is difficult due to the low water solubilities of NO_x gases.

8.7. ACID RAIN

Along with hydrogen chloride, HCl, emitted to the atmosphere by the combustion of chlorine-containing organic compounds, sulfur dioxide and nitrogen oxides react in the atmosphere to produce strongly acidic H_2SO_4 and HNO_3, respectively. Incorporated into rainwater, these acids fall to the ground as **acid rain**. A more general term, **acid deposition**, refers to the effects of atmospheric strong acids, acidic gases (SO_2), and acidic salts (NH_4NO_3 and NH_4HSO_4). Acid deposition is a major air pollution problem.

Figure 8.4 shows a typical distribution of acidic precipitation in the 48 contiguous U.S. states. This figure illustrates that acidic precipitation is a *regional* air pollution problem, not widespread enough to be a *global* problem, but spreading beyond *local* areas. (There have been some unfortunate cases where localized release of acid, usually as sulfur dioxide from metal ore smelting operations have affected local areas, often devastating vegetation within several kilometers of the source

Transport processes that move atmospheric acids and their precursor acid gases from their sources to downwind areas are very important in determining areas affected by acid rain. The northeastern U.S. and southeastern Canada are affected by acid originating from stack gas emissions carried by prevailing southwesterly winds from Missouri, Illinois, Kentucky and other regions to the southwest. Southern Norway, Sweden, and Finland receive acid precipitation originating farther south in Europe.

Numerous adverse effects have been reported as the result of acidic precipitation. These can be divided into the following major categories:

- Direct effects upon the atmosphere manifested by reduced and distorted visibility. These effects are due to the presence of sulfuric acid droplets and solutions or solid particles of acidic salts, such as NH_4HSO_4.
- Phytotoxicity (toxicity to plants) and destruction of sensitive forests. These effects can be direct, resulting from exposure of plant leaves and roots to acidic precipitation and to acid-forming gases, particularly SO_2 and NO_2. They can also be indirect, primarily by the liberation of phytotoxic Al^{3+} ion by the action of acidic rainfall on soil.
- Direct effects on humans and other animals. These are usually respiratory effects, and asthmatics are especially vulnerable.

- Effects upon plants and fish (especially fish fingerlings) in acidified lake water where the lake is not in contact with minerals, particularly $CaCO_3$, capable of neutralizing acid.
- Damage to materials. Stone (especially acid-soluble limestone and marble) and metal used in building can be corroded and etched by acidic precipitation. Electrical equipment, particularly relay contacts and springs can be corroded by acidic precipitation.

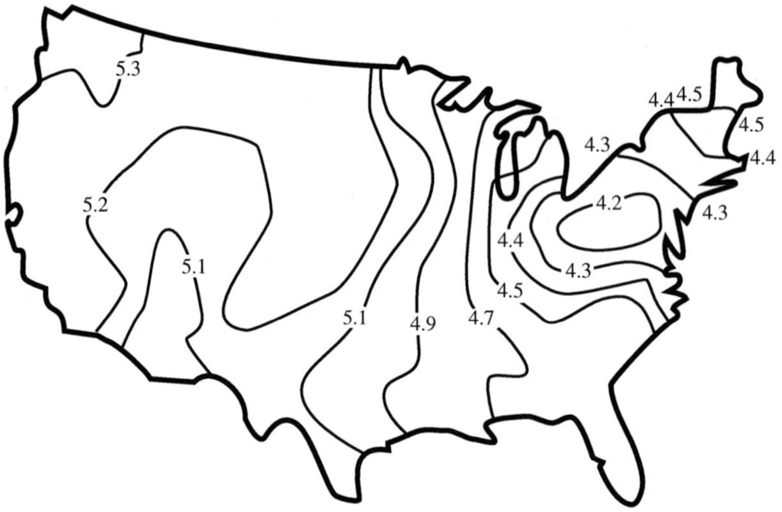

Figure 8.4. Lines showing precipitation with the same pH (isopleths) in the 48 contiguous United States. This is a hypothetical plot. The actual distribution of acidic precipitation varies constantly with time and climatic conditions, but is generally similar to the plot shown with lower values concentrating in the northeastern states.

Some measures can be taken to mitigate the effects of acid rain, although these are very limited once the pollutant has formed. Some success has been achieved with treating acidified lakes with pulverized limestone to neutralize acid. Corrosion-resistant materials can be used in applications where exposure to acid rain is likely. Protective coatings, such as corrosion-resisting paint primers on metals, can be applied to materials likely to be exposed to acidic precipitation. But the best protection is to prevent formation and release of SO_2 and NOx gases leading to acid rain formation by measures described in the preceding section.

8.8. MISCELLANEOUS GASES IN THE ATMOSPHERE

There are several inorganic gases other than oxides that can be significant atmospheric pollutants. One of the most common of these is ammonia, NH_3. In addition to industrial pollution, such as from heating coal to make coke for steel making, ammonia can be added to the atmosphere by bacterial sources, from sewage treatment, and from the decay of animal wastes. Accidental releases can occur from liquid anhydrous ammonia used as an agricultural nitrogen fertilizer.

Ammonia is strongly attracted to water, so it is normally present in the atmosphere in water droplets. It is the only significant gaseous base in the atmosphere, so that it reacts with atmospheric acids to produce corrosive ammonium salts as shown by the following reactions:

$$NH_3 + H_2SO_4 \rightarrow NH_4HSO_4 \quad (8.8.1)$$

$$NH_3 + HNO_3 \rightarrow NH_4NO_3 \quad (8.8.2)$$

Gaseous chlorine, fluorine, and volatile fluorides are uncommon air pollutants, but very serious where they occur. Elemental chlorine, Cl_2, is widely produced and distributed as a water disinfectant, bleach, and industrial chemical. It is very reactive and so toxic that it was the first poisonous gas used as a military poison in World War I. Most toxic exposures of chlorine occur as the result of transportation accidents leading to its release.

Hydrogen chloride, HCl, can get into the atmosphere from accidental releases of the gas, from reaction with atmospheric water of reactive chlorine-containing chemicals, such as $SiCl_4$,

$$SiCl_4 + 2H_2O \rightarrow SiO_2 + 4HCl \qquad (8.8.3)$$

and from the combustion of chlorine-containing polyvinylchloride (PVC) plastic. The strong affinity of HCl gas for water means that it exists as droplets of hydrochloric acid in the atmosphere. Atmospheric HCl is very irritating to mucous membrane tissue and damaging to corrodable materials.

Elemental fluorine (F_2) and hydrogen fluoride are both highly toxic. Fortunately, occurrences of these gases in the atmosphere are very rare. Gaseous silicon tetrafluoride, SiF_4, can be released during steel making and some metal smelting processes when fluorspar (CaF_2) reacts with sand (SiO_2):

$$2CaF_2 + 3SiO_2 \rightarrow 2CaSiO_3 + SiF_4 \qquad (8.8.4)$$

Sulfur hexafluoride, SF_6, is an astoundingly unreactive gaseous compound used to blanket and degas molten aluminum and magnesium and in gas-insulated electrical equipment. It lasts essentially forever in the atmosphere. The greatest concern with its release is that it is a powerful greenhouse warming gas with an effect per molecule about 24,000 times that of carbon dioxide.

Hydrogen sulfide, H_2S, enters the atmosphere from a number of natural sources including geothermal sources, the microbial decay of organic sulfur compounds and the microbial conversion of sulfate, SO_4^{2-}, to H_2S when sulfate acts as an oxidizing agent in the absence of O_2. Wood pulping processes can release hydrogen sulfide. Hydrogen sulfide is a common contaminant of petroleum and natural gas, and these sources are the most common source of poisoning by H_2S, which has about the same toxicity as hydrogen cyanide. A tragic incident of hydrogen sulfide poisoning occurred in Poza Rica, Mexico, in 1950 as the result of a process to recover H_2S from natural gas. Incredibly, the hydrogen sulfide byproduct was burned in a flare to produce sulfur dioxide. The flare became extinguished at night so that toxic hydrogen sulfide spread throughout the vicinity, killing 22 people and hospitalizing over 300. Atmospheric H_2S is phytotoxic, destroying immature plant tissue and reducing plant growth. It also affects some kinds of materials, forming a black coating of copper sulfide, CuS, on copper roofing. This coating weathers to a rather attractive green layer (patina) of basic copper sulfate, $CuSO_4 \cdot 3Cu(OH)_2$, which protects the copper from further attack. Hydrogen sulfide in the atmosphere becomes oxidized to sulfur dioxide, SO_2.

Carbonyl sulfide, COS, is another inorganic sulfur gas that can be detected in the atmosphere, though it is usually at very low levels. A related compound, carbon disulfide, CS_2, also occurs in the atmosphere.

8.9. CO_2: THE ULTIMATE AIR POLLUTANT?

Carbon dioxide, CO_2, is a normal constituent of the atmosphere, essential as a source of carbon for plant photosynthesis. Along with water vapor and other trace gases, atmospheric carbon dioxide absorbs outgoing infrared radiation from Earth, thus keeping the planet's surface temperature at a tolerable level. Levels of carbon dioxide gas in the atmosphere are now about 370 parts per million by volume. This represents an almost 40% increase over estimated pre-industrial concentrations of 260 ppm. Furthermore, as shown by the plot in Figure 8.5, global CO_2 levels are increasing by about 1 ppm per year, and ice core evidence indicates that these levels were only about 200 ppm at the peak of the last ice age around 18,000 years ago. So, humans are clearly increasing atmospheric carbon dioxide levels significantly, largely through the combustion of carbon-containing fossil fuels and as the result of destruction of forests. The importance of photosynthesis in determining atmospheric carbon dioxide is shown by the fact that there is an annual fluctuation of about 5 ppm CO_2 in the northern hemisphere attributed to photosynthesis. The minimum in this cycle occurs around September at the end of the summer growing season and the maximum occurs around April as photosynthesis is getting underway after winter.

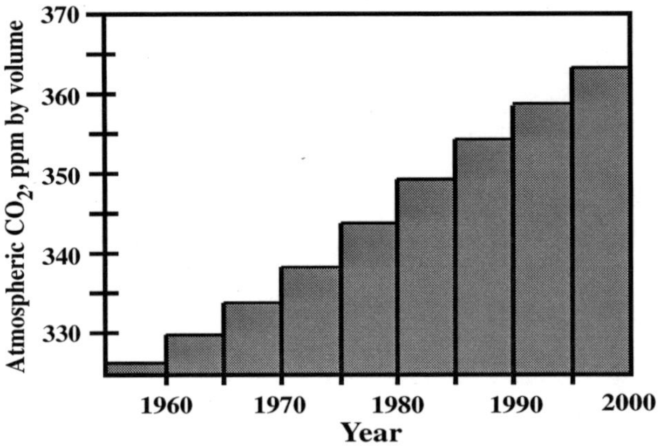

Figure 8.5. Concentrations of atmospheric CO_2 levels in recent decades. These levels are increasing by approximately 1 part per million by volume each year.

The concern with increasing carbon dioxide in the atmosphere is that it will lead to — indeed *is* leading to — an excess of a good thing, warming of the global atmosphere. This is the now well known **greenhouse effect**, which may not become as dire as some experts predict, but which has a real possibility of becoming the worst environmental problem so far created by humans.

Is Earth warming? Sophisticated computer models predict that it is backed by evidence from increasingly accurate temperature records over more than 100 years. Especially accurate temperature measurements taken by satellite show that the 1980s were the warmest decade on record globally. But this record decade was followed by the 1990s, which were warmer still. The 48 contiguous states of the U.S. had an average temperature of 3.6˚C for the 1999-2000 winter, the warmest winter on record, and records of 3.1˚C and 3.2˚C were set for the winters of 1997-1998 and 1998-1999, respectively. In New England and the Northern Plains, the first freezes and the first snowfalls were the latest on record during the 1999-2000 winter. February, 2000, was notable for setting record high temperatures in the northern regions of the contiguous 48 states. Both the first freeze and first snowfall came later than ever in colder regions of the 48 states, including New England and the Northern Plains. Numerous places across the northern tier of states set high temperature records during February, 2000. Continuing the trend, the spring of 2000 was the warmest on record in the 48 contiguous states of the U.S. Unfortunately, for those counting on the greenhouse effect to lower fuel bills, November and December of 2000-2001 turned out to be the coldest on record in the 48 contiguous states of the U.S.!

Gases other than carbon dioxide may be involved in greenhouse warming. These include chlorofluorocarbons and N_2O. The one most likely to cause a problem is methane, CH_4, which has increased from estimated pre-industrial atmospheric levels of 0.70 ppm, to present values of 1.8 ppm. Although these values are much lower than those of carbon dioxide, each methane molecule is 20–30 times more effective in trapping heat than is each CO_2 molecule. A number of human activities have contributed to the release of methane. Part of this is due to leakage of natural gas, which consists of methane, and from release as a byproduct of petroleum production. Bacteria growing in the absence of oxygen in municipal refuse landfills, in rice paddies, and in the stomachs of ruminant animals (cows, sheep, moose) release enormous quantities of methane. Green chemistry can also be applied to avoid the generation of ultrastable volatile compounds that have a high greenhouse gas potential. Sulfur hexafluoride, SF_6, was mentioned as such a gas in Section 8.8. Also included in this category are completely fluorinated hydrocarbons, such as CF_4.

Although there could be some benefits of mild global warming, the net effect would almost certainly be bad, perhaps catastrophic. Climate models predict an average global temperature increase of 1.5–5°C. That does not sound like much, but it is about as much again as the temperature increase that occurred from the last ice age until now. Especially if the warming is toward the high side of the projected range, it would greatly affect climate and rainfall. The melting of the polar and Greenland ice caps along with expansion of warmer ocean water would cause sea levels to rise as much as 0.5–1.5 meters. Decreased rainfall and increased water evaporation would contribute to severe drought and water shortages that could make some currently popular areas of the world virtually uninhabitable.

Can Green Chemistry Help Deal With Global Warming?

Green chemistry and the related area of industrial ecology can help deal with the problem of global warming in two major respects. The first approach is to provide means to prevent global warming from taking place. The second approach is in coping with global warming, if it occurs.

The prevention of global warming is best accomplished by avoiding the release of potential greenhouse gases. The most significant of these is carbon dioxide. One way to reduce the release of carbon dioxide is by using biomass as fuel or raw material for the manufacture of various products. Burning a biomass fuel does release carbon dioxide to the atmosphere, but an exactly equal amount of carbon dioxide was removed from the atmosphere in the photosynthetic process by which the biomass was made, so there is no net addition of CO_2. Unless or until biomass-derived materials used in feedstocks are burned, their use represents a net loss of carbon dioxide from the atmosphere.

Another potential use of green chemistry to prevent addition of carbon dioxide to the atmosphere is through **carbon sequestration** in which carbon dioxide is produced, but is bound in a form such that it is not released to the atmosphere. This approach has the greatest potential in applications where the carbon dioxide is produced in a concentrated form. In Section 6.7, reactions are shown by which carbon from coal is reacted with oxygen and water to produce elemental hydrogen and carbon dioxide. The net reaction for this production is the following:

$$2C + O_2 + 2H_2O \rightarrow 2CO_2 + 2H_2 \quad (8.9.1)$$

The hydrogen generated can be used as a pollution-free fuel in fuel cells or combustion engines. The carbon dioxide can be pumped into deep ocean waters, although this has the potential to lower ocean pH slightly, which would be detrimental to marine organisms. Another option is to pump the carbon dioxide deep underground. A side benefit of the latter approach is that in some areas carbon dioxide pumped underground can be used to recover additional crude oil from depleted oil-bearing formations.

An indirect green chemistry approach to the reduction of carbon dioxide emissions is to develop alternative methods of energy production. One thing that would be very beneficial is the development of more efficient photovoltaic cells. These devices have become marginally competitive for the generation of electricity, and even relatively small improvements in efficiency would enable their much wider use, replacing fossil fuel sources of electricity generation. Another device that would be extremely useful is a system for the direct photochemical dissociation of water to produce elemental hydrogen and oxygen, which could be used in fuel cells. An application of green biochemistry that would reduce carbon dioxide emissions is the development of plants with much higher efficiencies for photosynthesis. Plants now are only about 0.5% efficient in converting light energy to chemical energy. Raising this value to only 1% would make a vast difference in the economics of producing biomass as a substitute for fossil carbon.

Green chemistry can also be applied in the prevention of release of greenhouse gases other than carbon dioxide. This has been done, for example, in the replacement of chlorofluorocarbons (Freons) with analogous compounds having at least one C-H bond, that are rather readily destroyed in the troposphere. Both kinds of compounds act as greenhouse gases, but the latter last for much shorter times during which they are available to absorb infrared radiation. Another approach is to limit the emissions of methane, CH_4. Large quantities of methane are released by anaerobic bacteria growing in flooded rice paddies. By developing strains of rice and means of cultivation that enable the crop to be grown on unflooded soil, this source of methane can be greatly reduced. Methane collection systems placed in municipal waste landfills can prevent the release of methane from this source and provide a source of methane fuel.

Green chemistry, biochemistry, and biology can be used to deal with global warming when it occurs. Crops, fertilizers, and pesticides can be developed that enable plants to grow under the drought conditions that would follow global warming. Another approach is the development of salt-tolerant crops that can be grown on soil irrigated with saline water, where fresh water supplies are limited.

8.10. PHOTOCHEMICAL SMOG

One of the most common urban air pollution problems is the production of **photochemical smog**. This condition occurs in dry, stagnant air masses, usually stabilized by a temperature inversion (see Figure 8.1), that are subjected to intense sunlight. A smoggy atmosphere contains ozone, O_3, organic oxidants, nitrogen oxides, aldehydes, and other noxious species. In latter stages of smog formation visibility in the atmosphere is lowered by the presence of a haze of fine particles formed by the oxidation of organic compounds in smog.

The chemical ingredients of smog are nitrogen oxides and organic compounds, both released from the automobile, as well as from other sources. The driving energy force behind smog formation is electromagnetic radiation with a wavelength at around 400 nm or less, in the ultraviolet region, just shorter than the lower limit for visible light. Energy absorbed by a molecule from this radiation can result in the formation of active species, thus initiating *photochemical reactions*.

Although methane, CH_4, is one of the least active hydrocarbons in terms of forming smog, it will be used here to show the smog formation process because it is the simplest hydrocarbon molecule. Smog is produced in a series of chain reactions. The first of these occurs when a photon of electromagnetic radiation with a wavelength less than 398 nm is absorbed by a molecule of nitrogen dioxide,

$$NO_2 + h\nu \rightarrow NO + O \qquad (8.10.1)$$

to produce an oxygen atom, O. The oxygen atom is a very reactive species that can abstract a hydrogen atom from methane,

$$CH_4 + O \rightarrow H_3C\cdot + HO\cdot \qquad (8.10.2)$$

to produce a methyl radical, $H_3C\cdot$, and a **hydroxyl radical**, $HO\cdot$. In these formulas, the dot shows a single unpaired electron. A chemical species with such a single electron is a **free radical**. The hydroxyl radical is especially important in the formation of smog and in a wide variety of other kinds of photochemical reactions. The methyl radical can react with an oxygen molecule,

$$H_3C\cdot + O_2 \rightarrow H_3COO\cdot \qquad (8.10.3)$$

to produce a methylperoxyl radical, $H_3COO\cdot$. This is a strongly oxidizing, reactive species. One of the very important reactions of peroxyl radicals is their reaction with NO, produced in the photochemical dissociation of NO_2 (see Reaction 8.10.1 above),

$$NO + H_3COO\cdot \rightarrow NO_2 + H_3CO\cdot \quad (8.10.4)$$

to regenerate NO_2, which can undergo photodissociation, re-initiating the series of chain reactions by which smog is formed. Literally hundreds of other reactions can occur, leading eventually to oxidized organic matter that produces the small particulate matter characteristic of smog.

As the process of smog formation occurs, numerous noxious intermediates are generated. One of the main ones of these is ozone, O_3, and it is the single species most characteristic of smog. Whereas ozone is an essential species in the stratosphere, where it filters out undesirable ultraviolet radiation, it is a toxic species in the troposphere that is bad for both animals and plants. Another class of materials formed with smog consists of oxygen-rich organic compounds containing nitrogen of which peroxyacetyl nitrate, PAN,

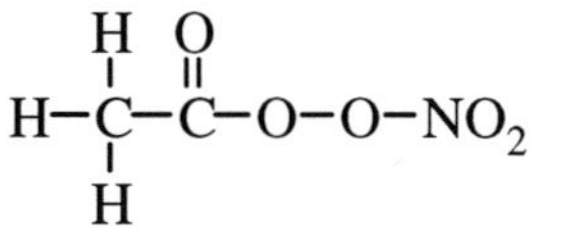

Peroxyacetyl nitrate (PAN)

is the most common example. This compound and ones similar to it are potent oxidizers and highly irritating to eyes and mucous membranes of the respiratory tract. Also associated with smog are aldehydes, which are irritants to eyes and the respiratory tract. The simplest aldehyde, and one commonly found in smoggy atmospheres, is formaldehyde:

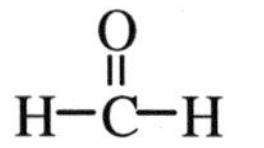

Formaldehyde

Harmful Effects of Smog

Smog adversely affects human health and comfort, plants, materials, and atmospheric quality. Each of these aspects is addressed briefly here. Ozone is the smog constituent that is generally regarded as being most harmful to humans, plants, and materials, although other oxidants and some of the noxious organic materials, such as aldehydes, are harmful as well.

People exposed to 0.15 parts per million of ozone in air experience irritation to the respiratory mucous tissues accompanied by coughing, wheezing, and bronchial constriction. These effects may be especially pronounced for individuals undergoing vigorous exercise because of the large amounts of air that they inhale. On smoggy days, air pollution alerts may advise against exercise and outdoor activities. Because of these effects, the U.S. Environmental Protection Agency has recommended an 8-hour standard limit for ozone of 0.08 ppm. In a smoggy atmosphere, the adverse effects of ozone are aggravated by exposure to other oxidants and aldehydes.

Plants are harmed by exposure to nitrogen oxides, ozone, and peroxyacetyl nitrate (PAN, see above), all oxidants present in a smoggy atmosphere. PAN is the most harmful of these constituents, damaging younger plant leaves, especially. Ozone exposure causes formation of yellow spots on leaves, a condition called chlorotic stippling. Some plant species, including sword-leaf lettuce, black nightshade, quickweed, and double-fortune tomato, are extremely susceptible to damage by oxidant species in smog and are used as bioindicators of the presence of smog. Costs of crop and orchard damage by smog run into millions of dollars per year in areas prone to this kind of air pollution, such as southern California.

Materials that are adversely affected by smog are generally those that are attacked by oxidants. The best example of such a material is rubber, especially natural rubber, which is attacked by ozone. Indeed, the hardening and cracking of natural rubber has been used as a test for atmospheric ozone.

Visibility-reducing atmospheric aerosol particles are the most common manifestation of the harm done to atmospheric quality by smog. The smog-forming process occurs by the oxidation of organic materials in the atmosphere, and carbon-containing organic materials are the most common constituents of the aerosol particles in an atmosphere afflicted by smog. Conifer trees (pine and cypress) and citrus trees are major contributors to the organic hydrocarbons that are precursors to organic particle formation in smog.

Preventing Smog with Green Chemistry

Smog is basically a chemical problem, which would indicate that it should be amenable to chemical solutions. Indeed, the practice of green chemistry and the application of the principles of industrial ecology can help to reduce smog. This is due in large part to the fact that a basic premise of green chemistry is to avoid the generation and release of chemical species with the potential to harm the environment. The best way to prevent smog formation is to avoid the release of nitrogen oxides and organic vapors that enable smog to form. At an even more fundamental level, measures can be taken to avoid the use of technologies likely to release such substances, for example, by using alternatives to polluting automobiles for transportation.

The evolution of automotive pollution control devices to reduce smog provide an example of how green chemistry can be used to reduce pollution. The first measures taken to reduce hydrocarbon and nitrogen oxide emissions from automobiles were very much command-and-control and "end-of-pipe" measures. These primitive measures implemented in the early 1970s did reduce emissions, but with a steep penalty in fuel consumption and in driveability of vehicles. However, over the last three decades, the internal combustion automobile engine has evolved into a highly sophisticated computer-controlled machine that generally performs well, emits few air pollutants, and is highly efficient. (And it would be much more efficient if those drivers who feel that they must drive "sport utility" behemoths would switch to vehicles of a more sensible size.) This change has required an integrated approach involving reformulation of gasoline. The first major change was elimination from gasoline of tetraethyllead, an organometallic compound that poisoned automotive exhaust catalysts (and certainly was not good for people). Gasoline was also reformulated to eliminate excessively volatile hydrocarbons and unsaturated hydrocarbons (those with double bonds between carbon atoms) that are especially reactive in forming photochemical smog.

An even more drastic approach to eliminating smog-forming emissions is the use of electric automobiles that do not burn gasoline. These vehicles certainly do not pollute as they are being driven, but they suffer from a very limited range between charges and the need for heavy lead/acid batteries. This inability to carry sufficient electrical energy to provide a convenient range of operation is probably an unsolvable problem. Furthermore, the use of so much lead in batteries comes with its own set of potential pollution problems from lead in the environment. It does appear, however, that hybrid automobiles using a small gasoline or diesel engine that provides electricity to drive electric motors propelling the automobile and to recharge relatively smaller batteries may be a viable solution to emission and fuel economy problems with automobiles. The internal combustion engine on these vehicles runs only as it is needed to provide power and, in so doing, can run at a relatively uniform speed that provides maximum economy with minimum emissions.

Another approach that is being used on vehicles as large as buses that have convenient and frequent access to refueling stations is the use of fuel cells that can generate electricity directly from the catalytic combination of elemental hydrogen and oxygen. Air pollution is virtually nonexistent because the only products are harmless water (see Section 6.7). There are also catalytic process that can generate hydrogen from liquid fuels, such as methanol, so that vehicles carrying such a fuel can be powered by electricity generated in fuel cells.

Green chemistry can be applied to devices and processes other than automobiles to reduce smog-forming emissions. This is especially true of organic solvents used for parts cleaning and other industrial operations, vapors of which are often released to the atmosphere. The substitution of water with proper additives or even the use of supercritical carbon dioxide fluid can eliminate such emissions.

QUESTIONS AND PROBLEMS

1. In addition to the oxygen that it provides, what are two vital protective functions served by the atmosphere?
2. In what respect is the composition of gases in the troposphere not uniform?
3. Other than avoiding turbulence due to lower altitude weather, suggest an advantage for commercial aircraft to cruise at a relatively high altitude of around 10 km.
4. After inspecting Figure 8.3, suggest why air in the troposphere mixes only slowly with that in the stratosphere.
5. What chemical species in the stratosphere is essential for life on Earth?
6. What starts a photochemical reaction? What is it called when a series of photochemical reactions continues?
7. What is a free radical?
8. In what two important respects may very small particles participate in atmospheric chemical processes?
9. In what respect does the radiation by which Earth loses energy differ from that by which it gets energy from the sun?
10. What are two catastrophic events that could cause a sudden cooling of Earth's atmosphere?
11. How is water vapor involved in moving energy through the atmosphere?
12. Distinguish among the terms meteorology, weather, and climate.
13. What do clouds consist of? What must happen before rain falls from clouds?
14. Why is there essentially no atmospheric chemistry involving elemental nitrogen gas in the atmosphere?
15. Cite an atmospheric chemical condition or phenomenon that shows that the O_2 molecule is easier to break apart than the N_2 molecule.
16. In what respect are elemental nitrogen and oxygen green elements?
17. Give a chemical reaction that produces oxygen that can be used for emergencies.
18. What are two major classes of atmospheric particles based upon how they are produced?
19. What is the source of fly ash?
20. Suggest why lead has become less of a problem as an atmospheric pollutant in recent years.
21. What is a radioactive element that can get into indoor spaces from underground sources?
22. What is an atmospheric phenomenon caused most prominently by particles 0.1 μm–1 μm in size? Why are very small particles especially dangerous to breathe?
23. List six means of controlling particle emissions.
24. What is the major health effect of carbon monoxide?
25. What is a serious air pollution phenomenon resulting from an atmospheric reaction of sulfur dioxide?

26. In what form may approximately half of the sulfur in coal be physically separated before combustion?

27. What is a method used to separate sulfur dioxide from furnace stack gas?

28. Name two ways in which green chemistry can be applied to reduce sulfur dioxide emissions.

29. What is an important health effect of nitrogen dioxide? Why is it particularly important in atmospheric chemistry?

30. What is the distinction between acid rain and acid deposition?

31. What are five categories of adverse effects from acid precipitation?

32. Chemically, what is distinctive about ammonia in the atmosphere?

33. What is the historic evidence for the toxicity of elemental chlorine in the atmosphere?

34. What are some sources of atmospheric hydrogen sulfide? Is it a health concern?

35. In what respect is atmospheric carbon dioxide essential to life on Earth? Why may it end up being the "ultimate air pollutant?"

36. What are some of the more harmful effects projected if global warming occurs to a significant extent?

37. What can green chemistry do about global warming?

38. What is a greenhouse gas other than carbon dioxide that is produced by microorganisms?

39. What are the ingredients and conditions leading to the formation of photochemical smog?

40. What substances are found in a smoggy atmosphere?

41. What are some harmful effects of smog?

42. What are some of the ways that green chemistry can help prevent smog?

43. The temperature of a specific number of moles of gas occupying initially 23.0 L was changed from 75˚C to -20˚C at constant pressure. Recalling the significance of 273 in such calculations, what was the volume of the gas after the temperature change?

44. The pressure on a specific number of moles of gas occupying initially 13.0 L was changed from 1.15 atm to 0.900 atm at constant pressure. What was the volume of the gas after the temperature change?

9. THE BIOSPHERE: HOW THE REVOLUTION IN BIOLOGY RELATES TO GREEN CHEMISTRY

9.1. GREEN CHEMISTRY AND THE BIOSPHERE

The **biosphere** consists of all living organisms and the materials and structures produced by living organisms. There is a very close connection between the biosphere and green chemistry including the following:

- Living organisms produce a wide range of materials that are used by humans for a variety of purposes.
- Large quantities of substances including pesticides and fertilizers are generated in the anthrosphere for use to control pests and enhance the growth and health of organisms in the biosphere.
- Reduction of the use and generation of toxic substances in the anthrosphere is designed to prevent harm to humans and other organisms in the biosphere.
- Environmental conditions largely determined by anthrospheric activities strongly affect organisms in the biosphere.

Individual organisms in the biosphere and organisms interacting in ecosystems can teach humans a lot about how to apply green chemistry. One important respect in which this is done is by the mild conditions under which organisms carry out complex chemical syntheses. Living things can function only within narrow temperature ranges that are close to those that humans find comfortable. (Even the 90-100˚C temperatures under which thermophilic bacteria function in hot springs and similar locations are not very far from room conditions.) Therefore, the enzyme-catalyzed reactions that organisms carry out occur under much milder conditions than the often high-temperature, high-pressure conditions of conventional chemical synthesis. Furthermore, organisms cannot tolerate highly toxic substances that are often used in chemical synthesis, the elimination of which is a primary objective of the practice of green chemistry.

Another lesson that living organisms provide for an efficiently operating anthrosphere is in the relationships between organisms with each other and with their environment in biological ecosystems. The wide variety of such ecosystems that have evolved over hundreds of millions of years of evolution have had to be sustainable to survive, completely recycling materials and preserving and enhancing their environment. This is in contrast to the way in which anthrospheric systems have evolved, especially during the last two centuries of the industrial revolution. In general, humans and their industrial systems have exploited nonrenewable resources and have polluted water, air, and land in a manner that simply cannot be sustained. Now, with nothing less at stake than the survival of the human race, it is imperative that humans develop sustainable industrial and economic systems. One way in which this can be done is through the practice of industrial ecology (see Chapter 11) in which various enterprises compose sustainable industrial ecosystems analogous to ecosystems in the biosphere.

9.2. BIOLOGY AND THE BIOSPHERE

Biology is the science of life and the organisms that comprise life. So what is life? Biologists define living organisms as those that share (1) constitution by particular classes of life molecules, (2) hierarchical organization, (3) capability to carry out metabolic processes, (4) ability to reproduce, (5) development, and (6) heredity. These areas are addressed in this section and in more detail later in the chapter.

The kinds of molecules that comprise living organisms were discussed in Chapter 5, Sections 5.6–5.10. Recall that these are *proteins* composed of polymers of nitrogen-containing amino acids, *carbohydrates* consisting of small molecules and polymers with an approximate simple formula of CH_2O, *lipids* defined by their property of solubility in organic solvents, and *nucleic acids* that are long polymers of sugars, nitrogen-containing bases, and phosphate. Two of these kinds of materials are sometimes bonded together as hybrid molecules. Along with water and some kinds of salts they make up living organisms. Literally thousands of kinds of structural and functional characteristics are due to the four kinds of molecules mentioned above. For example, proteins comprise muscle tissue and make up the enzyme molecules that act as catalysts to enable biochemical reactions to occur. A simple carbohydrate, glucose, $C_6H_{12}O_6$, is the primary organic product generated by plant photosynthesis and is present in animal bloodstreams. Large numbers of glucose molecules bonded together make up polymeric cellulose that is the structural material in plants. Lipids make up the crucial membranes that enclose living cells. And nucleic acids compose the genetic material that regulates cell function and reproduction.

Hierarchical organization applies to living organisms from the level of atoms all the way to the biosphere as a whole. Proteins, carbohydrates, lipids, and nucleic acids in living organisms are organized into distinct microscopic bodies contained in cells and called **organelles**. Cells are bodies of several micrometers (μm) in size that are the basic building blocks of organisms in that they are the smallest bodies of organisms that can exist independently (even cells of humans can be grown in cell cultures outside the body, given the appropriate nutrients and conditions). Bacteria, protozoa, and some other kinds of organisms are single-celled, but higher organisms are composed of many cells which have specialized functions. Cells with similar functions comprise **tissues** and tissues in turn make up **organs**, which may be organized into whole systems of organs. An **organism** is a collection of organs and organ systems. A group of organisms from the same species comprise a **population** and a group of populations existing in the same place make up a **community**. Numerous communities living in a particular environmental area, interacting with each other and with their environment, make up an **ecosystem**. Finally, all Earth's ecosystems comprise the entire **biosphere**.

The process of **metabolism** is what occurs when organisms mediate chemical (biochemical) processes to get energy, make raw materials required for tissues in organisms or modify raw materials for this purpose, and reproduce. Although there are thousands of different metabolic reactions, two stand out. The first of these is *photosynthesis* shown in Reaction 9.3.1 in which plants use light energy to convert inorganic CO_2 and H_2O to glucose sugar, $C_6H_{12}O_6$. The second major type of metabolic reaction is the mirror image of photosynthesis, **cellular respiration** in which glucose is oxidized to CO_2 and H_2O, yielding energy that is used by the organism. An interesting aspect of the conversion and utilization of energy in metabolism is that all organisms use the high-energy chemical species

adenosine triphosphate, ATP, (Figure 9.1) to transfer, convert, and store energy. Metabolism is addressed in more detail in Section 9.4.

All organisms carry out **reproduction** to produce offspring to continue the species. In addition to continuing a species, reproduction enables evolution to occur that results in new species. It is discussed further in Section 9.5.

Development is the process that occurs as an organism progresses from a fertilized egg to a juvenile and on to adulthood. Even single-celled bacteria that reproduce by cell division undergo development as the cells grow and produce additional organelles prior to further division. As anyone who has observed an infant grow into young adulthood knows, humans undergo development as well. With some species, development involves vastly different forms at different stages of development. An example of this is provided by multilegged crawling caterpillars that develop in a later stage into flying butterflies.

Heredity refers to the process by which traits characteristic of a species of organism are passed on to later generations. As discussed in Section 9.5, heredity occurs through the action of DNA. Heredity is the mechanism by which organisms have undergone evolution and adaptation to their environment.

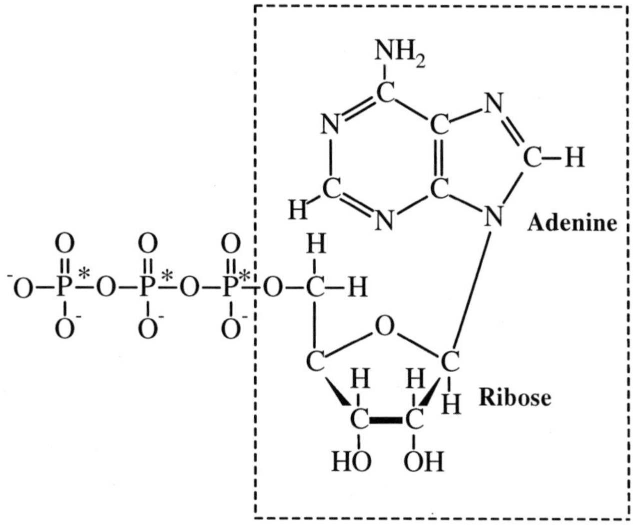

Figure 9.1. Adenosine triphosphate, ATP. The adenosine portion of the molecule is outlined by the dashed line. High-energy bonds are marked with asterices. Splitting off an ion of HPO_4^{2-} with the addition of H_2O (hydrolysis) forms adenosine diphosphate, ADP, releasing energy. Energy is required to produce ATP again. The ATP/ADP pair thus serves as a means of energy transfer in organisms.

Organisms that comprise living beings in the biosphere range in size and complexity from individual bacterial cells less than a micrometer in dimensions up to giant whales and human beings capable of thought and reasoning. Organisms comprising the biosphere belong to six kingdoms. **Archaebacteria** and **Eubacteria** are generally single-celled organisms without distinct, defined nuclei. **Protists** are generally single-celled organisms that have cell nuclei and may exhibit rather intricate structures. The three other kingdoms are **Plantae** (plants), **Animalia** (animals), and **Fungi** typified by molds and mushrooms.

Organisms are classified according to their food and energy sources and their utilization of oxygen. **Autotrophs** synthesize their food and biomass from simple inorganic substances, usually using solar energy to perform **photosynthesis**. **Chemautotrophs** mediate inorganic chemical reactions for their energy. **Heterotrophs**, including humans, derive their energy and biomass from the metabolism of organic matter, usually biomass from plants. **Aerobic** organisms require oxygen, whereas **anaerobic** organisms use alternate sources of oxidants. **Facultative** organisms can use oxygen or other oxidants depending upon conditions.

The biosphere is greatly influenced by the other environmental spheres. In an environment where temperatures are moderate, sunshine abundant, and nutrients readily available, the biosphere consists of diverse groups of organisms interacting and codependent within thriving **ecosystems**. Under extreme conditions, there may be only a few organisms composing the localized biosphere, specialized for existence at extreme temperatures, high acidities, high levels of pollutants, or other conditions that make life impossible for most organisms.

Just as the biosphere is strongly influenced by the environment in which the organisms are found, it has a strong influence upon its surroundings. Organisms act to break down inhospitable rock to form soil that supports a variety of plants. The oxygen in the atmosphere, upon which we all depend for our existence, was put there by photosynthesis performed by bacteria capable of photosynthesis. The nature of the anthrosphere that humans have constructed is influenced by the biosphere. The teepees that plains Native Americans used for portable dwellings were possible because of the abundant skins provided by large bison. The wood from which a large fraction of modern dwellings are constructed is obtained from trees in the biosphere.

9.3. CELLS: BASIC UNITS OF LIFE

As a fundamental unit of the biosphere, it is appropriate to choose the living **cell**. A single one of these very small entities visible only under a microscope may perform all the functions required for an organism to process nutrients and energy and to reproduce. Or cells may be highly specialized entities, such as human liver, brain, and red blood cells. There are two general classes of cells. **Prokaryotic cells** are those that make up bacteria and simple single-celled organisms that composed all of life on Earth for the first approximately 2 billion years of life on the planet. These cells are only about 1–2 micrometers in size, have only limited external appendages, and possess little differentiated internal structure. **Eukaryotic cells** compose all organisms other than bacteria, are typically 10 μm or more in size, often have external appendages, and generally show well differentiated internal structures with numerous distinct parts. These cells appeared only about 1.5 billion years ago in the estimated 3.5 billion years that life has existed on Earth. Figure 9.2 represents these two kinds of cells.

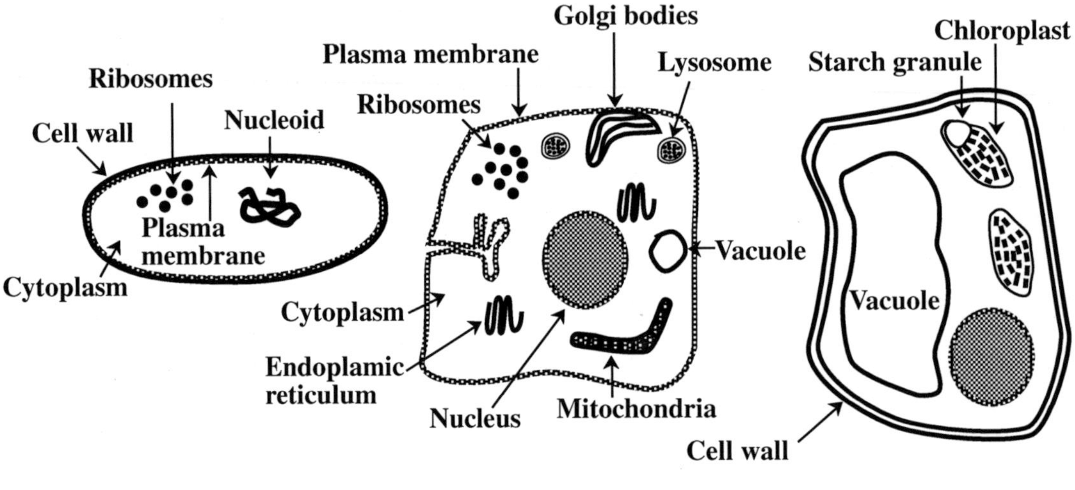

Figure 9.2. Representations of prokaryotic cells (left), eukaryotic animal cells (center), and eukaryotic plant cells (right). Eukaryotic cells are very complex and not all of the many organelles are shown.

Prokaryotic cells characteristic of bacteria are enclosed by strong **cell walls** composed largely of carbohydrates that hold the cells together. The plasma membrane controls passage of materials into and out of the cell and is the site of photosynthesis in photosynthetic bacteria. Gelatinous **cytoplasm** composed largely of protein and water fills the cell. There is not a defined nucleus, but the cell has a mass of genetic material (DNA) that composes a **nucleoid**. The DNA directs cell metabolism and reproduction. Proteins are made in the cell in **ribosomes** that are distributed around the cell interior. Ribosomes and other bodies in the prokaryotic cell are not enclosed by separate defined membranes as is the case with more complex eukaryotic cells.

Eukaryotic cells all are enclosed by **plasma membranes** (cell membranes), are filled with cytoplasm, and contain a variety of organelles that are enclosed by membranes. Figure 9.2 shows only a few of the most important organelles contained in eukaryotic cells. The genetic material in eukaryotic cells is contained in a **nucleus** that is enclosed by a membrane. This DNA is associated with proteins and RNA forming **chromosomes. Mitochondria** in eukaryotic cells are bodies in which oxidative metabolism, the process by which the cells use oxygen to gain energy from "burning" food nutrients, is carried out. **Lysosomes** are bodies that contain enzymes capable of breaking down cellular macromolecules (proteins, carbohydrates, nucleic acids, and lipids). This very important process destroys waste macromolecules that otherwise would accumulate and stop the cells from performing their necessary functions and in so doing recycles the small molecules that cells need for their metabolic processes. In a sense, therefore, lysosomes do green chemistry at the cellular level. The system of internal membranes composing the **endoplasmic reticulum** contains surface-bound enzymes that synthesize proteins, such as proteinaceous enzymes, that are exported from the cell. The endoplasmic reticulum is the feature that most clearly distinguishes eukaryotic from prokaryotic cells. **Golgi bodies** act to export materials from the cell.

Three features largely distinguish plant eukaryotic cells from animal cells in that the plant cells have a cell wall, a large central vacuole, and chloroplasts. The cell wall gives the plant cell strength and rigidity. The vacuole takes up most of the cell volume and allows contact with gases. The chloroplasts are sites in which chlorophyll uses light energy ($h\nu$) to synthesize carbohydrates as shown by the following reaction for the photosynthetic generation of glucose sugar:

$$6CO_2 + 6H_2O \text{ (light energy, } h\nu) \rightarrow C_6H_{12}O_6 \text{ (glucose)} + 6O_2 \qquad (9.3.1.)$$

Shown by the above reaction, **photosynthesis** was responsible for the greatest changes that the biosphere has ever caused in the atmosphere and geosphere. This occurred with the evolution of **cyanobacteria** (once thought to be algae and called "blue-green algae") about 3 billion years ago, the first organisms capable of carrying out photosynthesis and producing oxygen, which for them was a waste product. This raised the oxygen content of the atmosphere from virtually zero to the current value of 21% (by volume of dry air). The result was conversion of the atmosphere to an oxidizing medium. Vast deposits of solid iron minerals now used for iron ore were formed when atmospheric oxygen reacted with dissolved Fe in the oceans,

$$4Fe^{2+} + O_2 + 4H_2O \rightarrow 2Fe_2O_3 + 8H^+ \qquad (9.3.2)$$

to produce solid iron oxide. Part of the oxygen generated by photosynthesis dissolved in water, where it was available for the development of organisms that used oxygen to metabolize organic matter. Whereas Earth's surface had been a most inhospitable place for the existence of life, the oxygen released by photosynthesis enabled the formation of the ultraviolet radiation-filtering layer of ozone (O_3) in the stratosphere that made life possible outside the protective confines of water. Thus life became possible on Earth's land surface, soil was formed, aided by the weathering action of organisms that grew on rock surfaces, plants growing in soil became well established, and animals developed. The huge changes made possible by the action of single-celled cyanobacteria carrying out photosynthesis are obvious.

9.4. METABOLISM AND CONTROL IN ORGANISMS

Living organisms continually process materials and energy, a process called **metabolism**. *Photosynthesis*, which is discussed above, is the metabolic process that provides the base of the food chain for most organisms. Animals break down complex food materials to smaller molecules through the process of **digestion**. **Respiration** occurs as nutrients are metabolized to yield energy:

$$C_6H_{12}O_6 \text{ (glucose)} + 6O_2 \rightarrow 6CO_2 + 6H_2O + \text{energy} \qquad (9.4.1.)$$

Organisms assemble small molecules to produce biomolecules, such as proteins, by a **synthesis** process.

In addition to viewing metabolism as a phenomenon within an individual organism, it can be viewed as occurring within groups of organisms living in an ecosystem. Consider, for example, the metabolism of nitrogen within an ecosystem. Elemental nitrogen from the atmosphere may be fixed as organic nitrogen by bacteria living symbiotically on the roots of leguminous plants, then converted to nitrate when the nitrogen-containing biomass decays. The nitrate may be taken up by other plants and incorporated into protein. The protein may be ingested by animals and the nitrogen excreted as urea in their urine to undergo biological decay and return to the atmosphere as elemental nitrogen. Carbon from carbon dioxide in the atmosphere may be incorporated into biomass by plant photosynthesis, then eventually returned to the atmosphere as carbon dioxide as the biomass is used as a food source by animals.

Enzymes in Metabolism

In Section 4.5 *catalysts* were defined as materials that enable a reaction to occur without themselves being consumed. Living organisms have catalysts that are very important in metabolism. These catalysts are special proteins that enable biochemical reactions to take place called **enzymes**. Enzymes speed up metabolic reactions by as much as almost a billion-fold. In addition to making reactions go much more rapidly, enzymes are often highly specific in the reactions that they catalyze. The reason for the specificity of enzymes is that they have very specific structures that fit with the substances upon which they act.

Figure 9.3 illustrates the action of enzymes. The first step is the reversible formation of an enzyme/substrate complex that forms because of the complementary shapes of the enzyme (more specifically the active site on the enzyme) and the substrate. The second step is the formation of products accompanied by release of the unchanged enzyme molecule. This reaction implies that the substrate is split apart by enzyme action, a very common enzymatic process called hydrolysis when it is

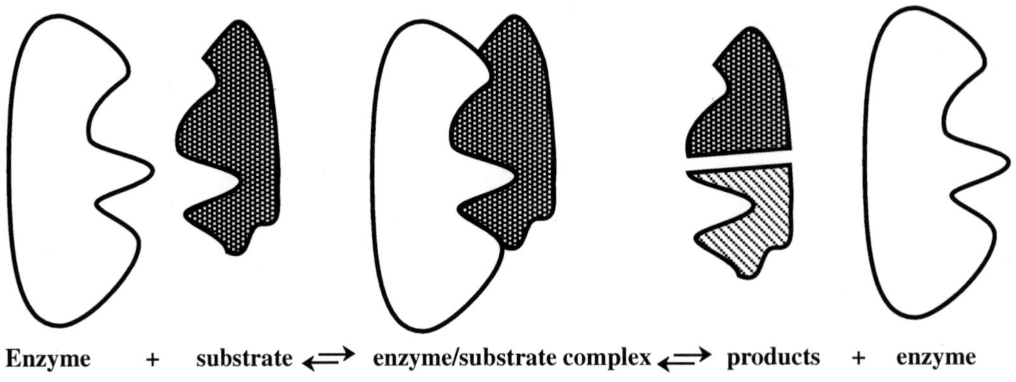

Figure 9.3. Enzyme action. The enzyme recognizes the substrate upon which it acts because of the complementary shapes of the enzyme and the substrate. The double arrows indicate that the processes are reversible.

accompanied by the addition of water with an H atom going to one of the products and an OH group to the other. Other types of enzyme-catalyzed reactions occur, including the joining of two molecules, modifications of functional groups (see Section 5.4) on substrate molecules, and rearranging the structures of molecules.

The names of enzymes, usually ending in "*-ase*" often reflect their functions and may also indicate where they operate. An example is *gastric proteinase*, a name that indicates the enzyme acts in the stomach (*gastric*) and hydrolyzes proteins (*proteinase*). The enzyme released by the pancreas that hydrolyzes fats is called *pancreatic lipase*.

A number of factors can affect enzyme action. One important factor is temperature. Organisms without temperature-regulating mechanisms have enzymes that increase in activity as temperature increases up to the point where the heat damages the enzyme, after which the activity declines precipitously with increasing temperature. Enzymes in mammals function optimally at body temperature (37˚C for humans) and are permanently destroyed by about 60˚C. There is particular interest in enzymes that function in bacteria that live in hot springs and other thermal areas where the water is at or near boiling. These enzymes may turn out to be very useful in commercial biosynthesis operations where the higher temperature enables reactions to occur faster. Acid concentration also affects enzymes, such as those that function well in the acidic environment of the stomach, but stop working when discharged into the slightly basic environment of the small intestine (were this not the case, they would tend to digest the intestine walls).

A significant concern with potentially toxic substances is their adverse effects upon enzymes. As an example, organophosphate compounds, such as insecticidal parathion and military poison sarin "nerve gas" bind with acetylcholinesterase required for nerve function, causing it not to act and stopping proper nerve action. Some substances cause the intricately wound protein structures of enzymes to come apart (denaturation) which stops enzyme action. The active sites of enzymes which recognize substrates have a high population of -SH groups. Heavy metals, such as lead and cadmium, have a strong affinity for -SH and may bind at enzyme active sites destroying the function of the enzymes.

Enzymes are of significant concern in the practice of green chemistry. One obvious relationship is that between enzymes and chemicals that are toxic to them. In carrying out green chemical processes, such chemicals should be avoided wherever possible. Another obvious relationship has to do with the use of biological processes to perform chemical operations, which are usually done under much milder and environmentally friendly conditions biologically than chemically. Biochemical processes are all carried out by enzymes. For example, several enzymes, starting with hexokinase, are involved in the multistepped biochemical fermentation synthesis of ethyl alcohol from carbohydrate glucose. With recombinant DNA technology (see Section 9.8) it is now possible to invest bacteria with enzyme systems from other organisms designed to carry out desired biochemical processes. Bacteria are much more amenable to handling and usually much more efficient than the organisms from which the genes for the desired enzyme systems are taken. Another approach is to use isolated enzymes immobilized on a solid support to carry out biochemical processes without the direct involvement of an organism.

Nutrients

The raw materials that organisms require for their metabolism are **nutrients**. Those required in larger quantities include oxygen, hydrogen, carbon, nitrogen, phosphorus, sulfur, potassium, calcium, and magnesium and are called **macronutrients**. Plants and other autotrophic organisms use these nutrients in the form of simple inorganic species, such as H_2O and CO_2, which they obtain from soil, water, and the atmosphere. Heterotrophic organisms obtain much of the macronutrients that they need as carbohydrates, proteins and lipids (see Chapter 5) from organic food material.

An important consideration in plant nutrition is the provision of **fertilizers** consisting of sources of nutrient nitrogen, phosphorus, and potassium. A large segment of the chemical manufacturing industry is involved with fixing nitrogen from the atmosphere as ammonia, NH_3, and converting it to nitrate (NO_3^-), urea (CON_2H_4), or other compounds that are applied to the soil as nitrogen fertilizer. Phosphorus is mined as mineral phosphate that is converted to biologically available phosphate ($H_2PO_4^-$ and HPO_4^{2-} ions) by treatment with sulfuric or phosphoric acid. Potassium is mined as potassium salts and applied directly as fertilizer. The ongoing depletion of sources of phosphorus and potassium fertilizer is a sustainability issue of significant concern.

Organisms also require very low levels of a number of micronutrients, which are usually used by essential enzymes that enable metabolic reactions to occur. For plants, essential micronutrients include the elements boron, chlorine, copper, iron, manganese, sodium, vanadium, and zinc. The bacteria that fix atmospheric nitrogen required by plants require trace levels of molybdenum. Animals require in their diet elemental micronutrients including iron and selenium as well as micronutrient vitamins consisting of small organic molecules.

Control in Organisms

Organisms must be carefully regulated and controlled in order to function properly. A major function of these regulatory functions is the maintenance of the organism's homeostasis, its crucial internal environment. The most obvious means of control in animals is through the **nervous system** in which messages are conducted very rapidly to various parts of the animal as **nerve impulses**. More advanced animals have a brain and spinal cord that function as a **central nervous system** (CNS). This sophisticated system receives, processes, and sends nerve impulses that regulate the behavior and function of the animal. Effects on the nervous system are always a concern with toxic substances. For example, exposure to organic solvents that dissolve some of the protective lipids around nerve fibers can lead to a condition in which limbs do not function properly called **peripheral neuropathy**. Therefore, a major objective of green chemistry is to limit the use of and human exposure to such solvents.

Both animals and plants employ **molecular messengers** that move from one part of the organism to another to carry messages by which regulation occurs. Messages sent by these means are much slower than those conveyed by nerve impulses. Molecular messengers are often **hormones** that are carried by a fluid medium in the organism, such as the bloodstream, to cells where they bind to **receptor proteins** causing some sort of desired response. For example, the process may cause the cell to synthesize a protein to counteract an imbalance in homeostasis. Some hormones called **pheromones** carry messages from one organism to another. They commonly serve as sex attractants. Some biological means of pest control use sex pheromones to cause sexual confusion in pesticidal insects, thus preventing their reproduction. Figure 9.4 shows a common plant hormone and a common animal hormone.

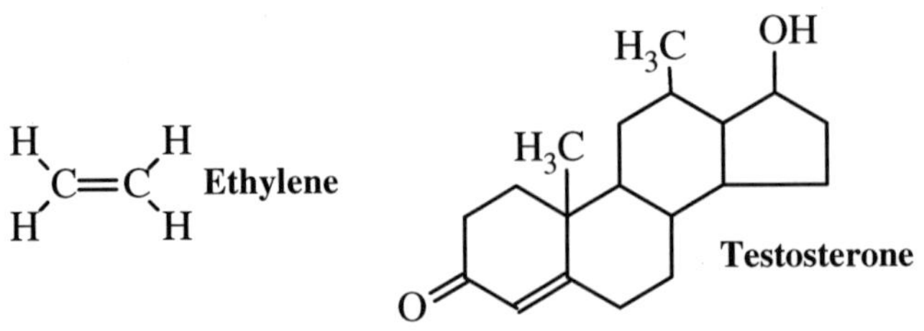

Figure 9.4. A simple molecule that acts as a plant hormone to promote maturation processes (ethylene) and a common animal hormone, testosterone, the male sex hormone. See Chapter 5, Section 5.3, for the meaning of line formulas, such as those shown for testosterone.

In animals, regulatory hormones are commonly released by **endocrine glands**, such as the anterior pituitary gland that releases human growth hormone, the parathyroid gland that releases a hormone to stimulate uptake of calcium into the blood from bones and the digestive tract, and the pancreas that releases insulin to stimulate glucose uptake from blood. These hormones are carried to target cells in fluids external to the cells. Some toxic substances interfere with the function of endocrine glands. Another concern is that some toxic substances may mimic the action of hormones. For example, evidence exists to suggest that premature sexual development in some young female children can be caused by ingestion of synthetic chemicals that mimic the action of the female sex hormone estrogen.

9.5. REPRODUCTION AND INHERITED TRAITS

As noted in the preceding section, one of the major activities of organisms is metabolism by which organisms process materials and energy. The other major activity of all organisms is reproduction. Most organisms are capable of reproducing a large excess of their species because throughout time predators and hostile conditions have required large numbers of juveniles to ensure survival of enough members to continue the species. Unrestrained reproduction, especially by humans, poses a strong threat of overpopulation and is a major concern related to reproduction and the environment. A second major concern is the potential effect of environmental chemicals upon reproduction and the potential that such chemicals have to cause birth defects. Therefore, chemicals that may affect reproduction are given strong consideration in the practice of green chemistry.

Primitive single-celled organisms, particularly bacteria, undergo **asexual** reproduction in which a cell simply splits to form two cells. Humans and most other multicelled organisms undergo **sexual reproduction** requiring that male sperm cells fertilize female egg cells to produce cells capable of dividing and producing new individuals.

Reproduction is directed by **genes** which occur in molecules of deoxyribonucleic acid, DNA, discussed in Chapter 5, Section 5.5. The DNA of an individual, which in sexual reproduction has contributions from both parents, determines the physical, biochemical, and behavioral traits of the organism. The DNA can be altered resulting in changes called **mutations**. A miniscule fraction of mutations are desirable and convey advantages to an individual that are passed along as heritable characteristics in offspring. This is the process of *natural selection* that has resulted in literally millions of different species of organisms.

Some chemicals are capable of producing mutations. Control of production and exposure to these **mutagens** is a major thrust of green chemistry. This is particularly so because substances that cause mutations are generally regarded as being capable of causing cancer as well.

9.6. STABILITY AND EQUILIBRIUM OF THE BIOSPHERE

In order for an organism to survive and thrive, it must reach a state of stability and equilibrium with its environment. The term given to such a state is **homeostasis** ("same status"). In maintaining homeostasis, an organism must interact with its surroundings and other organisms in its surroundings and must balance flows and processing of matter (including nutrients) and energy. On an individual basis, organisms do a remarkably good job of keeping their internal levels of water, materials such as calcium in blood, and temperature within a range conducive to their well being. Mammals have developed extraordinary capabilities of homeostasis; a healthy individual maintains its internal temperature within a few tenths of a degree. The concept of homeostasis applies to entire groups of organisms living together in ecosystems and, ultimately, to the entire biosphere. Therefore, a major objective of environmental science, including the practice of green chemistry, is to maintain and enhance conditions of homeostasis in the biosphere.

Ecology describes the interaction of organisms with their surroundings and each other. An important consideration in ecology is the manner in which organisms process matter and energy. An **ecosystem** describes a segment of the environment and the organisms in it with all of the interactions and relationships that implies. An ecosystem has means of capturing energy, almost always by plants or algae that perform photosynthesis. Light, temperature, moisture, and nutrient supplies are critical aspects of an ecosystem. Ecosystems recycle essential nutrient carbon, oxygen, nitrogen, phosphorus, sulfur, and trace elements. An important part of any ecosystem is the **food chain**, or more complicated **food webs**, in which food generated by photosynthesis is utilized by different organisms at different levels. An important aspect of the food chain in respect to persistent, poorly degradable organic chemicals that are soluble in lipid (fat) tissue occurs through the sequence of animals eaten in the food chain (small creatures in water are fed upon by small fish that are eaten by large fish that are eaten by large birds). Thus, aquatic pollutants become more concentrated in lipid tissue at the top of the food chain, a process called biomagnification (see Section 9.9). An objective of the practice of green chemistry is to avoid the generation and use of chemicals with strong tendencies to undergo biomagnification in the environment.

The surroundings over a relatively large geographic area in which a group of organisms live constitute a **biome**. There are a number of different kinds of biomes. Regions near the equator may support **tropical rain forest** biomes that stay warm all of the year and in which nutrients remain largely in the organisms (rain forest soil is often notably poor in nutrients). Temperate regions may support **temperate deciduous forests** in which the trees grow new leaves for a warm, wet summer season and shed them for cold winters. Temperate regions may also have **grassland biomes** in which grass grows from a tough mass of dense roots called **sod**. **Tundra** are treeless arctic regions in which during summer only a layer of wet soil thaws above a permanently frozen foundation of permafrost.

Different kinds of biomes pose a variety of environmental challenges. Some of these have come about from the conversion of biomes to cropland. Grasslands in which the sod has been broken to support wheat and other crops have proven susceptible to wind erosion, which gave rise to the catastrophic Dust Bowl on the U. S. Great Plains during the 1930s. Climate changes resulting from global warming could change the distribution of biomes, giving rise to much larger areas of hot deserts that humans might have to learn how to utilize.

Response of Life Systems to Stress

Organisms and the ecosystems in which they exist are subject to a number of threats that can result in loss of populations and even total destruction of the system. Natural threats include drought, flooding, fire, landslide, and volcanic eruption. Humans threaten life systems with cultivation, deforestation, mining, and severe pollution. The ability of a community of organisms to resist alteration and damage from such threats, sometimes called **inertia**, depends upon several factors and provides important lessons for the survival of the human community in the face of environmental threats. One of the basic factors involved in providing resistance of a community to damage is its overall rate of photosynthesis, its **productivity**. Another important factor is **diversity** of species so that if one species is destroyed or seriously depleted, another species may take its place. **Constancy** of numbers of various organisms is desirable; wide variations in populations can be very disruptive to a biological community. Finally, **resilience** is the ability of populations to recover from large losses. The ability of a biological system to maintain high levels of these desirable factors is commonly determined by factors other than the organisms present. This is clearly true of productivity, which is a function of available moisture, suitable climate, and nutrient-rich soil. Since all organisms depend upon the availability of good food sources, diversity, constancy, and resilience tend to follow high productivity.

Relationships Among Organisms

In a healthy, diverse ecosystem, there are numerous, often complex relationships among the organisms involved. Species of organisms strongly influence each other. And organisms strongly influence the physical portion of the system in which they live. An example of such an influence is the tough, soil-anchoring sod that develops in grassland biomes.

In most ecosystems there is a **dominant plant species** that provides a large fraction of the biomass anchoring the food chain in the ecosystem. This might be a species of grass, such as the bluestem grass that thrives in the Kansas Flint Hills grasslands. Herbivores feed upon the dominant plant species and other plants and, in turn, are eaten by carnivores. At the end of the food cycle are organisms that degrade biomass and convert it to nutrients that can nourish growth of additional plants. These organisms include earthworms that live in soil and bacteria and fungi that degrade biological material.

In a healthy ecosystem different species compete for space, light, nutrients, and moisture. Much of agricultural chemistry is devoted to trying to regulate the competition of weeds with crop plants. Large quantities of herbicides are applied to cropland each year to kill competing weeds. In this never-ending contest, green chemistry has an important role in areas such as the synthesis of herbicides that have maximum impact on target pests with minimum impact on the environment. In an undisturbed ecosystem the **principle of competitive exclusion** applies in which two or more potential competitors exist in ways that minimize competition for nutrients, space, and other factors required for growth.

Within ecosystems there are large numbers of **symbiotic relationships** between organisms which exist together to their mutual advantage. The classic case of such a relationship is that of lichen consisting of algae and fungi growing together. The fungi anchor the system to a rock surface and produce substances that slowly degrade the rock and extract nutrients from it. The algae are photosynthetic, so they produce the biomass required by the system, which is utilized in part by the fungi. Another important symbiotic relationship is that in which nitrogen-fixing bacteria grow in nodules on leguminous plant roots. The bacteria receive nutrients from the plants in exchange for chemically fixed nitrogen required for plant nutrition.

9.7. DNA AND THE HUMAN GENOME

In Chapter 5, Section 5.10, *deoxyribonucleic acid, DNA*, was discussed and it was noted that this macromolecule stores and passes on genetic information that organisms need to reproduce and synthesize proteins. Recall that DNA is composed of repeating units called *nucleotides* each consisting of a molecule of the sugar 2-deoxy-β-D-ribofuranose, a phosphate ion, and one of the four nitrogen-containing bases, adenine, cytosine, guanine, and, thymine (conventionally represented by the letters A, C, G, and T, respectively). DNA is one of two *nucleic acids*, the other one of which is *ribonucleic acid, RNA*. Like DNA, RNA consists of repeating nucleotides but the sugar in RNA is β-D-ribofuranose and it contains uracil instead of thymine in its bases. The structural formulas of segments of DNA and RNA are shown in Figure 5.10.

The structure of DNA is a key aspect of its function, and its elucidation by Watson and Crick in 1953 was a scientific insight that set off a revolution in biology that is going on to this day. The huge DNA molecules consist of two strands counterwound with each other and held together by hydrogen bonds (discussed in Section 7.1 and illustrated for water molecules in Figure 7.1). A representation of this structure is shown in Figure 9.5. In this structure, the hydrogen bonds connecting complementary bases on the two strands are represented by dashed lines. Because of their structures that make hydrogen bonding possible, adenine on one strand is always hydrogen-bonded to thymine on the opposite strand and guanine to cytosine. During cell division the two strands of DNA unwind and each

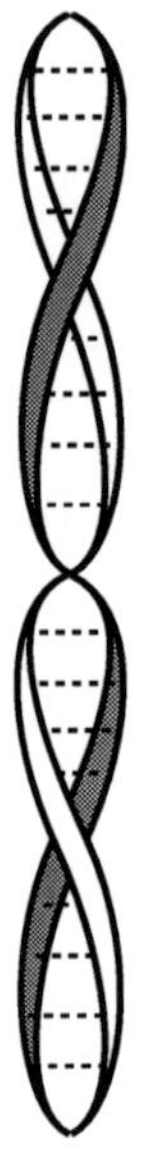

Figure 9.5. Representation of the double helix structure of DNA. Hydrogen bonds between complementary bases on the two strands are shown by dashed lines.

generates a complementary strand for the DNA of each new cell. Figure 9.6 shows a representation of two complementary strands of DNA in which thymine and adenine from the two strands are hydrogen-bonded together. When guanine is opposite cytosine, the two bases are bonded by three hydrogen bonds.

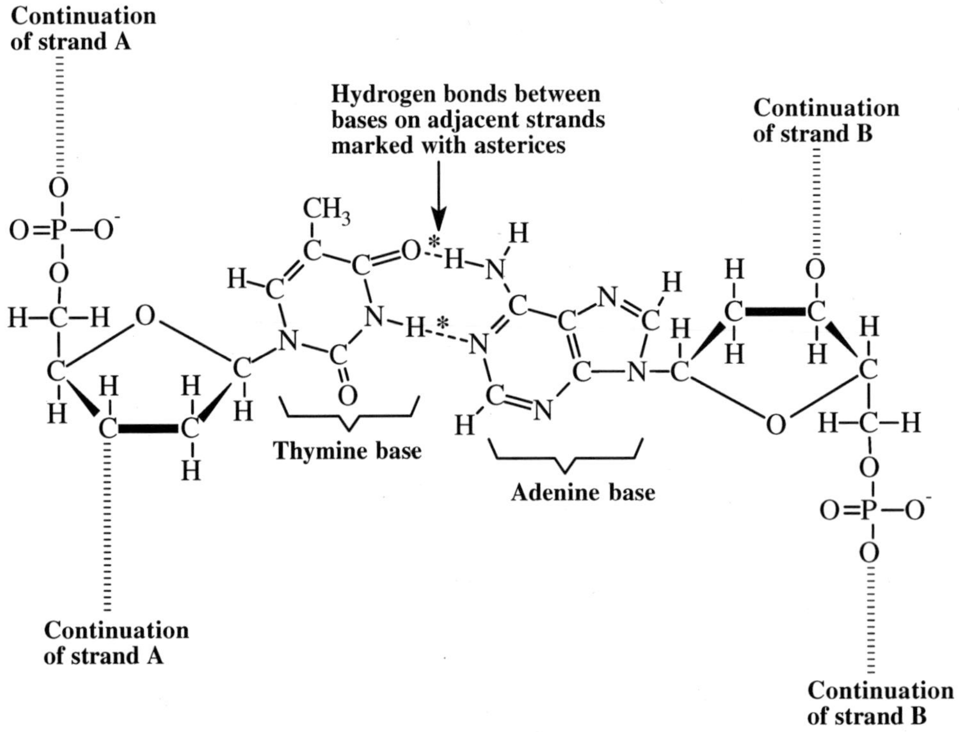

Figure 9.6. Representation of two nucleotides in two adjacent strands of DNA showing hydrogen bonding between the bases thymine and adenine. These two bases bonded together by hydrogen bonds constitute a base pair.

In organisms with eukaryotic cells, DNA is divided into units associated with protein molecules called **chromosomes**. The number of these varies with the organism; humans have 23 pairs of chromosomes, a total of 46. The strands of DNA in chromosomes, in turn, are divided into sequences of nucleotides, each distinguished by the nitrogen-containing base in it. These sequences of nucleotides give directions for the synthesis of a specific kind of protein or polypeptide. (Polypeptide is a general term for polymers of amino acids; proteins are the relatively long-chain polypeptides.) These specific groups of nucleotides, each of which has a specific function, are called **genes**. When a particular protein is made, DNA produces a nucleic acid segment designated *m*RNA, which goes out into the cell and causes the protein to be formed through a process called **transcription** and **translation** (the gene is said to be **expressed**). Whereas the number of chromosomes in an organism is now easy to determine, the number of genes is a matter of debate and is still not known for most organisms. Original estimates of the number of genes in human DNA were in excess of 100,000, but extensive investigation of human DNA now indicates that it may be only 30,000, or even fewer (much to the disappointment of venture capitalists who financed human DNA studies with the prospect of patenting large numbers of human genes.)

Proteins are the biological molecules that make up much of the structure of cells and that perform most of the key functions of living organisms. Proteins are made according to directions provided by cellular DNA. The steps in protein synthesis are the following:

1. The DNA in a gene that is specific for a particular protein transfers information for the protein synthesis to RNA.

2. The RNA links with a cell ribosome, which is the protein-synthesizing entity of the cell.

3. Using directions provided by the RNA, the ribosome assembles amino acids into a protein.

4. The protein performs the function for which it is designed in the organism; for example, it may function as an enzyme to carry out metabolic processes.

As the entities that give the directions for protein synthesis, genes are obviously of the utmost importance in living organisms. As discussed in Section 9.8, genes can now be transferred between different kinds of organisms and will direct the synthesis of the protein for which they are designed in the recipient organism. It is now known that a number of human diseases are the result of defective genes, and there is a genetic tendency toward getting other kinds of diseases. For example, certain gene characteristics are involved in susceptibility to breast cancer.

Because of the known relationship of gene characteristics to disease, the decision was made in the mid-1980s to map all the genes in the human body. This collective body of genes is called the **human genome** and the project to map it is called the Human Genome Project. The original impetus for this project in the U. S. arose because of interest in the damage to human DNA by radiation, such as that from nuclear weapons. But, from the beginning, it was recognized that the project had enormous commercial potential, especially in the pharmaceutical industry, and could be very valuable in human health.

The sequencing of the human genome has been done on individual chromosomes. Each chromosome consists of about 50 million base pairs (see Figure 9.6). However, it is possible to sequence only about 500-800 base pairs at one time, so the DNA has to be broken into segments for sequencing. There are two approaches to doing this. The publicly funded consortium working on the human genome project identified short marker sequences on the DNA that could be recognized in reassembling the information from the sequencing. The private concern involved in the effort used a process in which the DNA was broken randomly into fragments, each of which was sequenced. The data from the sequencing were then analyzed using powerful computer programs to show overlap, and the complete gene sequence was then assembled.

In 2001, a joint announcement from the parties involved in the Human Genome Project revealed that the genome had been sequenced. This accomplishment is leading to a vast effort to understand genetically-based diseases in humans, to develop pharmaceutical agents based upon genetic information, and other areas that can use information about the genome. The benefits and consequences of mapping the human genome will be felt for many decades to come.

Genome Sequencing and Green Chemistry

The Human Genome Project and related genome sequencing of other organisms have a number of implications for green chemistry. One of the key goals of green chemistry is to use chemicals that have maximum effectiveness for their stated purpose with minimum side effects. This certainly applies to pharmaceuticals in which a knowledge of the human genome may enable development of drugs that do exactly what they are supposed to do without affecting nontarget systems. This means that drugs can be made very efficiently with little waste material.

Some of the most important effects of DNA sequencing as it relates to green chemistry has to do with a wide variety of organisms other than humans. With an exact knowledge of DNA and the genes that it contains, it is possible to deal with organisms on a highly scientific basis in areas such as pest control and the biosynthesis of raw materials. An accurate map of the genetic makeup of insects, for example, should result in the synthesis of precisely targeted insecticides which kill target pests without affecting other organisms. Such insecticides should be effective at very low doses, thus minimizing the amount of insecticide that has to be synthesized and applied, consistent with the goals of green chemistry.

An exact knowledge of the genomes of organisms is extremely helpful in the practice of genetic engineering in which genes are transferred between species to enable production of desired proteins and to give organisms desirable characteristics, such as pest resistance. A number of medically useful proteins and polypeptides are now produced by genetically engineered microorganisms, most commonly genetically modified *Escherichia coli* bacteria. Perhaps the greatest success with this technology has been the biosynthesis of human insulin, a lack of which causes diabetes in humans. Two genes are required to make this relatively short polypeptide which consists of only 51 amino acids. Other medically useful substances produced by genetically engineered organisms include human growth hormone, tissue plasmogen activator that dissolves blood clots formed in heart attacks and strokes, and various vaccine proteins to inoculate against diseases such as meningitis, hepatitis B, and influenza. Genetic engineering is discussed in more detail in Section 9.8 and in Chapters 10 and 12.

9.8. GENETIC ENGINEERING

Ever since humans started raising crops (and later animals) for food and fiber they have modified the genetic makeup of the organisms that they use. This is particularly evident in the cultivation of domestic corn which is physically not at all like its wild ancestors. Until now, breeding has been a slow process. Starting with domestication of wild species, selection and controlled breeding have been used to provide desired properties, such as higher yield, heat and drought tolerance, cold resistance, and resistance to microbial or insect pests. For some domesticated species these changes have occurred over thousands of years. During the 1900s, increased understanding of genetics greatly accelerated the process of breeding different varieties. The development of high-yielding varieties of wheat and rice during the "green revolution" of the 1950s has prevented (or at least postponed) starvation of millions of people. A technology that enabled a quantum leap in productivity of domestic crops was the development of **hydrids** from crossing of two distinct lines of the same crop, dating in a practical sense from the mid-1900s.

This section discusses the genetic modification of organisms to enhance their value. It addresses plants primarily because more effort has been made and more things have been accomplished in plant breeding than with other kinds of organisms. However, the general principles discussed apply to animals and other kinds of organisms as well.

Traditional breeding normally takes a long time and depends largely upon random mutations to generate desirable characteristics. One of its greatest limitations has been that it is essentially confined to the same species, whereas more often than not, desired characteristics occur in species other than those being bred. Since about the 1970s, however, the possibility has arisen of using **transgenic technology** to transfer genes from one organism to an entirely different kind. This has raised a vast array of possibilities for greatly modified species that could be applied to many different purposes. And it has led as well to a number of concerns regarding unintended consequences of the technology. Ideally, transgenic technology can be used beneficially in plant breeding to increase tolerance to stress, increase yield, enhance the value of the end product by enriching it in desired biochemicals such as essential amino acids, and otherwise make plants more useful.

Transgenic technology is possible because of the existence within cells of deoxyribonucleic acid, DNA. This long-chain biological polymer directs cell reproduction and metabolism as discussed in Section 9.7. Transgenic technology is possible because a gene in DNA will make the protein for which it is designed in an organism that is quite different from the one in which the gene originated. So a gene transferred from one organism to another as a segment of DNA will often perform the function for which it was developed in the recipient organism. The details of how segments of DNA are transferred between organisms are beyond the scope of this work. Enzymes are used in the process, with restriction enzymes cutting out desired regions of DNA and ligase enzymes joining the ends of DNA together. Enzymes are used to further manipulate and amplify the DNA.

Perhaps the most difficult aspect of transgenic technology is identifying the genes responsible for desired characteristics and locating them among the millions of repeating units comprising the DNA strand. In addition to identifying specific genes, it is necessary to learn how they interact with other genes and the mechanisms by which they are regulated and expressed, the process by which a gene generates a specific protein.

After a specific gene is isolated, it is cloned by insertion into a bacterium, which reproduces the gene many times. In order for a gene to generate a desired protein at the appropriate time and location in a plant, a **promoter** must be added that functions as a switch. The easiest promoter to use is a **constitutive promoter** that causes the gene to be expressed in most of the plant's tissues and throughout its lifetime. The most successful promoter for this purpose is designated CaMV35S, which is isolated from the cauliflower mosaic virus. Other more specific promoters have also been used, such as those that are induced by light and function during photosynthetic processes. Much of the current effort in transgenic technology is devoted to the use of specific promoters that cause the gene to be expressed only where and when its protein product is needed.

So far, two major methods have been used to insert genes into a plant cell. The **gene gun** uses a very small projectile to literally shoot genetic information into cells. This method has been used with monocot ("grassy") species including corn and wheat. It suffers from a low percentage of "hits."

The most widely employed method of gene insertion is the *Agrobacterium* method widely used on dicot (broadleafed) species, such as potatoes and soybeans, and more recently adapted to monocots as well. This method uses a bacterium that thrives in soil called *Agrobacterium tumafaciens* (the cause of crown gall disease in plants), which infects plants, using the plants' metabolic processes for its own reproduction. The mechanism by which *Agrobacterium* is used to insert genetic information into plant cells is complicated and not completely understood. The bacterium enters a plant through a wound in the plant stem or leaves. Somehow the DNA incorporated into the bacterium is transferred through plant cell protoplasm to the plant DNA. This process may occur when the plant DNA becomes uncoiled during cell reproduction.

Only a few percent of plant cells targeted for gene insertion actually incorporate and express the gene. Therefore, it is necessary to have some means of knowing if the gene insertion has been successful. This is accomplished by the insertion of **marker genes** that make plant cells resistant to herbicidal compounds or antibiotics that kill normal plant cells. Plant cells are placed in media containing the toxic materials, and those cells that reproduce are the ones into which the desired genes have been successfully inserted. Following selection of the viable cells that presumably contain the desired transplanted genes, the cells are grown in tissue cultures in the presence of growth-promoting hormones and nutrients required for growth. This leads to the production of whole plants that produce seeds. Additional plants are grown from these seeds and evaluated for the desired characteristics.

Once plants containing desired transgenes have been produced, an exhaustive evaluation process occurs. This process has several objectives. The most obvious of these is an evaluation of the transplanted gene's activity to see if it produces adequate quantities of the protein for which it is designed. Another important characteristic is whether or not the gene is passed on reliably to the plant's progeny through successive generations. It is also important to determine whether the modified plant grows and yields well and if the quality of its products are high.

Only a few strains of plants are amenable to the insertion of transgenes, and normally their direct descendants do not have desired productivity or other characteristics required for a commercial crop. Therefore, transgenic crops are crossbred with high-yielding varieties. The objective is to develop a cross that retains the transgene while having desired characteristics of a commercially viable crop. The improved variety is subjected to exhaustive performance tests in greenhouses and fields for several years and in a number of locations. Finally, large numbers of genetically identical plants are grown to produce seed for commercial use.

Many kinds of genetically modified plants have been developed and more are being marketed commercially every year. These are discussed in more detail in Chapter 10.

9.9. BIOLOGICAL INTERACTION WITH ENVIRONMENTAL CHEMICALS

Organisms in the environment interact significantly with xenobiotic materials (those foreign to living systems) in their surroundings. The uptake of such materials by organisms is discussed in this section. The biodegradation of xenobiotic substances, primarily through the action of bacteria, is discussed in Section 9.10.

Bioaccumulation is the term given to the uptake and concentration of xenobiotic materials by living organisms. The materials may be present in water in streams or bodies of water, sediments in bodies of water, drinking water, soil, food, or even the atmosphere. Bioaccumulation can lead to **biomagnification** in which xenobiotic substances become successively more concentrated in the tissues of organisms higher in the food chain. This usually occurs with poorly degradable, lipid-soluble organic compounds. Suppose, for example, that such a compound contacts lake water, accumulates in solid detritus in the water, sinks to the sediment, is eaten by small burrowing creatures in the sediment, which are eaten by small fish. The small fish may be eaten by larger fish, which in turn are consumed as food by birds. At each step, the xenobiotic substance may become more concentrated in the organism and may reach harmful concentrations in the birds at the top of the food chain. This is basically what happened with DDT, which almost caused the extinction of eagles and hawks.

Fish that bioaccumulate poorly degradable, lipid-soluble organic compounds from water will lose them back to water if they are placed in an unpolluted environment. The process by which this occurs is called **depuration**. The time required to lose half of the bioaccumulated xenobiotic material is called the **half-life** of the substance.

The most straightforward case of bioaccumulation is **bioconcentration**, which occurs when a substance dissolved in water enters the body of a fish or other aquatic organism by passive processes (basically, just "dissolves" in the organism), and is carried to bodies of lipid in the organism in the blood flow. The model of bioconcentration assumes that the organism taking up the compound does not metabolize the compound, a good assumption for refractory organic compounds such as DDT or PCBs. It also assumes that uptake is by nondietary routes, including diffusion through the skin and especially through the gills of fish. This model applies especially to substances that have low water solubilities (though high enough to make the compound available for uptake) and high lipid solubilities. This model of bioconcentration assumes a dynamic equilibrium between the xenobiotic substance dissolved in water and the same substance dissolved in lipid tissue. It is called the **hydrophobicity model** because of the hydrophobic ("water-hating") nature of the substance being taken up.

The degree of bioconcentration depends upon a number of factors. The most important of these are the relative water and lipid solubility of the compound. The size and shape of the xenobiotic molecule also seem to be factors, as is temperature. In addition, bioconcentration depends upon the species of fish and their age, size, and lipid contents. Bioconcentration may be expressed by **bioconcentration factors** defined as

$$\text{Bioconcentration factor} = \frac{\text{Concentration of xenobiotic in lipid}}{\text{Concentration of xenobiotic in water}} \quad (9.9.1)$$

The bioconcentration factor can also be regarded as the ratio of the solubility of the compound in lipid to its solubility in water. Typical bioconcentration factors for PCBs and hexachlorobenzene in sunfish, trout, and minnows range from somewhat more than 1,000 to around 50,000, reflecting the high lipid solubility of these compounds.

9.10. BIODEGRADATION

Bacteria, fungi, and protozoa in the environment play an important role in biodegrading both natural materials and synthetic substances. These processes occur predominantly in water, in sediments in bodies of water, and in soil. Biodegradation is the process by which biomass from deceased organisms is broken down to simple inorganic constituents, thus completing the cycle in which biomass is produced from atmospheric carbon dioxide and from water by photosynthesis.

The biodegradation of substances in the environment by the action of enzymes in microorganisms can be divided metabolically into two categories. The first of these is the utilization by microorganisms of organic matter that can be metabolized for energy and as material to synthesize additional biomass. This is the route taken by microorganims degrading biomass from other organisms, and to a lesser extent in the biodegradation of some xenobiotic materials. The second way in which microorganisms metabolize environmental chemicals is through **cometabolism** in which the organism's enzymes act upon the substances as a "side-line" of their normal metabolic processes. The substances that are cometabolized are called secondary substrates because they are not the main compounds for which the enzymatic processes are designed.

A commonly cited example of cometabolism occurs with the action of *Phanerochaete chrysosporium* on organochlorine compounds, including PCBs and dioxins. Commonly known as the white rot fungus, this organism has an enzyme system that normally breaks down lignin, the degradation-resistant "glue" that holds cellulose together in wood and woody plants. Under certain stressed conditions, however, the enzyme will act to cometabolize synthetic organochlorine compounds and was once widely promoted as a means of remediating hazardous waste sites contaminated with such compounds.

The degree of biodegradation varies over a wide range. In the simplest case, the change to the substrate molecule is relatively minor, such as addition, deletion, or modification of a functional group. Complete biodegradation to simple inorganic species — CO_2 for carbon, NH_4^+ or NO_3^- for nitrogen, HPO_4^{2-} for phosphorus, SO_4^{2-} for sulfur — is the process of **mineralization**, which is crucial in completing elemental cycles in the environment.

An important step in biodegradation is the modification of a substance to reduce its toxicity. This process is called **detoxication**. An example of detoxication is given in Reaction 9.10.2 below for the conversion of insecticidal paraoxon, a potent nerve poison, to *p*-nitrophenol, which is only about 0.005 × as toxic. In some cases, however, action of microorganisms in the environment may produce a much more toxic material. An example of this is the generation of highly toxic, mobile methylmercury species, $Hg(CH_3)_2$ and $HgCH_3^+$ from insoluble, relatively harmless inorganic mercury species.

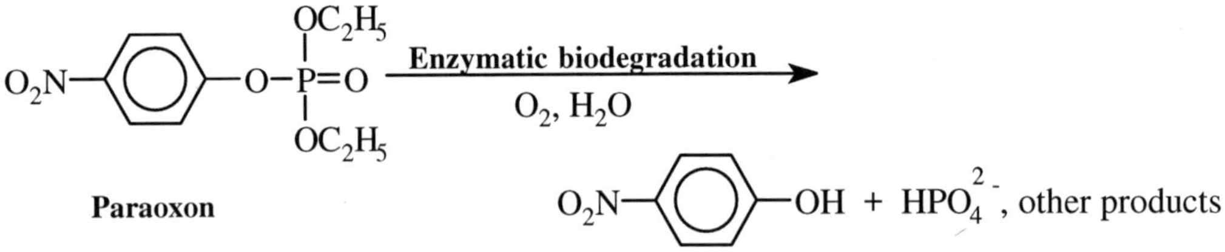

(9.10.2)

Paraoxon **p-nitrophenol**

A number of factors are involved in determining the effectiveness and rate of biodegradation. The compound in question has to be biodegradable. Biodegradability is influenced by both physical properties, such as water solubility, and chemical characteristics including the presence of functional groups amenable to microbial attack. As illustrated by the example, below,

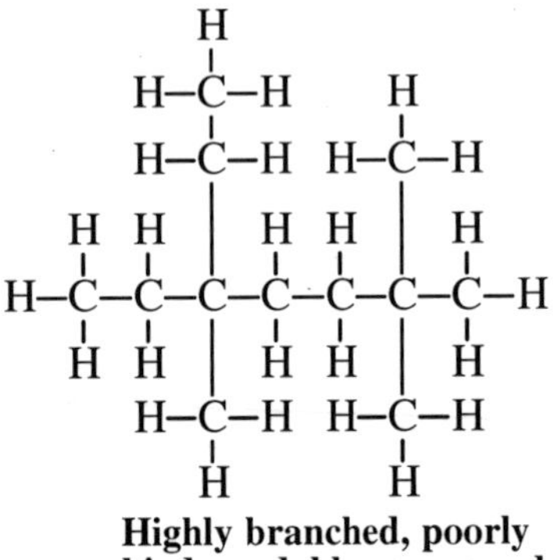

Highly branched, poorly biodegradable compound

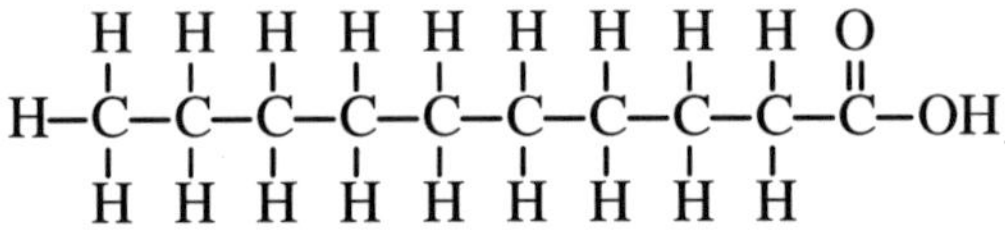

Highly biodegradable compound with a straight hydrocarbon chain and a functional group amenable to biological attack

branched-chain hydrocarbons are very resistant to biodegradation whereas straight-chain hydrocarbons, especially those with a suitable functional group, are readily metabolized by microorganisms. It should be noted that even very poorly biodegradable compounds can often be degraded under suitable conditions. As an example, phenol,

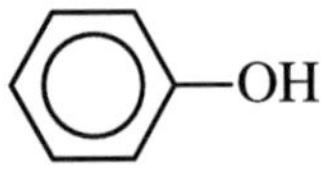

Phenol

is a biocidal compound that kills bacteria and was once the most commonly used disinfectant. However, in dilute solution and under the appropriate conditions, phenol can be destroyed by bacteria. An important aspect of biodegradation of resistant compounds is to use microorganisms acclimated to the particular kind of compounds. Populations of such microorganisms are found in locations where the kinds of compounds to be treated have been spilled, such as in petroleum spills on soil.

Biodegradability of compounds is an important consideration in green chemistry. This is especially true of "consumable" materials that are dissipated to the environment. As discussed in Section 7.13, the most common example of the use of a biodegradable material as a consumable material is the use of readily degraded straight-chain LAS surfactant in place of nonbiodegradable ABS surfactant in household detergents.

9.11. THE ANTHROSPHERE IN SUPPORT OF THE BIOSPHERE

Although humans are commonly blamed — usually with ample justification — for adverse effects upon the biosphere, human activities and modern technology have a high potential for benefitting the biosphere. This is especially true now that there is such an explosion in the understanding of biological sciences. The realization that the biosphere can contribute immeasurably to the benefit of humankind through such things as the provision of renewable feedstocks as raw materials provides a strong incentive to use technology to the benefit of the biosphere.

The most direct interface between the biosphere and technology occurs in agriculture. The production of biomass per unit area of land has increased in a spectacular fashion in recent decades with the use of fertilizers, herbicides, insecticides, and sophisticated means of cultivation and harvesting. Now the application of recombinant DNA technology (see Section 9.8) to agriculture promises even greater advances. In the past, the ways in which techniques for improved agricultural productivity were applied were largely divorced from considerations of the natural ways in which plants and animals grow on land. Fortunately, there is a growing realization of the important information that nature can provide in maintaining agricultural productivity. For example, in the prevention of water erosion, terraces constructed on land are designed to funnel excess water runoff onto grassed waterways. By planting these waterways to native grasses, a tough, erosion-resistant sod can be established that stands up under the punishment of occasional deluges of runoff water while surviving intermittent severe droughts. On a larger scale, in place of cultivating drought-prone prairie land to grow grain to feed to cattle, a better approach may be to reseed these lands to tough native grasses and allow bison to feed upon the grass as a source of meat (less fat and more healthy than beef from cattle).

The restoration and development of "natural" areas has become an important endeavor commonly termed **restoration ecology**. This often is advised with farmland that is too marginal to support profitable agricultural operations. The example of restoring native grasslands was mentioned above. Much of the rocky, hilly, unproductive farmland in New England is now reverting to forests. In such restoration efforts, modern construction machinery with the capacity to move enormous quantities of dirt have proven useful. One example in which such machinery is used is in leveling large areas for the construction of wetlands. Rivers that were once straightened to facilitate water flow — with catastrophic results in the form of flooding and erosion — are now being restored with the bends and meanders that characterize a healthy river. Following the catastrophic 500-year floods on the Missouri and Mississippi Rivers in 1993, large areas of cropland in the river bottoms were purchased by the Federal Government, river dikes designed to prevent flooding were breached, and the land was allowed to to revert to a wild state. Land disrupted by strip mining has been smoothed over to reduce erosion, topsoil applied, and trees planted to produce natural areas and wildlife habitat.

A significant amount of restoration ecology has been devoted to restoring game animals, some of which had been driven virtually to extinction by overhunting and habitat destruction. Animals that have come back in significant numbers include wild turkeys, wood ducks, snowy egrets, and American bison. Some of these efforts have been almost too successful. Once endangered Canadian geese have greatly increased in numbers and now populate many suburban areas where they often show their displeasure with sharing their new habitats with humans by hissing and pinching their fellow two-legged animals. In many areas deer now destroy crops and are a traffic hazard. Sophisticated captive breeding techniques are now used to reproduce endangered species of animals, and animal cloning may reach a point at which these efforts are routine.

In the area of green chemistry, sophisticated chemical analysis techniques can now be used to find and eliminate the sources of chemical hazards to wildlife. The classic example of this occurred in the 1960s when it was found that insecticidal DDT, biomagnified through the food chain, was preventing reproduction of endangered eagles and hawks at the top of the food chain. In 1970 a newly developed technique for the determination of mercury showed that large fish were contaminated by this heavy metal released from sediments by bacterial methylation. Analysis of lipid tissue in humans, caribou, and polar bears now indicate a global distillation mechanism by which persistent organic compounds evaporate into the atmosphere in warmer regions of Earth and condense in the polar regions, leading to significant contamination of food supplies. One of the major objectives of green chemistry is the elimination of the generation and use of such materials.

As the projected effects of global warming become more pronounced during the next century, technology will be employed to a greater extent to deal with these effects upon the biosphere. Increasingly sophisticated genetic engineering techniques will be used to develop plant varieties that can withstand the heat and drought resulting from global warming. Another possibility is the development of plants that can grow in saltwater. Using renewable solar and wind energy, vast water desalination projects will be developed to provide fresh water to irrigate high-value crops where the costs can be justified.

QUESTIONS AND PROBLEMS

1. Fill in the blanks in the following statement: Living organisms carry out chemical processes in ____ ____________, which are bodies of the order of a ________________ in size.
2. Name several constituents of living cells. What are two constituents typical of plant cells?
3. What are hormones? Where are they produced?
4. To what general class of kinds of biomolecules do enzymes belong? What is their function?
5. How are enzymes named?
6. How does temperature relate to enzyme action, particularly as it pertains to enzymes acting in the body?
7. What is recombinant DNA? Why is it produced?
8. What is metabolism? How is respiration related to metabolism?
9. What can natural ecosystems teach humans about how to manage their industrial systems?
10. Distinguish between bioaccumulation and biomagnification as they apply to lipid-soluble organic compounds.
11. What are six characteristics shared by all living organisms?
12. What are the four basic kinds of molecules that comprise living organisms?
13. Photosynthesis is the reaction that provides food and energy to keep biological systems going. What important biological process is the opposite of photosynthesis?
14. What are the six kingdoms of living organisms?
15. Justify the statement that human beings are aerobic heterotrophs.

16. What is the general effect of good living conditions, such as mild climate and abundant nutrients, on ecosystems?
17. Give four organelles found in eukaryotic cells.
18. What huge effect did photosynthesis by cyanobacteria have upon the atmosphere in the early years of life on earth?
19. In order to use the food that humans eat, it must undergo digestion and respiration. What is the distinction between these two phenomena?
20. One of two major characteristics of enzymes is their ability to greatly speed up metabolic reactions. What is the other characteristic?
21. Describe the overall process of enzyme action.
22. Name two important aspects of enzymes as they relate to green chemistry.
23. Which three elements are commonly regarded as plant fertilizers?
24. Name two ways in which messages are sent in the body of an animal. Do they have about the same speed?
25. Give examples of organs that produce hormones.
26. Describe a concern pertaining to synthetic chemicals and hormones.
27. Name a concern other than mutations regarding exposure to mutagens.
28. In order for an organism to live and thrive, its internal environment must be maintained within acceptable limits. Give the name of this state.
29. Name some of the critical factors in an ecosystem. How do functioning ecosystems handle materials?
30. Tropical rain forests, temperate deciduous forests, and grasslands are all examples of ___________.
31. Name four characteristics that enable life systems to respond well to stress.
32. What are symbiotic relationships between organisms? Give an example.
33. Name several factors that have contributed to the increased agricultural productivity of biomass and another factor that promises such increases in the future.
34. Define the human genome and give reasons why it is so important.
35. What are the relationships among cell DNA, RNA, and ribosomes in protein synthesis?
36. Name some substances that are beneficial to human health that are made by genetically engineered organisms. Name the organism often used for this purpose
37. What was the "green revolution?" What technology will likely result in a "second green revolution?"
38. How is transgenic technology distinguished from conventional breeding techniques? What kind of biomolecule makes transgenic technology possible?
39. Give the function of *Agrobacterium* in transgenic technology.
40. What are xenobiotic substances? What kind of metabolic process is largely responsible for their biodegradation?

41. Distinguish between bioaccumulation and biomagnification.

42. What physical property of a xenobiotic compound largely determines the magnitude of its bioconcen tration factor in an organism?

43. Discuss the range of changes that can occur on a xenobiotic compound as a consequence of biodegradation. What is perhaps the most important such change that can occur with respect to effects of xenobiotic substances on organisms?

44. Describe an important consideration of biodegradation related to consumable consumer products.

45.What is the explanation of high levels of persistent organic compounds in nonindustrialized polar regions?

10. THE GEOSPHERE, SOIL, AND FOOD PRODUCTION: THE SECOND GREEN REVOLUTION

10.1. THE SOLID EARTH

The next environmental sphere to be discussed is the **geosphere**. The geosphere consists of all the rocks, minerals, soil and sediments that compose the solid earth. Most of our food is grown on the geosphere and humans extract from it metals, fossil fuels, fertilizers for plants, and a variety of minerals used in construction and for other purposes. Over the years huge quantities of waste products have been discarded to the geosphere, in some cases very carelessly in a manner that poses threats to humans and other organisms. A thin layer on top of the geosphere — in places only a few centimeters deep — composes topsoil which supports the plant life upon which humans and most other land-dwelling organisms depend for their food.

The geosphere interacts strongly with the other environmental spheres. Streams and rivers flow through channels in the geosphere, lakes and reservoirs occupy cavities on the surface of the geosphere, and groundwater occurs in aquifers underground that are part of the geosphere. Gases are exchanged between the geosphere and the atmosphere, light and infrared radiation transmitted through the atmosphere warm the surface of ground, and it in turn radiates back to the atmosphere the infrared radiation by which Earth loses the energy it absorbs from the sun.

The geosphere is tied to green chemistry in many important respects, including the following:

- Plants that provide most food for humans and animals grow on the geosphere.
- Plants growing on the geosphere already provide, and have the potential to provide much more, biomass for use as renewable materials, such as wood, fiber, raw materials, and fuel.
- The geosphere is the source of nonrenewable minerals, ores, fossil fuels, and other materials used by modern industrialized societies.
- Modifications and alterations of the geosphere have profound effects upon the environment.
- Sources of fresh water are stored in lakes and rivers on the surface of the geosphere, move by means of streams, rivers, and canals on the geosphere, and occur in aquifers underground.
- The geosphere is the ultimate sink for disposal of a variety of wastes.

This chapter briefly addresses the nature of the geosphere, and resource utilization from the geosphere. Waste disposal on land or underground are considered in later chapters. Because of the special importance of soil and the plants that grow on it as sources of food and fiber, emphasis is placed upon soil and agriculture.

Physical Nature of the Geosphere

At the center of Earth is an iron-rich inner core, hot enough to be molten under normal pressures, but compressed to a solid by the enormous pressures at such great depths. Surrounding this core is an outer core consisting of molten rock. Earth's solid outer layer consists of the mantle and the **crust**, a layer that is only 5-40 km thick. Only the upper layers of the crust are accessible to humans.

For the most part, the crust consists of **rocks**, which in turn are made up of **minerals** characterized by a definite chemical composition and crystal structure. Only about 25 of the approximately 2000 known minerals compose most rocks. Because most of the crust consists of chemically combined oxygen (49.5%) and silicon (25.7%), the most abundant minerals are **silicates** composed of various silicon oxides, examples of which are quartz, SiO_2, and potassium feldspar, $KAlSi_3O_8$. Other elements in Earth's crust are aluminum (7.4%, commonly occurring as Al_2O_3), iron (4.7% as Fe_3O_4 and other iron oxides), calcium (3.6% in limestone, $CaCO_3$, and dolomite, $CaCO_3 \cdot MgCO_3$), sodium (2.8%), potassium (2.6%), and magnesium (2.1%). That leaves only 1.6% of the crust to serve as a source of other important mineral substances, including metals other than iron and aluminum, phosphorus required for plant growth, and sulfur widely used in industrial applications.

Igneous rock is rock that has solidified from molten rock that has penetrated to near Earth's surface. Exposed to water, atmospheric oxygen, and various organisms, igneous rock becomes highly altered, reaching a state of greater physical and chemical equilibrium with the atmosphere. This a process called **weathering**. Weathering products end up as soil and are carried by water to be deposited as sediments. Sediments that become buried and compressed produce **secondary minerals**, among the most abundant of which are **clays**, consisting of silicon and aluminum oxides, produced by the weathering of minerals such as potassium feldspar, $KAlSi_3O_8$. A common clay is **kaolinite**, $Al_2Si_2O_5(OH)_4$.

Although the earth's crust is very thin compared to its total diameter, there is an even much thinner, fragile and vitally important layer covering the crust — **soil**. Soil is the finely divided mixture of mineral and organic matter upon which plants grow, providing the food that humans and most other animals eat. Productive soil may be only a few centimeters thick, and rarely is more than a few meters in thickness. Because of its unique importance and fragility, soil is the subject of much of the rest of this chapter.

Geochemistry is the branch of chemistry that deals with rocks and minerals and the chemical interactions of the geosphere with other environmental spheres. The specialized branch of geochemistry relating to environmental influences and interactions of the geosphere is **environmental geochemistry**. Weathering by chemical processes is a particularly important aspect of geochemistry. Almost imperceptible under dry conditions, weathering proceeds at a much more rapid rate in the presence of water. The rate of weathering is also increased by the action of microorganisms, some of which secrete chemical species that attack rock and leach nutrients from it. Particularly important to weathering are **lichens**, which are algae and fungi living together synergistically. The algae utilize solar energy to convert atmospheric carbon dioxide to plant biomass and the fungi utilize the biomass and anchor the organisms to the rock surface and extract nutrients from it.

Human Influences

Human activities have a tremendous influence on the geosphere as evidenced by hills leveled, valleys filled in, and vast areas paved to make freeways, parking lots, and shopping centers. One such influence is on **surface albedo**, defined as the percentage of impinging solar energy reflected back from Earth's surface. The surface albedo of an asphalt paved surface is only about 8%. A more alarming effect is **desertification** in which normally productive soil is converted to unproductive desert in areas where rainfall is marginal. This phenomenon is discussed in more detail in Section 10.11.

10.2. ENVIRONMENTAL HAZARDS OF THE GEOSPHERE

There are a number of geospheric hazards that can be very damaging and destructive. Some of these are essentially beyond human influence. Volcanoes and earthquakes fall into these categories. Even in these cases, however, human activities can significantly influence the degree of damage done. As examples, structures constructed on poorly consolidated fill dirt are much more susceptible to earthquake damage than are those attached firmly to bedrock, and the construction of dwellings in areas known to be subject to periodic volcanic eruptions simply means that unstoppable lava flows and other volcanic effects will be much more damaging when they occur. Other, less spectacular, but very destructive geospheric phenomena can be greatly aggravated by human activities. Destructive and sometimes life-threatening landslides, for example, often result from human alteration of surface soil and vegetation.

Earthquakes

Earthquakes consist of violent horizontal and vertical movement of Earth's surface resulting from tectonic plates moving relative to each other. Plates move along fault lines. Huge masses of rock may be locked relative to each other for as long as centuries, then suddenly move along fault lines. This movement and the elastic rebound of rocks that occurs as a result causes the earth to shake, often violently and with catastrophic damage.

History provides many examples of astoundingly damaging earthquakes. Over 1 million lives (out of a much lower global population than now) were lost by an earthquake in Egypt and Syria in 1201 A.D. The Tangshan, China, earthquake of 1976 killed approximately 650,000. During the latter 1990s and early 2000s, a number of fatalities resulted from earthquakes in Turkey, Greece, Taiwan, Iran, and India. Financial costs of earthquakes in highly developed areas are enormous; the 1989 Loma Prieta earthquake in California cost about 7 billion dollars. Phenomena caused by earthquakes can add to their destructiveness. In addition to their direct shaking effects, earthquakes can cause ground to rupture, subside, or rise. **Liquefaction** of poorly consolidated ground, especially where groundwater levels are shallow, occurs when soil particles disturbed by an earthquake separate and behave like a liquid, causing structures to sink and collapse. One of the more terrifying effects of earthquakes are giant ocean waves called **tsunamis** that can reach heights of as much as 30 meters. An April 1, 1946, earthquake off the coast of Alaska produced a Tsunami that reached Hilo, Hawaii, about 5 hours after the quake, killing 159 people.

Earthquakes have defied all efforts to predict them, a fact that makes them all the more frightening. However, earthquake-prone areas, such as southern California, are well known, and loss of life and property can be minimized by taking appropriate measures. Buildings can be constructed to resist the effects of earthquakes using practices that have been known for some time. For example, some buildings in Niigata, Japan, were constructed to be earthquake-resistant in the 1950s. When a destructive earthquake hit that city in 1964, some buildings tipped over on the liquified soil but remained structurally intact! (Current practice calls for the construction of more flexible structures designed to dissipate the energy imparted to them by an earthquake.) The construction of buildings, roadways, railroads, and other structures to withstand the destructive effects of earthquakes provides an excellent example of designing the anthrosphere in a manner that is as compatible as possible with the geosphere.

Volcanoes

A **volcano** results due to the presence of liquid rock magma near the surface. In addition to liquid rock lava at temperatures ranging from 500°C to 1400°C that flows from volcanoes, these often very destructive phenomena are manifested by discharges of gases, steam, ash, and particles. Volcanic disasters have plagued humankind forever. The 79 A.D. eruption of Mount Vesuvius in ancient Rome buried the city of Pompei in ash, preserving a snapshot of life in Rome at that time. The astoundingly massive eruption of Indonesia's Tambora volcano in Indonesia in 1815 was caused when water infiltrated the hot magma beneath the volcano resulting in an explosion equivalent to 100 million tons of TNT explosive and blasting an estimated 30 cubic kilometers of solid material into the atmosphere. The May 18, 1980, Mount St. Helens eruption in Washington State blew about 1 cubic kilometer of material into the atmosphere, killed 62 people, and caused about $1 billion in damage.

In addition to their immediate effects upon surrounding areas, volcanoes can affect the atmosphere and climate. The Tambora volcano blasted enough particulate matter in the atmosphere to cause a very pronounced cooling effect. The following "year without a summer" caused global crop failures and starvation, and perceptible global cooling was observed for the next 10 years. Huge quantities of water vapor, dense carbon dioxide gas, carbon monoxide, hydrogen sulfide, sulfur dioxide, and hydrogen chloride may be emitted to the atmosphere. People may suffocate in the carbon dioxide or be poisoned by the toxic carbon monoxide and hydrogen sulfide. Hydrogen chloride along with hydrogen sulfide and sulfur dioxide oxidized in the atmosphere to sulfuric acid can contribute to acidic rainfall. Volcanic emissions differ in their atmospheric chemical effects. The 1982 El Chichón eruption in Mexico generated little particulate mineral matter but vast amounts of sulfur oxides that were oxidized to sulfuric acid in the atmosphere. The tiny droplets of sulfuric acid suspended in the atmosphere effectively reflected enough sunlight to cause a perceptible cooling in climate.

Massive, atmospheric-damaging eruptions of volcanoes in recorded history have caused catastrophic crop failures. These will happen again. And since the world as a whole carries little food surplus from year to year, the certainty of food supply disruptions due to volcanic activity point to the desirability of storing substantial amounts of food for emergency use.

Surface Effects

Though less spectacular than major earthquakes or volcanic eruptions, surface earth movement causes enormous damage and significant loss of life. Furthermore, surface earth movement is often strongly influenced by human activities. Surface phenomena result from the interaction of forces that act to thrust earth upward countered by weathering and erosion processes that tend to bring earth masses down. Both of these phenomena are influenced by the exposure of earth masses to water, oxygen, freeze-thaw cycles, alternate saturation with water and drying, organisms and human influences. **Weathering** is the physical and chemical breakdown of rock to fine, unconsolidated particles. **Erosion** occurs when weathered materials are moved by the action of wind, liquid water, and ice. Weathered and eroded rock can be deposited as sediments and may be buried and converted to sedimentary rocks.

Landslides occur when finely divided (unconsolidated) earthen material slides down a slope. The results can be devastating. The 1970 earthquake-initiated landslide of dirt, mud, and rocks on the slopes of Mt. Huascaran in Peru may have killed 20,000 people. A 1963 landslide on slopes surrounding a reservoir held by the Vaiont Dam in Italy suddenly filled the reservoir causing a huge wall of water to overflow the dam killing 2600 people and destroying everything in its path.

Along with weather and climate, human activities can influence the likelihood and destructiveness of landslides. Roads and structures constructed on sloping land can weaken the integrity of earthen material or add mass to it, increasing its tendency to slide. In some cases, strong root structures of trees and brush anchor sloping land in place. However, some plants have roots that destabilize and add mass to soil and increase the accumulation of water underground, resulting in a greater tendency for the earth to slide. Fortunately, predicting a tendency for landslides to occur is relatively straightforward based upon the nature and slope of geological strata, climate conditions, and observations of evidence of a tendency toward landslides, such as movement of earth and evidence of cracked foundations in buildings built on slopes. In some cases remedial actions may be taken, but more important are the indications that structures should not be built on slide-susceptible slopes.

Less spectacular than landslides is **creep** characterized by a slow, gradual movement of earth. Creep is especially common in areas where the upper layers of earth undergo freeze/thaw cycles. A special challenge is **permafrost** which occurs in northern Scandinavia, Siberia, and Alaska. Permafrost refers to a condition in which ground at a certain depth never thaws, and thawing occurs only on a relatively thin surface layer. Structures built on permafrost may end up on a pool of water-saturated muck resting on a mixture of frozen ice and soil. One of the greater challenges posed by permafrost in recent times has been the construction of the Trans-Alaska pipeline in Alaska.

Expansive clays that alternately expand and contract when saturated with water, then become dried out, can cause enormous damage to structures, making the construction of basements virtually impossible in some areas. **Sinkholes** occur in areas where rock formations are dissolved by chemical action of water (particularly dissolved carbon dioxide acting on limestone). Earth can fall into a cavity generated by this phenomenon causing huge holes in the ground that can swallow several houses at a time.

10.3. WATER IN AND ON THE GEOSPHERE

The geosphere is the repository of virtually all the world's fresh water. As shown in Figure 10.1, this water may be in underground aquifers as groundwater; on the surface as streams, rivers, lakes and impoundments; or as deposits of ice (glaciers) resting on Earth's surface. Water collected by the geosphere constitutes virtually all Earth's fresh water resources. This water is susceptible to pollution which in extreme cases can render the water sources virtually useless. One of the greater water pollution problems is when water in underground aquifers becomes contaminated with hazardous waste material improperly discarded in the geosphere.

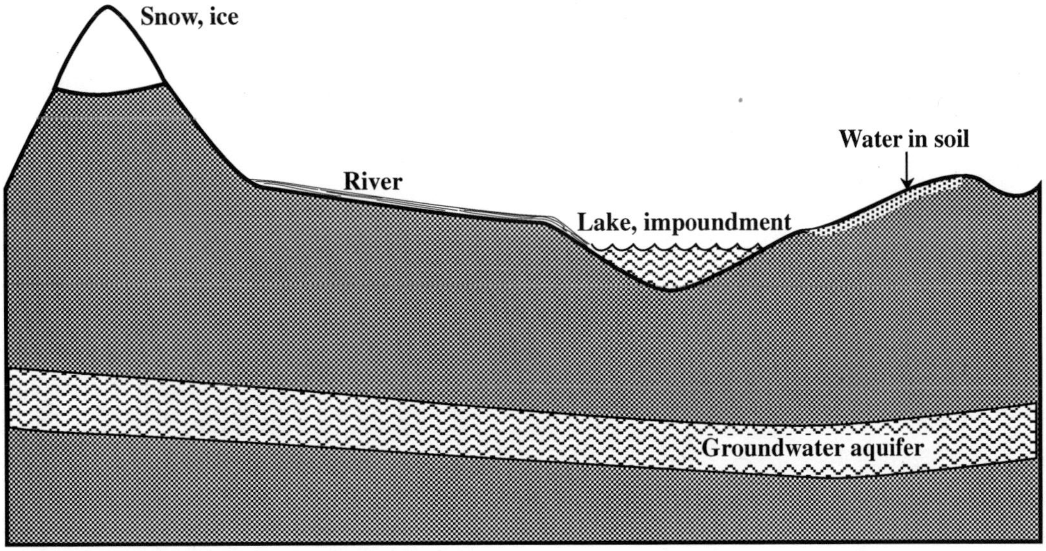

Figure 10.1. Major aspects of water on and in the geosphere.

Water commonly moves on the geosphere in **streams** or **rivers** consisting of channels through which water flows. Rivers collect water from drainage basins or watersheds. In order to protect water quality in rivers, pollution and pollution-causing agricultural practices in drainage basins must be avoided. Rivers continually erode the geosphere over which they flow and leave deposits of sediments. Over time, a river will erode earth away and create valleys. An undisturbed river continually cuts curving patterns known as **meanders** in a river valley. The flat area of a valley formed by erosion and sediment deposition in the valley and susceptible to periodic flooding is the river's floodplain.

Floods are the phenomena associated with river flow that are most likely to cause damage to human structures. Despite their destructiveness, floods are normal phenomena by which a river does much of its work of shaping the surface of the geosphere. However, by unwisely building in floodplains, humans have made themselves susceptible to the damaging effects of floods. Human activities on the geosphere surface can make the effects of floods much worse. For example, flash floods following intense rainfall in urban areas are made much worse by the removal of vegetation from watersheds and its replacement with paving. Concrete and asphalt surfaces do not slow down the flow of water like well-rooted plants do and such surfaces prevent the infiltration of water into the ground.

Attempts to control water flow and flooding provide interesting examples of how humans can interact with their natural environment. Control measures have concentrated on the downstream end on the rivers themselves by construction of levees to confine rivers to their banks, straightening and deepening river channels to increase the velocity and flow of the water in an effort to move it quickly downstream away from the potentially flooded area, and by building dams to contain floodwater until it can be safely released. Such measures can be deceptively successful, sometimes for many decades, until a massive flood overwhelms them. When a contained river carrying vast amounts of water flowing at a high velocity eventually breaks through the levees and dams designed to contain it, the resulting damage can be catastrophic.

An approach to flood control based upon the principles of industrial ecology provide a means of minimizing flood damage. Such an approach tends to concentrate more on the upstream end, the watersheds from which water produced by rainfall flows into the river. With the proper kind of vegetation cover, such as forests, and with terraces and small dams designed to temporarily slow the flow of water into the river from the watershed, extremes of high water (flood crests) can be greatly reduced. With regard to protection of dwellings and agricultural land in the river's flood plain, a fundamental question has to be asked whether houses should even be located in these areas and whether the land should be cultivated. In many cases, the answer is no, and the least costly alternative overall is to pay for removal of the structures and conversion of the land back to an uncultivated state, simply allowing the flooding that comes naturally to the river.

10.4. ANTHROSPHERIC INFLUENCES ON THE GEOSPHERE

The urge to "dig in the dirt" and alter Earth surfaces seems to be innate in humans. During recent decades the potential of humans to alter the geosphere has been greatly increased by the development of massive earth-moving equipment. Flooding of rivers caused by human activities was discussed above. Other geospheric disturbances that can be detrimental include landslides on mounds of waste mine tailings, adverse effects resulting from exposure of minerals during mining (production of acid mine water from exposure of pyrite, FeS_2, in coal mining), and filling and destruction of wetlands upon which many forms of wildlife depend for breeding grounds.

Human effects upon the geosphere can be both direct and indirect. Construction of dams and reservoirs, flattening whole mountain tops to get to underground coal seams, and plowing natural prairies to grow crops are obvious direct effects. Indirect effects include pumping so much water from underground aquifers that the ground subsides, or exposing minerals by strip mining that weather to produce polluted acidic water. In extracting minerals from the earth, it is disturbed and rearranged in ways that can cause almost irreversible damage to the environment. A major objective of the practice of green chemistry and industrial ecology is to minimize these detrimental effects and, to the extent possible, eliminate them entirely.

Many of the effects of human activities on the geosphere have to do with the extraction of resources of various kinds from Earth's crust. These may range from gravel simply scooped from pits on Earth's surface to precious metals at such low concentrations that tons of ore must be processed to get a gram or less of the metal. The most straightforward means of obtaining materials from Earth's crust is surface mining. This often involves removing unusable material in the form of the overburden of soil and rock that covers the desired resource. This may leave a pit that fills with water alongside a pile of the overburden. This kind of mining practice caused many environmental problems in the past. With the modern practice of surface mining, however, topsoil is first removed and stored, rock removed to get to the resource is either placed back in the pit or on contoured piles, and the topsoil placed over it for revegetation. In favorable cases, the result can be attractive lakes that support fish life and vegetated, gently sloping artificial hills.

Underground mining usually does not leave the visible scars that may be inflicted by surface mining. However, it can have profound environmental effects. Collapse of underground mines can cause surface subsidence. Water flowing through and from underground mines can pick up water pollutants. Most ores require a degree of beneficiation in which the usable portion of the ore is concentrated, leaving piles of tailings. These may collapse, and materials leached from them can pollute water. Examples of the latter include acidic water produced by the action of bacteria on iron pyrite, FeS_2, removed from coal and radium leached from the tailings remaining from uranium mining operations.

10.5. THE GEOSPHERE AS A WASTE REPOSITORY

As discussed above, mineral processing produces large quantities of waste solids. Other sources of waste solids include ash from coal combustion, municipal garbage, and solid wastes from various industrial processes. Ultimately, these wastes are placed on or in the geosphere. Such measures have an obvious potential for pollution.

One of the most common waste materials that ends up as part of the geosphere is **municipal refuse**, the "garbage" generated by human activities. This material is largely disposed in **sanitary landfills** made by placing the solid wastes on top of the ground or in depressions in the ground and covering it with soil to minimize effects such as windblown waste paper and plastic, emission of odorous materials to the atmosphere, and water pollution. Although "garbage dumps" used to be notably unsightly and polluting, modern practice of sanitary landfilling can result in areas that can be used as parkland, golf courses, or relatively attractive open space. The unconsolidated nature of decaying garbage and the soil used to cover it make municipal landfills generally unsuitable for building construction. Biological decay of degradable organic material ($\{CH_2O\}$) in the absence of oxygen generates methane gas by a process represented as

$$2\{CH_2O\} \rightarrow CO_2 + CH_4 \qquad (10.5.1)$$

Methane is a powerful greenhouse gas in the atmosphere, so it is undesirable to release it to the atmosphere. However, modern sanitary landfills may be equipped with pipes and collection systems so that the methane can be collected and used as a fuel.

Whereas the release of gases, particularly methane, to the atmosphere is a potential air pollution problem with sanitary landfills, contaminated **leachate** consisting of water seeping through the landfilled wastes can pollute water, especially groundwater. This water may contain heavy metals, organic acids, odor-causing organics, and other undesirable pollutants. There are two general approaches to minimizing problems from contaminated landfill leachate. One of these is to construct the landfill in a manner that minimizes water infiltration, thus reducing the amount of leachate produced to lowest possible levels. To prevent the leachate from getting into groundwater, it is desirable to locate the landfill on a layer of poorly permeable clay. In some cases, the bottom of the landfill may be lined with impermeable synthetic polymer liners that prevent leachate from getting into groundwater. In cases where significant quantities of badly polluted leachate are generated, it is best to collect and treat the leachate, usually by biological treatment processes.

Minimization of the quantities of materials requiring sanitary landfill disposal using the best practice of industrial ecology and green chemistry is highly desirable. The best way to do that is by reducing quantities of materials at the source, simply using less material that ultimately will require disposal. Wherever possible, materials, such as packing materials, that ultimately get into landfills should be biodegradable. Recycling programs in which glass, plastic, paper, and food cans are removed from refuse prior to disposal are effective in reducing quantities of material requiring disposal in landfills. Burning of garbage with proper pollution control measures can reduce it to a low-volume ash that can be placed in a landfill. Although not practiced to a significant extent, anaerobic digestion of macerated wet refuse in an oxygen-free digester has the potential to produce methane for fuel use and greatly reduce the mass of the degradable wastes.

Sanitary landfills are not suitable for the disposal of hazardous substances. These materials must be placed in special **secure landfills**, which are designed to contain the wastes and leachate, thus preventing pollution of water, air, and the geosphere. One way in which this is accomplished is with impermeable synthetic membranes that prevent water from seeping into the fill and prevent leachate from draining into groundwater. These landfills are often equipped with water treatment systems to treat leachate before it is released from the system. Unfortunately, many hazardous chemicals **never** degrade and a "secure" chemical landfill leaves problems for future generations to handle. One of the major objectives of green chemistry is to prevent the generation of any hazardous materials that would require disposal on land. The best way to do that is to avoid making or using such materials. In cases where that is not possible and hazardous materials are generated, they should be treated in a way that renders them nonhazardous prior to disposal.

10.6. HAVE YOU THANKED A CLOD TODAY?

A common bumper sticker is one that asks the question, "Have you thanked a green plant today?" this is an obvious reference to plants whose photosynthesis produces the food that we and most other animals depend upon for our existence. An even more fundamental question is whether we have thanked the soil — the clods of dirt — upon which green plants depend for their existence. Good, productive soil combined with a suitable climate and adequate water is the most valuable asset that a nation can have. Vast areas of the world lack this fundamental asset, and the people living in areas with poor soil often suffer poverty and malnutrition as a result. Furthermore, areas that once had adequate soil have seen it abused and degraded to the extent that it is no longer productive. One of the central challenges faced by the practice of green chemistry and industrial ecology is to retain and enhance the productive qualities of soil.

It should be kept in mind that soil receives significant amounts of many pollutants. Some of these, such as herbicides used to control weed growth, are applied deliberately as part of agricultural practice. Others, such as particulate matter produced as an end product of photochemical smog formation, are deposited upon soil. Some of these materials contribute to water and air pollution so that soil is very much involved with environmental chemical cycles. As examples, herbicide runoff from soil can pollute water and N_2O released by bacteria acting upon nitrogen fertilizers is a potential air pollutant.

Soil is a term that actually describes a wide range of finely divided mineral matter containing various levels of organic matter and water that can sustain and nourish the root systems of plants growing on it. Soil is largely a product of the weathering of rock by physical, chemical, and biochemical processes that produces a medium amenable to support of plant growth. A healthy soil contains water available to plants, has a somewhat loose structure with air spaces, and supports an active population of soil-dwelling organisms, including fungi and bacteria that degrade dead plant biomass and animals, such as earthworms. Although the solids in a typical soil are composed of about 95% inorganic matter, some soils contain up to 95% organic matter, and some sandy soils may have only about 1% organic matter.

Figure 10.2 shows the major aspects of the physical structure of soil. Soil is divided into layers called **horizons** formed by weathering of parent rock, chemical processes, biological processes, and the action of water including leaching of colloidal matter to lower horizons. The most important of these for plant growth is **topsoil**. Plant roots take water and plant nutrients from topsoil, and it is the layer of

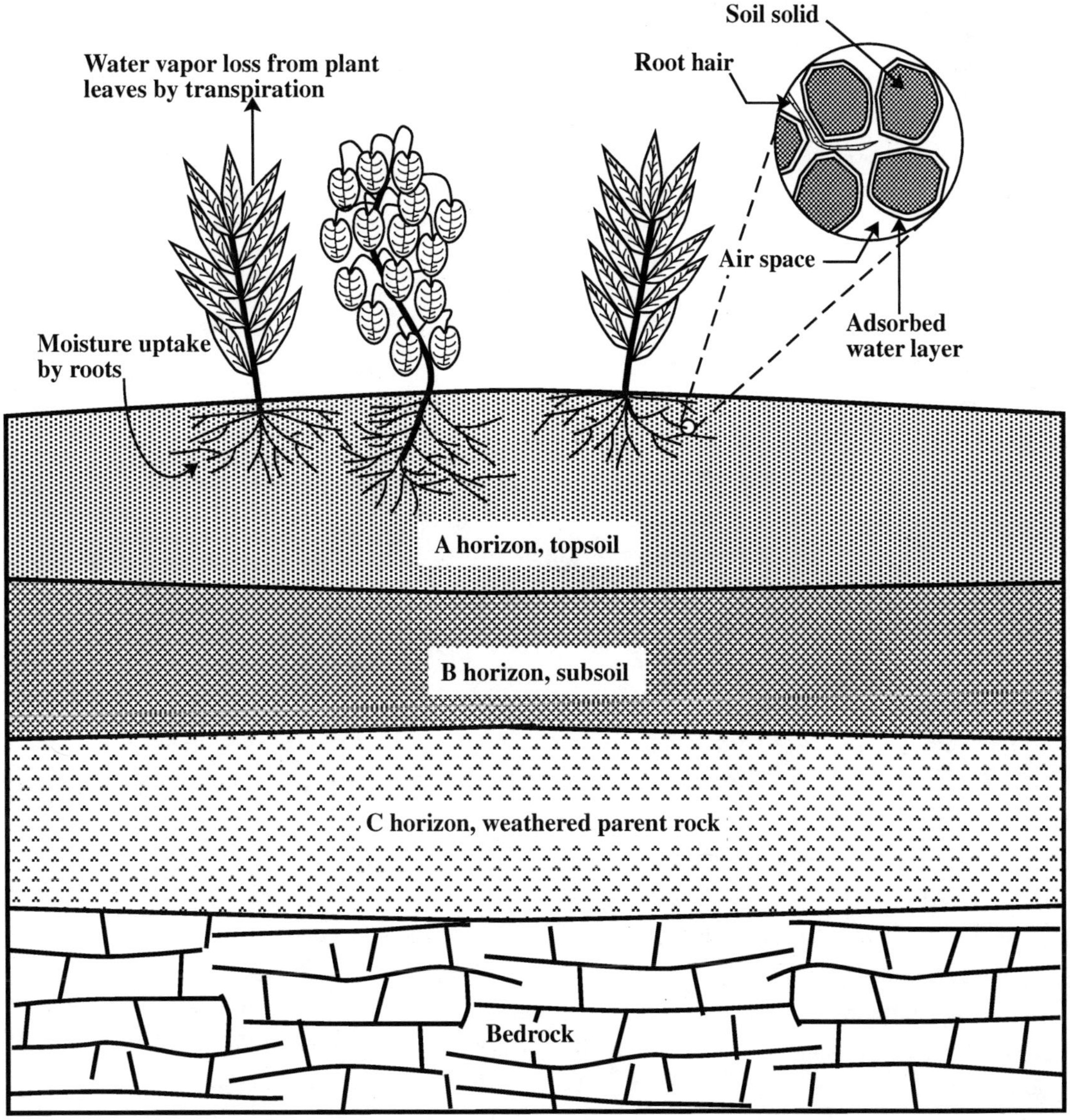

Figure 10.2. Major aspects of soil structure showing various soil horizons. The inset shows aspects of soil microstructure.

maximum biological activity. The **rhizosphere** is the part of topsoil in which plant roots are especially active and in which the elevated levels of biomass are composed of plant roots and microorganisms associated with them. There are strong synergistic relationships between plant root systems and microorganisms in the rhizosphere. The surfaces of root hairs are commonly colonized by microorganisms, which thrive upon carbohydrates, amino acids, and root-growth-lubricant mucigel secreted from the roots.

Inorganic Solids in Soil

Reflecting the fact that the two most common elements in the earth's crust are oxygen and silicon, silicates are the most common mineral constituents of soil. These include finely divided quartz (SiO_2), orthoclase ($KAlSi_3O_8$), and albite ($NaAlSi_3O_8$). Other elements that are relatively abundant in Earth's crust are aluminum, iron, calcium, sodium, potassium, and magnesium, and their abundance is reflected by minerals such as epidote ($4CaO{\bullet}3(AlFe)_2O_3{\bullet}\ 6SiO_2{\bullet}H_2O$), geothite (FeO(OH)), magnetite (Fe_3O_4), calcium and magnesium carbonates ($CaCO_3$, $CaCO_3{\bullet}MgCO_3$), and oxides of manganese and titanium in soil. Soil parent rocks undergo weathering processes to produce finely divided colloidal particles, particularly clays. These secondary minerals hold moisture and mineral nutrients, such as K^+ required for plant growth, that are accessible by plant roots and are repositories of plant nutrients. Inorganic soil colloids can absorb toxic substances in soil, thus reducing the toxicity of substances that would harm plants. It is obvious that the abundance and nature of inorganic colloidal material in soil are important factors in determining soil productivity.

Soil Organic Matter

The few percent of soil mass consisting of organic matter has a strong influence upon the physical, chemical, and biological characteristics of soil. Organic matter is effective in holding soil moisture and it holds and exchanges with plant roots some of the ions that are required as plant nutrients. Temperature, moisture, and climatic conditions significantly affect the kinds and levels of soil organic matter. Cold, wet conditions in which soil stays saturated with moisture preventing access of microorganisms to oxygen tend to prevent complete biodegradation of plant residues that compose soil organic matter allowing it to accumulate. This is clearly illustrated by accumulation of peat in Ireland and other locales with similar climatic conditions such that most of the solid soil is composed of organic matter. Tropical conditions, especially with alternate wet and dry seasons, can result in loss of soil organic matter. One reason that the soil supporting tropical rain forests degrades so quickly when the trees are removed is that the organic matter in the soil undergoes rapid biodegradation when the forest cover is removed.

The plant biomass residues that form soil organic matter undergo a biodegradation process by the action of soil bacteria and fungi in which the cellulose in the biomass is readily degraded leaving modified residues of the lignin material that binds the cellulose to the plant matter. This is the process of **humification** and the residue is **soil humus**, a black organic material of highly varied chemical structure. A fraction of soil humus is soluble in water, especially when base is present in the water. Another fraction called **humin** does not dissolve and stays in the solid soil.

Though composing usually not more than a few percent of soil, soil humus has a very strong influence on the characteristics of soil. It has a strong affinity for water and holds much of the water in a typical soil. Largely because of their carboxylic acid (CO_2H) groups, soil humic molecules exchange H^+ ion and act to buffer the pH of water in soil (the soil solution). Humic substances bind metal ions and other ionic plant nutrients. Soil humus also binds and immobilizes organic materials, such as herbicides applied to soil.

Water in Soil and The Soil Solution

Water in soil is required for plants. This water is taken up by plant root hairs, transferred through the plant, and evaporated from the leaves, a process called **transpiration** (see Figure 10.2). The quantities of water involved are enormous; for example, the water transpired to produce a kilogram of dry hay can amount to several hundred kg. Most of the water in normal soils is not present as visible liquid, but is absorbed to various degrees upon the soil solids. In fact, a condition in which all the spaces in soil are filled with water—waterlogging—slows the growth of most plants. The water that is available in soil is called the **soil solution** and contains a number of dissolved materials, including plant nutrients. It plays an essential role in transferring substances, such as dissolved metal cations, between roots and the soil solid. Cations commonly present in the soil solution include H^+, Ca^{2+}, Mg^{2+}, Na^+, and K^+ along with very low levels of Fe^{2+}, Mn^{2+}, and Al^{3+}. Common anions present are HCO_3^-, CO_3^{2-}, HSO_4^-, SO_4^{2-}, Cl^-, and F^-.

10.7. PRODUCTION OF FOOD AND FIBER ON SOIL — AGRICULTURE

The most important use of soil is **agriculture**, the production of food and fiber by growing crops and livestock. Agriculture is very closely tied with the practice of green chemistry in many ways. Agricultural chemicals, including fertilizers, herbicides, and insecticides are produced and applied to crops and land in enormous quantities. Annual production of millions of kilograms of these chemicals demands the proper practice of green chemistry and engineering. The judicious use of relatively small quantities of herbicides enables planting of crops in soil covered with residues of the previous year's crops with little or no cultivation of soil. This practice of low-tillage agriculture, now called conservation tillage (see Section 10.11), is in keeping with the best practice of green chemistry and industrial ecology. Organic biomass produced by plants can be used as a renewable source of raw material and fuel. Some plants are now being genetically engineered to produce specific chemicals.

The practice of agriculture is absolutely essential for the survival of humankind. In order to continue to feed growing world populations while maintaining and even enhancing the ability of soil to produce food, it is necessary that the practice of agriculture be as green as possible. In the past and still today, this has often not been true. Cultivation of soil by humans has displaced native plants, destroyed wildlife habitat, contaminated soil with pesticides, filled rivers and bodies of water with sediments, and otherwise perturbed and damaged the environment. Agricultural practices arguably represent the greatest incursion of the anthrosphere into the other environmental spheres. On the positive side, growth of domestic crops temporarily removes greenhouse gas carbon dioxide from the atmosphere and provides organic raw materials and biomass fuel without any net addition of carbon dioxide to the atmosphere.

The basis of agriculture is the development of domestic plants from their wild ancestors. (The same can be said of animals, but only a handful of animal species have been domesticated, although each consists of many different breeds.) Our prehistoric ancestors learned to select plants with desired characteristics for the production of food and fiber, developing new species that often require the careful efforts of expert botanists to relate them to their wild ancestors. Only around 1900 were the scientific principles of heredity applied to plant breeding, with excellent results. Using scientific methods, agriculturists accomplished a "green revolution" in the 1950s and 1960s that resulted in varieties of rice and wheat, especially, that had vastly increased yields. The techniques used included selective breeding, hybridization, cross-pollination, and back-crossing to develop grain varieties which, combined with chemical fertilizers and pesticides, lead to much higher crop yields. India, a country on the verge of starvation in the 1940s, increased its grain output by 50%. Developments such as higher yielding and faster maturing dwarf varieties of rice enabled better nutrition for an increasing world population, at least postponing the inevitable problems that will result from population growth. By breeding plants resistant to cold, drought, and insects, overall crop productivity has been further increased. Increased nutritional values for grain have been achieved, such as the development of corn varieties that have higher levels of lysine amino acid.

One of the major advances in plant breeding has been the development of **hybrids** produced by crossing **true-breeding** strains of plants. So-called "hybrid vigor" is well known, and many hybrids have vastly greater yields than their parent strains. Corn, a remarkably productive photosynthesizer, has proven most amenable to the production of hydrids, in part because of the separation of male flowers which grow on the tops of plants from female flowers attached to the budding corn ears. By planting rows of corn that alternate between two different strains and cutting the tassels from the tops of the plants that are to produce the corn seed, hybrid corn varieties are readily produced. More recently, techniques have been developed for growing hybrids of other kinds of plants.

There are, of course, many factors other than the genetic strains of plants that are involved in high crop productivity. The effects of weather have been mitigated by the development of crop varieties that resist heat, cold, and drought. The provision of water by artificial irrigation has greatly increased crop productivity and is essential for crop productivity in some regions, such as the vegetable-growing areas of California. Irrigation practices continue to become more efficient with the replacement of wasteful spray irrigators by systems that apply water directly to soil, or even directly to plant roots. Computerized control of irrigation can make it much more efficient. Environmentally, widespread use of herbicides has had some excellent benefits, along with some harmful effects, by enabling greater crop productivity with less tillage of land.

10.8. PLANT NUTRIENTS AND FERTILIZERS

Plant biomass is composed largely of carbon, hydrogen, and oxygen, which plants extract from water and atmospheric carbon dioxide. Other nutrients that plants require in relatively large quantities are calcium, magnesium, and sulfur, which are usually in sufficient abundance in soil, and nitrogen, phosphorus, and potassium, which are commonly added to soil as fertilizers.

Soil acidity in the form of H^+ ion builds up as plant roots exchange H^+ for other cationic nutrients in soil. When acidity reaches excessive levels, the soil is no longer productive. Acidity can be neutralized by the addition of lime ($CaCO_3$), which neutralizes acidity according to the following reaction:

$$\mathrm{Soil\}(H^+)_2 + CaCO_3 \rightarrow Soil\}Ca^{2+} + CO_2 + H_2O} \qquad (10.8.1)$$

This process also adds calcium to soil.

Essential plant nutrient nitrogen is very much involved with nature's nitrogen cycle, which is significantly modified by human activities. Major aspects of this cycle are the following:

- At 79% N_2, Earth's atmosphere constitutes an inexhaustible nitrogen resource, although, because of the extreme stability of the N_2 molecule, it is difficult to extract nitrogen from air in a chemically combined form.
- *Rhizobium* bacteria growing on the roots of leguminous plants, such as clover and soybeans, convert atmospheric nitrogen to nitrogen chemically bound in biomolecules. This nitrogen is converted to ammonium ion, NH_4^+, when plant residues and animal feces, urine, and carcasses undergo microbial decay.
- Lightning and combustion processes convert atmospheric nitrogen to nitrogen oxides, and ammonia manufacturing plants produce NH_3 from atmospheric elemental nitrogen and elemental hydrogen produced by natural gas.

- Soil microbial processes oxidize ammoniacal nitrogen (NH_4^+) to nitrate ion, NO_3^-, the form of nitrogen most readily used by plants. Microbial processes also produce gaseous N_2 and NO_2 which are released to the atmosphere, a process called denitrification that completes the nitrogen cycle.

Natural processes usually do not produce sufficient nitrogen to allow maximum plant growth, so that artificial means are used to extract nitrogen in a chemically combined form from the atmosphere. This is done by the Haber process combining elemental N_2 and H_2 over a catalyst at very high pressures of about 1000 times atmospheric pressure and an elevated temperature of 500°C. The reaction is

$$N_2 + 3H_2 \rightarrow 2NH_3 \qquad (10.8.2)$$

producing ammonia that is 82% chemically bound N. This anhydrous ammonia can be applied directly below the soil surface where its tremendous attraction to soil moisture binds it to the soil. It can also be applied as a 30% solution of NH_3 in water, and is sometimes added directly to irrigation water. Ammonia, which is held in soil as ammonium ion, NH_4^+, is not well assimilated directly by most plants. But it is slowly oxidized by the action of soil bacteria using atmospheric O_2 oxidant to nitrate ion, NO_3^-, which is used directly by plants.

A solid form of nitrogen fertilizer can be made by reacting ammonia with oxygen over a platinum catalyst to make nitric acid, HNO_3, and reacting the acid with basic ammonia to make ammonium nitrate, NH_4NO_3. This molten material is solidified into small pellets that can be applied to soil as fertilizer. Ammonium nitrate mixed with fuel oil is used for blasting to quarry rock, and it was the explosive used in the bombing of the Oklahoma City Federal Building in 1995. A safer alternative to ammonium nitrate as a solid nitrogen fertilizer is urea,

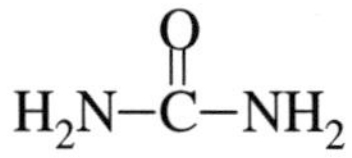

which is made by a process that, overall, involves the reaction of carbon dioxide and ammonia:

$$CO_2 + 2NH_3 \rightarrow CO(NH_2)_2 + H_2O \qquad (10.8.3)$$

Phosphorus is an essential plant nutrient required for cellular DNA and other biomolecules. It is utilized by plants as $H_2PO_4^-$ and HPO_4^{2-} ions. Phosphate minerals that can be used to manufacture phosphorus-containing fertilizers occur in a number of places throughout the world. In the United States, Florida has especially abundant phosphate resources, largely as fluorapatite, $Ca_5(PO_4)_3F$, as well as hydroxyapatite, $Ca_5(PO_4)_3OH$. These phosphate minerals are too insoluble to serve directly as fertilizers and are treated with phosphoric acid and sulfuric acid to make superphosphates that are much more soluble and available to plants:

$$2Ca_5(PO_4)_3F(s) + 14H_3PO_4 + 10H_2O \rightarrow 2HF(g) + 10Ca(H_2PO_4)_2{\bullet}H_2O \qquad (10.8.4)$$

$$2Ca_5(PO_4)_3F(s) + 7H_2SO_4 + 3H_2O \rightarrow 2HF(g) + 3Ca(H_2PO_4)_2{\bullet}H_2O + 7CaSO_4 \qquad (10.8.5)$$

Potassium as the potassium ion, K^+, is required by plants to regulate water balance, activate some enzymes, and enable some transformations of carbohydrates. Potassium makes up 2.6% of Earth's crust, but much of it is tied up in relatively unavailable forms in minerals such as potassium feldspar, $KAlSi_3O_8$. Potassium is a common cation held in an exchangeable form by clay minerals in soil from which it is readily available.

Potassium is one of the most abundant elements in the earth's crust, of which it makes up 2.6%; however, much of this potassium is not easily available to plants. For example, some silicate minerals such as leucite, $K_2O{\bullet}Al_2O_3{\bullet}4SiO_2$, contain strongly bound potassium. Exchangeable potassium held by clay minerals is relatively more available to plants. Potassium for fertilizer is simply mined from the ground as salts, particularly KCl, or pumped from beneath the ground as potassium-rich brines. Large potassium deposits occur in the Canadian province of Saskatchewan.

Plants require several **micronutrients**, largely elements that occur only at trace levels, for their growth. These include boron, chlorine, copper, iron, manganese, molybdenum (for N-fixation), and zinc. Some of these are toxic at levels above those required for optimum plant growth. Most of the micronutrients are required for adequate function of essential enzymes. Photosynthetic processes use manganese, iron, chlorine, and zinc. Since the micronutrients are required at such low levels, soil normally provides sufficient amounts.

10.9. PESTICIDES AND AGRICULTURAL PRODUCTION

Pesticides and related substances such as growth regulators are widely used in agriculture. The most common agricultural pesticides are insecticides and herbicides, which are an integral part of modern agricultural production. The Federal Insecticide, Fungicide, and Rodenticide (FIFRA) act regulates agricultural pesticides in the United States. This act, first passed in 1947 and amended several time since then, requires registration of agricultural pesticides, considering toxicity, degradability, and other aspects of their use.

Recombinant DNA technology (see Section 10.12) is having some significant effects upon pesticide use. One of the more interesting developments has been the splicing of genetic material into cotton, corn, and other crops that cause them to produce an insecticide that is generated by some kinds of bacteria. Another interesting aspect of recombinant DNA applied to herbicides is the breeding of genetically modified plants that are not affected by specific herbicides. These kinds of measures significantly reduce levels of chemical production needed for pesticides as well as levels of pesticide residues on soil and in water and are, therefore, very much in keeping with the practice of green chemistry.

10.10. SOIL AND PLANTS RELATED TO WASTES AND POLLUTANTS

Soil is a repository of large quantities of wastes and pollutants, and plants act as filters to remove significant quantities of pollutants from the atmosphere. Sulfates and nitrates from the atmosphere, including acid-rain-causing H_2SO_4 and HNO_3 deposit largely on the land and the plants growing on it. Gaseous atmospheric SO_2, NO, and NO_2 are absorbed by soil and oxidized to sulfates and nitrates. Soil bacteria and fungi are known to convert atmospheric CO to CO_2. When leaded gasoline was widely used, soil along highways became contaminated with lead, and lead mines and smelters were significant sources of this toxic element. Organic materials, such as those involved in photochemical smog formation, are removed by contact with plants and are especially attracted by the waxy organic-like surfaces of the needles of pine trees.

A number of materials that can be considered as pollutants are deliberately added to soil. The most obvious of these consists of insecticides and herbicides added to soil for pest and weed control. Chemicals from hazardous waste disposal sites can get onto soil or below the soil surface by leaching from landfill or drainage from waste lagoons. Some kinds of wastes, especially petroleum hydrocarbons, are disposed on soil where adsorption and microbial processes immobilize and degrade the wastes. Soil can be effective for the treatment of sewage. Leakage from underground storage tanks of organic liquids, such as gasoline and diesel fuel, have created major soil contamination problems.

Soils in parts of New York State have been contaminated with polychlorinated biphenyls (PCBs) discarded from the manufacture of industrial capacitors. Analyses of PCBs in United Kingdom soils archived for several decades have shown levels of these pollutants that parallel their production. Starting with very low levels around 1940 before PCBs were manufactured in large quantities, concentrations of PCBs increased markedly, peaking around 1970, when PCB manufacture was ceased. More recent soil samples have shown PCB concentrations near the pre-1940 levels. It is believed that these results reflect evaporation of PCBs and their condensation onto soil. They are consistent with observations of high PCB levels in remote Arctic and sub-Arctic regions believed to be due to the condensation of these compounds from the atmosphere onto soil in very cold regions.

The degradation and eventual fates of the enormous quantities of herbicides and other pesticides applied to soil are very important in understanding the environmental effects of these substances. Many factors are involved in determining pesticide fate. One of the main ones of these is the degree of adsorption of pesticides to soil, strongly influenced by the nature and organic content of the soil surface as well as the solubility, volatility, charge, polarity, and molecular structure and size of the pesticides. Strongly adsorbed molecules are less likely to be released and thus harm organisms, but they are less biodegradable in the adsorbed form. The leaching of adsorbed pesticides into water is important in determining their water pollution potential. The effects and potential toxicities of pesticides to soil bacteria, fungi, and other organisms have to be considered. It must be kept in mind that pesticides may be converted to more toxic products by microbial action.

10.11. SOIL LOSS — DESERTIFICATION AND DEFORESTATION

Soil erosion refers to the loss and relocation of topsoil by water and wind action. About a third of U.S. topsoil has been lost to erosion since cultivation began on the continent and at present about a third of U.S. cropland is eroding at a rate sufficient to lower productivity. About 10% of U.S. land is eroding at an unacceptable rate in excess of 14 tons of toposil per acre annually. Soil erosion is largely a product of cultivation. Except in cases of extreme slopes, very high winds, and torrential rains, uncultivated soils undergo little erosion. Erosion was recognized as a problem in the central United States within a few years after forests and prairie grasslands were first plowed to raise crops, particularly in the latter 1800s. The recognition that precious topsoil was being lost at an unsustainable rate lead to soil conservation measures going back to 1900, or even earlier. In that sense, soil conservation was the first environmental movement, predating efforts to alleviate water and air pollution by many decades.

Water erosion is responsible for greater loss of soil than is wind erosion. Whereas wind erosion tends to move soil around and deposit it in areas where it can still be used for growing crops, water erosion normally moves greater quantities of soil and carries them into streams and rivers and ultimately to the oceans. The overall pattern of soil erosion in the Central Continental United States is shown in Figure 10.3. This figure shows that erosion is especially bad in agricultural areas draining into the Missouri and Mississippi Rivers; millions of tons of soil are carried by these rivers into the Gulf of Mexico each year. These are areas of relatively high rainfall, which can sometimes come as very intense rainstorms, especially during the spring. A high proportion of the farmlands in these areas are devoted to rowcrops, which are crops such as corn, soybeans, and sorghum grains planted in rows with bare soil in between. This mode of cultivation leaves soil that is especially susceptible to water erosion.

The ultimate result of soil erosion and other unsustainable agricultural practices in relatively dry areas is **desertification**. This condition occurs when permanent plant cover is lost from soil, it loses its capacity to retain moisture, dries out, and loses fertility so that plants no longer grow on it. Among the interrelated factors involved in desertification are wind erosion, water erosion (which occurs during sporadic cloudbursts even in arid areas), development of adverse climate conditions, depletion of underground water aquifers, lack of water for irrigation, accumulation of salt in water supplies, loss of soil organic matter, and deterioration of soil physical and chemical properties. Eventually the land becomes unable to support agriculture, grazing, or even significant human populations. Desertification is one of the most troublesome results of global warming caused by greenhouse warming. It is actually a very old problem and is a serious concern in many parts of the world, such as the Mideast, the southern boundary of Africa's Sahara, and regions of the southwestern U. S. Formerly productive areas of the Middle East and North Africa, "lands of milk and honey" described in biblical terms, have turned into desert, largely due to human agricultural activities. The growth of domestic grazing animals on these areas — especially goats, which tend to pull vegetation up by its roots — has been a particularly strong contributor to desertification. Much of the productive capacity of arid grasslands in the western and southwestern United States has been drastically diminished by overgrazing.

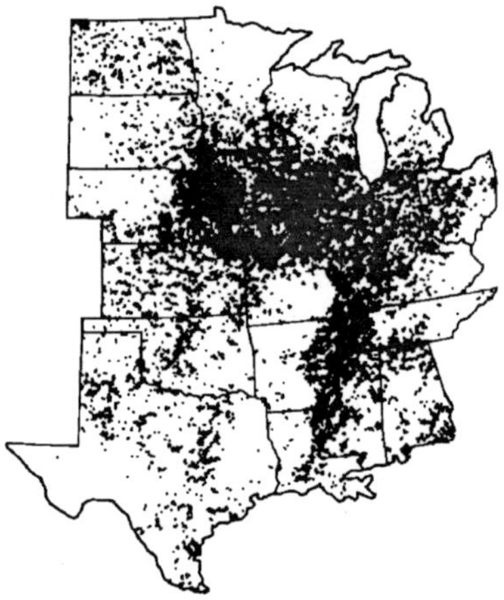

Figure 10.3 Pattern of soil erosion (dark areas) in the major agricultural states of the central continental U.S. It is seen that erosion is particularly severe in the highly cultivated Missouri and Mississippi River watersheds.

Fortunately, human ingenuity and technological tools can be used to prevent or reverse desertification. For example, water, which upon occasion falls as torrential rain upon normally dry desert lands, can be collected and used to recharge underground water aquifers. Advanced cultivation and irrigation techniques can be used to establish perennial plant cover on erosion-prone desert soils. Potentially, plants can be genetically engineered to grow under severe conditions of temperature, drought, and salinity. Environmentally friendly mining practices can be employed, and land surfaces damaged by harmful strip mining practices can be restored.

The loss of forest growth to cultivated land — **deforestation** — has occurred extensively in the United States. However, much of the colonial U.S., particularly in New England, which was deforested for cultivation of crops, is now undergoing largely spontaneous reforestation as unprofitable farmlands are abandoned and trees become established again. Deforestation is a particularly severe problem in tropical regions. Rich tropical forests contain most known plant and animal species many of which are becoming extinct as the forests are destroyed. Once destroyed, tropical forests are almost impossible to restore. This is because tropical forest soil has been leached of nutrients by the high annual rainfalls in tropical regions. When forest cover is removed, the soil erodes rapidly, loses the plant roots and other biomass that tends to hold it together, loses nutrients, and becomes unable to sustain either useful crops or the kinds of forests formerly supported.

The key to preventing soil loss from erosion as well as preventing desertification from taking place lies in a group of practices that agriculturists term **soil conservation**. A number of different approaches are used to retain soil and enhance its quality. Some of these are old, long-established techniques such as construction of terraces and planting crops on the contour of the land (see Figure 10.4). Crop rotation and occasional planting of fields to cover crops, such as clover, are also old practices. A relatively new practice involves minimum cultivation and planting crops through the residue of crops from the previous year. This practice, now commonly called **conservation tillage**, is very effective in reducing erosion because of the soil cover of previous crops and the roots that are left in place. Conservation tillage makes minimum use of herbicides to kill competing weeds until the desired crop is established enough to shade out competing plants. There is some concern that fungi (molds) will thrive in old crop residues and cause problems with new crops.

The ultimate in no-till agriculture is the use of perennial plants that do not have to be planted each year. Trees in orchards and grape vines in vinyards are such plants. The roots of perennial plants are very effective in holding soil in place. Efforts to develop perennial plants that produce grain have not been successful to date. This is because a successful grain-producing plant is one that dedicates its metabolic processes to the production of large quantities of seed that can be used for grain, whereas perennial plants put their energy into the development of large, bulbous root structures that store food for the next growing system. It is possible that sometime in the future genetic engineering may be applied to the development of perennial plants capable of producing high grain yields.

Figure 10.4. Construction of terraces on the contour of land and planting crops on the contour are practices that have been very effective in reducing soil erosion.

Among the most successful plants at stopping erosion are trees. These plants grow for many years and some tree varieties will grow back from their root structures when the wood is harvested. Wood and wood products are probably the most widely used renewable resources. Hybrid tree varieties have been developed that are outstanding producers of biomass.

Wood is a renewable resource used for construction in place of steel, aluminum, and cement, all produced by very energy-intensive processes. Wood is about 50% cellulose, a carbohydrate polymer that is used directly to make paper. Although humans and many other animals cannot use cellulose directly for fuel, it can be broken down chemically or biochemically to glucose sugar. This material serves as a food source for yeasts (a form of fungi) that generate ethanol, an alcohol that can be used as fuel and to make other chemicals. In the process, the yeasts produce protein that can be fed to animals.

Water and Soil Conservation

Conservation of soil and conservation of water go together very closely. Soil is normally the first part of the geosphere that water contacts, and contaminated soil yields contaminated water. Most fresh water falls initially on soil, and the condition of the soil largely determines the fate of the water and how much is retained in a usable condition. Soil in a condition that retains water allows rainwater to infiltrate into groundwater. If water drains too rapidly from soil, the soil erodes and the water runoff is badly contaminated with soil sediments. Measures taken to conserve soil usually conserve water as well. Terraces, contour cultivation, constructed waterways, and water-retaining ponds (Figure 10.4) prevent water from washing soil away, but also retain water and help prevent flash floods. Some of these measures involve modification of the contour of the soil, particularly terracing, construction of waterways, and construction of water-retaining ponds. Bands of trees can be planted on the contour to retain both soil and water. Avoiding practices, such as overgrazing, that tend to lead to desertification and reforestation of land unsuitable for growing crops conserves water as well as land.

10.12. AGRICULTURAL APPLICATIONS OF GENETICALLY MODIFIED ORGANISMS

As discussed in Sections 9.7 and 9.8 of Chapter 9, genes composed of deoxyribonucleic acid, DNA, located in the nuclei of cells direct cell reproduction and synthesis of proteins and generally direct the organism activities. Plant scientists are now able to modify DNA by processes called *recombinant DNA technology*. (Recombinant DNA technology is also being applied to animals, but to a lesser extent than with plants.) Recombinant DNA technology normally involves taking a single characteristic from one organism — the ability to produce a bacterially synthesized insecticide, for example — and splicing it into another organism. By so doing, for example, corn and cotton have been genetically engineered to produce their own insecticide. Plants produced by this method are called **transgenic** plants. During the 1970s, the ability to manipulate DNA through genetic engineering became a reality, and during the 1980s, it became the basis of a major industry. This technology promises some exciting developments in agriculture and, indeed, is expected to lead to a "second green revolution." Direct manipulation of DNA can greatly accelerate the process of plant breeding to give plants that are much more productive, resistant to disease, and tolerant to adverse conditions. In the future, entirely new kinds of plants may even be engineered.

Plants are particularly amenable to recombinant DNA manipulation. In part this is because huge numbers of plant cells can be grown in appropriate media and mutants can be selected from billions of cells that have desired properties, such as virus resistance. Individual plant cells are capable of generating whole plants, so cells with desired qualities can be selected and allowed to grow into plants which may have the qualities desired. Ideally, this accomplishes in weeks what conventional plant breeding techniques would require years to do.

There are many potential green chemistry aspects from genetic engineering of agricultural crops. One promising possibility is to increase the efficiency of photosynthesis, which is only a few tenths of a percent in most plants. Doubling this efficiency should be possible with recombinant DNA techniques, which might significantly increase the production of food and biomass by plants. For example, with some of the more productive plant species, such as fast-growing hybrid poplar trees and sugarcane, biomass is almost economical as a fuel source. A genetically engineered increase in photosynthesis efficiency could enable biomass to economically replace expensive petroleum and natural gas for fuel and raw material. A second possibility with genetic engineering is the development of the ability to support nitrogen-fixing bacteria on plant roots in plants that cannot do so now. If corn, rice, wheat, and cotton could be developed with this capability it could save enormous amounts of energy and natural gas (a source of elemental hydrogen) now consumed to make ammonia synthetically.

Transgenic crops have many detractors, and demonstrations have broken out and test plots of crops destroyed by people opposed to what they call "Frankenfoods." There is some evidence to suggest that bacterial insecticide produced by transgenic corn kills beautiful Monarch butterflies that have contacted the corn pollen. In year 2000 a lot of concern was generated over the occurrence of transgenic corn in taco shells made for human consumption, and a large recall of the product from supermarket shelves occurred. Opposition has been especially strong in Europe, and the European Commission, the executive body of the European Union, has disallowed a number of transgenic crops. Despite these concerns, transgenic crops are growing in importance and there is a lot of interest in them in highly populated countries, particularly China, where they are seen as a means of feeding very large populations.

The Major Transgenic Crops and their Characteristics

The two characteristics most commonly developed in transgenic crops is tolerance for herbicides that kill competing weeds and resistance to pests, especially insects, but including microbial pests (viruses) as well. The most common transgenic crop grown in the U. S. is the soybean, of which about 53 million acres, or 47% of the total, consisted of transgenic varieties in 1999. This increased to 54% of the soybean acreage in 2000 and 68% in 2001. About 25 million acres of transgenic corn were grown in 1999, but somewhat less transgenic corn was planted in 2000 because of reduced infestations of the European corn borer. The percentage of corn that was transgenic in 2001 has been estimated at 26%. In 1999 about 9 million acres each of transgenic cotton and canola (grown as a source of canola oil) were grown. In 2001, it was estimated that 69% of the cotton grown in the U. S. was transgenic. Only about 2–3% of the potato crop was transgenic during the 1998–2000 period, and small fractions of the squash and papaya crops were transgenic as well. In 2001, 72 million acres of U. S. farmland were planted to transgenic crops.

The overwhelming majority of characteristics spliced into transgenic crops consist of herbicide tolerance and resistance to insects. Insect resistance has been imparted by addition of a gene from *Bacillus thuringiensis* (Bt) that causes the plant to produce a natural insecticide in the form of a protein that damages the digestive systems of insects, killing them. Of the acreages of crops planted in 1999, 70.2% were herbicide tolerant, 22.2% were Bt insect resistant, 7.3% were both herbicide tolerant and insect resistant, and 0.3% were virus resistant.

The disruption of natural ecosystems by cultivation of land and planting agricultural crops provides an excellent opportunity for opportunistic plants — weeds — to grow in competition with the desired crops. To combat weeds, farmers use large quantities of a variety of herbicides. The heavy use of herbicides poses a set of challenging problems. In many cases, to be effective without causing undue environmental damage, herbicides must be applied in specified ways and at particular times. Collateral damage to crop plants, environmental harm, and poor biodegradation leading to accumulation of herbicide residues and contamination of water supplies are all problems with herbicides. A number of these problems can be diminished by planting transgenic crops that are resistant to particular herbicides. The most common such plants are those resistant to Monsanto's Roundup herbicide (glyphosate, structural formula below):

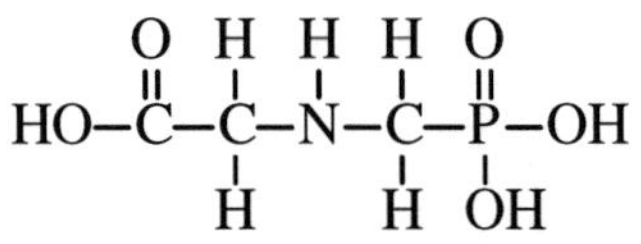

Glyphosate, Roundup herbicide

This widely used compound is a broad-spectrum herbicide, meaning that it kills most plants that it contacts. One of its advantages from an environmental standpoint is that it rapidly breaks down to harmless products in soil, minimizing its environmental impact and problems with residue carry-over. By using "Roundup Ready" crops, of which by far the most common are transgenic soybeans, the herbicide can be applied directly to the crop, killing competing weeds. Application when the crop plants are relatively small, but after weeds have had a chance to start growing, kills weeds and enables the crop to get a head start. After the crop has developed significant size, it deters the growth of competing weeds by shade that deprives the weeds of sunlight.

Aside from weeds, the other major class of pests that afflict crops consists of a variety of insects. Two of the most harmful of these are the European corn borer and the cotton bollworm, which cost millions of dollars in damage and control measures each year and can even threaten an entire year's crop production. Even before transgenic crops were available, *Bacillus thuringiensis* (Bt) was used to control insects. This soil-dwelling bacterium produces a protein called delta-endotoxin. Ingested by insects, delta-endotoxin partially digests the intestinal walls of insects causing ion imbalance, paralyzing the system, and eventually killing the insects. Fortunately, the toxin does not affect mammals or birds. Bt has been a popular insecticide because as a natural product it degrades readily and has gained the acceptance often accorded to "natural" materials (many of which are deadly).

Genetic engineering techniques have enabled transplanting genes into field crops that produce Bt. This is an ideal circumstance in that the crop being protected is generating its own insecticide, and the insecticide is not spread over a wide area. There are several varieties of insecticidal Bt, each produced by a unique gene. Several insecticidal pests are well controlled by transgenic Bt. In addition to the European corn borer mentioned above, these include the Southwestern corn borer and corn earworm. Cotton varieties that produce Bt are resistant to cotton bollworm. Bt-producing tobacco resists the tobacco budworm. Potato varieties have been developed that produce Bt to kill the Colorado potato beetle, although this crop has been limited because of concerns regarding Bt in the potato product consumed directly by humans. Although human digestive systems are not affected adversely by Bt, there is concern over its being an allergen because of its proteinaceous nature.

The greatest success to date with Bt crops has occurred with cotton, which has saved as much as a half million kilograms of synthetic insecticides in the U. S. each year. The benefits of Bt corn are less certain. One of the concerns with Bt corn is the production of the insecticide on pollen, which spreads from the corn plants. Some studies have suggested that this pollen deposited on milkweed that is the natural source of food for Monarch butterflies is a serious threat to this beautiful migratory insect. Another concern with all Bt crops is the potential to develop resistance in insects through the process of natural heredity. To combat resistance, farmers are required to plant a certain percentage of each field to non-Bt crops with the idea that insects growing in these areas without any incentive to develop resistance will crossbreed with resistant strains, preventing them from becoming dominant.

Virus resistance in transgenic crops has concentrated on papaya. This tropical fruit is an excellent source of Vitamins A and C and is an important nutritional plant in tropical regions. The papaya ringspot virus is a devastating pest for papaya, and transgenic varieties resistant to this virus are now grown in Hawaii. One concern with virus-resistant transgenic crops is the possibility of transfer of genes responsible for the resistance to wild relatives of the plants that are regarded as weeds, but are now kept in check by the viruses. For example, it is possible that virus-resistant genes in transgenic squash may transfer to competing gourds, which would crowd out the squash grown for food.

Future Crops

The early years of transgenic crops can be rather well summarized by soybeans, corn, and cotton resistant to herbicides and insects. In retrospect, these crops will almost certainly seem rather crude and unsophisticated. In part, this lack of sophistication is due to the fact that the genes producing the desired qualities are largely expressed by all tissues of the plants and throughout their growth cycle, giving rise to problems such as the Bt-contaminated corn pollen that may threaten Monarch butterflies or Bt-containing potatoes that may not be ideal for human consumption. It is anticipated that increasingly sophisticated techniques will overcome these kinds of problems and will lead to much improved crop varieties in the future.

A wide range of other transgenic crops are under development. One widely publicized crop is "golden rice" which incorporates β-carotene in the grain, which is therefore yellow, rather than the normal white color of rice. The human body processes β-carotene to Vitamin A, the lack of which impairs vision and increases susceptibility to maladies including respiratory diseases, measles, and diarrhea. Since rice is the main diet staple in many Asian countries, the widespread distribution of golden rice could substantially improve health. As an example of the intricacies of transgenic crops, two of the genes used to breed golden rice were taken from daffodil and one from a bacterium! Some investigators contend that humans cannot consume enough of this rice to provide a significant amount of Vitamin A.

One of the first transgenic crops designed for human consumption was a variety of tomato that ripened slowly and could be left on the vine longer than conventional tomatoes, thus developing a better flavor than other varieties, which are normally picked while still green. Unfortunately the genetically engineered variety, which was given the brand name of FlavrSavr, did not have other desirable characteristics and failed. Work is continuing on delayed-ripening tomatoes and on improving the nutritional value of tomatoes, such as by raising the content of lycopene, which is involved with the production of Vitamin A.

Work continues on improved transgenic oilseed crops. The one getting the most attention is canola, the source of canola oil. Efforts are underway to modify the distribution of oils in canola to improve the nutritional value of the oil. Another possibility is increased Vitamin E content in transgenic canola. Sunflower, another source of vegetable oils, is the subject of research designed to produce improved transgenic varieties. Herbicide tolerance and resistance to white mold are among the properties that are being developed in transgenic sunflowers.

Decaffeinated coffee and tea have become important beverages. Unfortunately, the processes that remove caffeine from coffee beans and tea leaves also remove flavor, and some such processes use organic solvents that may leave undesirable residues. The genes that produce caffeine in coffee and tea leaves have now been identified, and it is possible that they may be removed or turned off in the plants to produce coffee beans and tea leaves that give full-flavored products without the caffeine. Additional efforts are underway to genetically engineer coffee trees in which all the beans ripen at once, thereby eliminating the multiple harvests that are now required because of the beans ripening at different times.

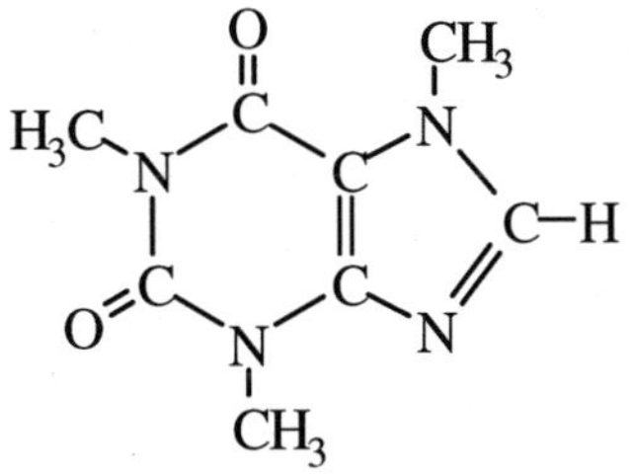

The caffeine molecule

Although turf grass for lawns would not be regarded as an essential crop, enormous resources in the form of water and fertilizers are consumed in maintaining lawns and grass on golf courses and other locations. Healthy grass certainly contributes to the "green" esthetics of a community. Furthermore, herbicides, insecticides, and fungicides applied to turf grass leave residues that can be environmentally harmful. So the development of improved transgenic varieties of grass and other groundcover crops can be quite useful. There are many desirable properties that can benefit grass. Included are tolerances for adverse conditions of water and temperature, especially resistance to heat and drought. Disease and insect resistance are desirable. Reduced growth rates can mean less mowing, saving energy. For grass used on waterways constructed to drain excess rain runoff from terraced areas (see Figure 10.4) a tough, erosion resistant sod composed of masses of grass roots is very desirable. Research is underway to breed transgenic varieties of grass with some of these properties. Also, grass is being genetically engineered for immunity to the effects of Roundup herbicide, which is environmentally more benign than some of the herbicides such as 2,4-D currently used on grass.

An interesting possibility for transgenic foods is to produce foods that contain vaccines against disease. This is possible because genes produce proteins that resemble the proteins in infectious agents, causing the body to produce antibodies to such agents. Diseases for which such vaccines may be possible include cholera, hepatitis B, and various kinds of diarrhea. The leading candidate as a carrier for such vaccines is the banana. This is because children generally like bananas and they are readily grown in some of the tropical regions where the need for vaccines is the greatest.

QUESTIONS AND PROBLEMS

1. Suggest the main contributions made by the geosphere to the biosphere.
2. Distinguish between rocks and minerals.
3. How does igneous rock turn into secondary minerals?
4. What is the branch of chemistry that deals with rocks and minerals and their chemical characteristics and interactions?
5. Suggest ways in which improved materials, some made by green chemical processes, can reduce the effects of earthquakes.
6. In what respect do volcanoes have the potential to drastically affect global climate. Is there any evidence for such an effect?
7. How may human activities lead to landslides?
8. What are the formations called that contain water under ground? What is a major threat to groundwater in such formations?
9. What is FeS_2? Why is its exposure from mining a potential problem?
10. Give an example of an air pollutant and of a water pollutant that may be generated by sanitary landfills.
11. How is soil divided? Which is the top one of these divisions?
12. What is humification, and what does it have to do with soil?
13. What is water in soil called? Give the name of the process by which this water enters the atmosphere by way of plants.
14. In what respects is conservation tillage consistent with the practice of green chemistry?
15. Explain why corn is especially amenable to the production of hybrids.
16. How do human activities affect the nitrogen cycle.
17. Name a gaseous form and two solid forms of fixed nitrogen used as fertilizer.
18. How are phosphate minerals treated to make the phosphorus more available to plants?
19. Name a pollutant that was once commonly transferred from the atmosphere through the atmosphere to soil. Why is this pollutant no longer such a problem?
20. Explain relatively high levels of PCBs and similar persistent organic compounds in remote arctic regions.

21. Explain what is meant by desertification.
22. What is the good news in the U. S. regarding deforestation?
23. What is the potential use of perennial plants in grain production?
24. Give the meaning of transgenic.
25. What are the two main qualities currently developed in transgenic field crops? What are some other possibilities?
26. Explain the importance of *Bacillus thuringiensis* and glyphosate in transgenic crops.
27. Why is the potato not a very good candidate for Bt insecticide?
28. What is a potential environmental problem with Bt corn?
29. Name a concern with transgenic crops, such as squash, that are virus resistant.
30. How might transgenic crops be used to produce vaccines?

11. TOWARD A GREENER ANTHROSPHERE THROUGH INDUSTRIAL ECOLOGY

11.1. INDUSTRIAL ECOLOGY AND INDUSTRIAL ECOSYSTEMS

Recall that the *anthrosphere* has been defined as a fifth sphere of the environment, the one made and modified by human activities. As such, it has developed in ways that are often in conflict with other spheres of the environment, including even the human denizens of the biosphere who have constructed the anthrosphere. This has given rise to the many environmental, resource, and sustainability problems that afflict the world today.

It is crucial for humankind and, indeed, the Earth as a whole, that the anthrosphere be brought into a state of compatibility with the other environmental spheres and with Earth. Green chemistry has a key role to play in this endeavor. In a sense, green chemistry is all about the greening of the anthrosphere. In order to understand how this may occur, it is necessary to introduce and explain the key concept of **industrial ecology**. Industrial ecology integrates the principles of science, engineering, and ecology in industrial systems through which goods and services are provided in a way that minimizes environmental impact and optimizes utilization of resources, energy, and capital. In so doing, industrial ecology considers every aspect of the provision of goods and services from concept, through production, and to the final fate of products remaining after they have been used. It is above all a **sustainable** means of providing goods and services. It is most successful in its application when it mimics natural ecosystems, which are sustainable by nature.

Industrial ecology works through groups of industrial concerns, distributors, and other enterprises functioning to mutual advantage, using each others' products, recycling each others' potential waste materials, and utilizing energy as efficiently as possible. By analogy with natural ecosystems, such a system is an **industrial ecosystem**. Successful industrial ecosystems achieve the maximum possible degree or recycling. To quote Kumar Patel of the University of California at Los Angles, "The goal is *cradle to reincarnation*, since if one is practicing industrial ecology correctly there is no grave." As has been the case with natural ecosystems, the best means of assembling industrial ecosystems is through natural selection in which the various interests involved work out mutually advantageous relationships. However, with knowledge of the feasibility of such systems, external input and various kinds of incentives can be applied to facilitate the establishment of industrial ecosystems. A key measure of the success of such a system can be given by the following relationship:

$$\frac{\text{Market value of product}}{\text{Consumption of material and energy}}$$

Just as organisms in natural ecosystems develop strong symbiotic relationships — the inseparable union of algae and bacteria in lichens growing on rock surfaces, for example — concerns operating in industrial ecosystems develop a high degree of **industrial symbiosis**. It is the development of such mutually advantageous interactions between two or more industrial enterprises that cause an industrial ecosystem to develop in the first place. The recycling components of an industrial ecosystem are absolutely dependent upon symbiotic relationships with their sources of supply.

Figure 11.1 outlines a general industrial ecosystem. The major inputs to such a system are energy and virgin raw materials. A successful system minimizes use of virgin raw materials and maximizes efficiency of energy utilization. The materials processing sector produces processed materials such as sheet steel or synthetic organic polymers. These in turn go to a goods fabrication sector in which the processed materials are formed and assembled or, in the case of consumables such as detergents, formulated to give the desired product. Scrap materials, rejected components, and off-specification consumables generated during goods fabrication may go to recycling and remanufacturing. From goods fabrication, manufactured items or formulated substances are taken to a user sector, which includes consumers and industrial users. In a successful system of industrial ecology, waste materials from the user sector are minimized and, ideally, totally eliminated. Spent goods from the user sector are taken to recycling and remanufacturing to be introduced back into the materials flow of the system. Such items may consist of automobile components that are cleaned, have bearings replaced, and otherwise refurbished for the rebuilt automobile parts market. Another typical item is paper, which is converted back to pulp that is made into paper again. In some cases the recycling and remanufacturing sector salvages materials that go back to materials processing to start the whole cycle over. An example of such a material is scrap aluminum that is melted down and recast into aluminum for goods fabrication. Communications are essential to a successful industrial ecosystem, as is a reliable, rapid transportation system. It is especially important that these two sectors work well in modern manufacturing practice which calls for "just in time" delivery of materials and components to avoid the costs of storing such items.

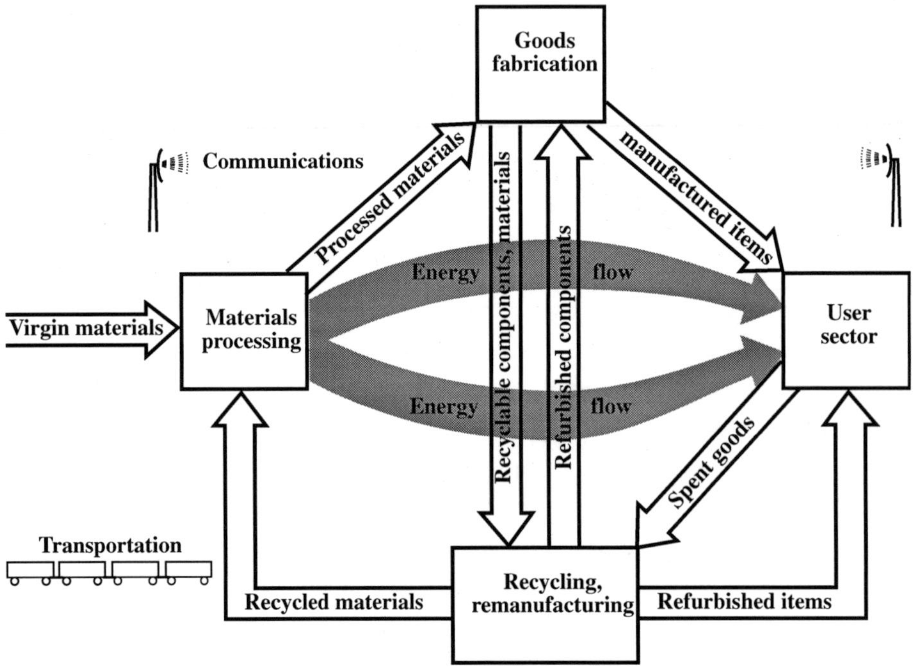

Figure 11.1. Outline of the major components of an industrial ecosystem.

An important characteristic of an industrial ecosystem is its **scope**. A regional scope large enough to encompass several industrial enterprises, but small enough for them to interact with each other on a constant basis is probably the most satisfactory scale to consider. Frequently such systems are based around transportation systems. Segments of interstate highways over which goods and materials move between enterprises by truck may constitute industrial ecosystems.

11.2. METABOLIC PROCESSES IN INDUSTRIAL ECOSYSTEMS

Industrial metabolism refers to the processes to which materials and components are subjected in industrial ecosystems. It is analogous to the metabolic processes that occur with food and nutrients in biological systems. Like biological metabolism, industrial metabolism may be addressed at several levels. A level of industrial metabolism at which green chemistry, especially, comes into play is at the molecular level where substances are changed chemically to give desired materials or to generate energy. Industrial metabolism can be addressed within individual unit processes within a factory, at the factory level, at the industrial ecosystem level, and even globally.

A significant difference between industrial metabolism as it is now practiced and natural metabolic processes relates to the wastes that these systems generate. Natural ecosystems have developed such that true wastes are virtually nonexistent. For example, even those parts of plants that remain after biodegradation of plant materials form soil humus (see "Soil Organic Matter" in Chapter 10, Section 10.6) that improves the conditions of soil on which plants grow. Anthropogenic industrial systems, however, have developed in ways that generate large quantities of wastes, where a waste may be defined as *dissipative use of natural resources*. Furthermore, human use of materials has a tendency to dilute and dissipate materials and disperse them to the environment. Materials may end up in a physical or chemical form from which reclamation becomes impractical because of the energy and effort required. A successful industrial ecosystem overcomes such tendencies.

Organisms performing their metabolic processes degrade materials to extract energy (catabolism) and synthesize new substances (anabolism). Industrial ecosystems perform analogous functions. The objective of successful industrial metabolism is to make desired goods with the least amount of byproduct and waste. This can pose a significant challenge. For example, to produce lead from lead ore for the growing battery market in hybrid gasoline/electric automobiles requires mining large quantities of ore, extracting the relatively small fraction of the ore consisting of lead sulfide mineral, and roasting and reducing the mineral to get lead metal. The whole process generates large quantities of lead-contaminated tailings left over from mineral extraction and significant quantities of byproduct sulfur dioxide, which must be reclaimed to make sulfuric acid and not released to the environment. The recycling pathway, by way of contrast, takes essentially pure lead from recycled batteries and simply melts it down to produce lead for new batteries; the advantages of recycling in this case are obvious.

There are some interesting comparisons between natural ecosystems and industrial systems as they now operate. The basic unit of a natural ecosystem is the organism, whereas that of an industrial system is the firm or, in the case of large corporations, the branch of a firm. Natural ecosystems handle materials in closed loops; with current practice, materials traverse an essentially one-way path through industrial systems. It follows that natural systems completely recycle materials, whereas in industrial systems the level of recycling is often very low. Organisms have a tendency to concentrate materials. For example, carbon in carbon dioxide that is only about 0.04% of atmospheric air becomes concentrated in organic carbon through photosynthesis. Industrial systems in contrast tend to dilute materials to a level where they cannot be economically recycled, but still have the potential to pollute. Aside from maintaining themselves during their limited lifetime, the major function of organisms is reproduction. Industrial enterprises do not have reproduction of themselves as a primary objective; their main function is to generate goods and services.

Unlike natural ecosystems in which reservoirs of needed materials are essentially constant (oxygen, carbon dioxide, and nitrogen from air as examples) industrial systems are faced with largely depleting reservoirs of materials. For example, the lead ore cited above is a depleting resource; more may be found, but only a finite amount is ultimately available. Fossil energy resources are also finite. For example, much more fossil energy from coal may be available, but it would come at an unacceptable cost of global warming from carbon dioxide emissions. Again, industrial metabolic processes that emphasize recycling are desirable because recycling gives essentially constant reservoirs of materials in the recycling loop. Ideally, even in the case of energy, renewable energy resources such as wind and solar power provide an essentially constant, nondepleting energy source.

As discussed under "Control in Organisms" under Section 9.4, biological systems have elaborate systems of control. Considering the metabolism that occurs in an entire natural ecosystem, it is seen to be **self-regulating**. If herbivores that consume plant biomass become too abundant and diminish the stock of the biomass, their numbers cannot be sustained, the population dies back, and their food source rebounds. The most successful ecosystems are those in which this self-regulating mechanism operates continuously without wide variations in populations. Industrial systems do not inherently operate in a self-regulating manner that is advantageous to their surroundings, or even to themselves in the long run. Examples of the failure of self-regulation of industrial systems abound in which enterprises have wastefully produced large quantities of goods of marginal value, running through limited resources in a short time, and dissipating materials to their surroundings, polluting the environment in the process. Despite these bad experiences, within a proper framework of laws and regulations designed to avoid wastes and excess, industrial ecosystems can be designed to operate in a self-regulating manner. Such self-regulation operates best under conditions of maximum recycling in which the system is not dependent upon a depleting resource of raw materials or energy.

Obviously, recycling is the key to the successful function of industrial metabolism. Figure 11.2 illustrates the importance of the **level of recycling**. In low-level recycling, a material or component is taken back to near the beginning of the steps through which it is made. For example, an automobile engine block might be melted down to produce molten metal from which new blocks are then cast. With high-level recycling, the item or material is recycled as close to the final product as possible. In the case of the automobile engine block, it may be cleaned, the cylinder walls rehoned, the flat surfaces replaned, and the block used as the platform for assembling a rebuilt engine. In this example and many others that can be cited, high-level recycling uses much less energy and materials and is inherently more efficient. The term given to the value represented to an item or material recycled near the top of the energy/materials pyramid shown in Figure 11.2 is called its **embedded utility**.

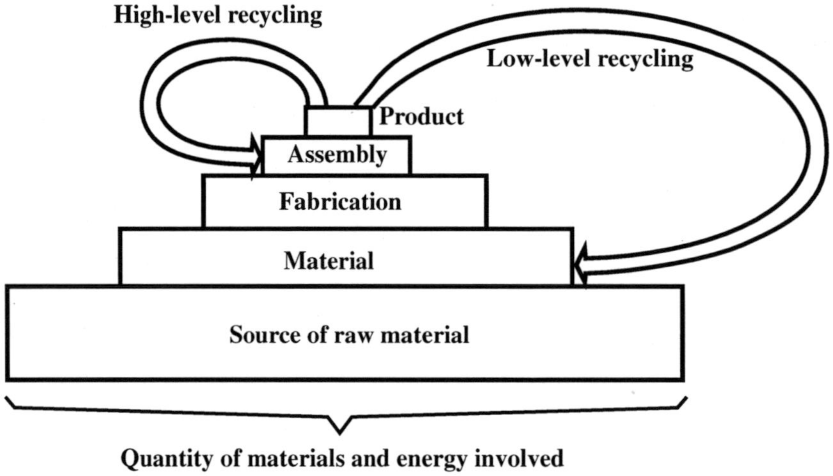

Figure 11.2. The level at which recycling occurs strongly influences the amount of energy required and the quantity of materials that must be processed.

11.3. LIFE CYCLES IN INDUSTRIAL ECOSYSTEMS

In conventional industrial systems, a product is manufactured and marketed after which the vendor forgets about it (unless some product defect, such as treads coming off of tires forces a recall). In a system of industrial ecology, however, the entire **life cycle** of the product is considered. An important aspect of such a consideration is the **life-cycle assessment**. The overall goal of a life-cycle assessment is to determine, measure, and minimize environmental and resource impacts of products and services.

An important decision that must be made at the beginning of a life-cycle assessment is determination of the **scope** of the assessment. Parameters included in the scope include the time period to be considered, the area (space) to be considered, and the kinds of materials, processes, and products that will go into the assessment. As an example, consider the chemical synthesis of an insecticide that releases harmful vapors and generates significant quantities of waste material. A narrowly focussed life-cycle assessment might consider control measures to capture released vapors and the best means of disposing of the waste byproducts. A broader scope would consider a different synthetic process that might not cause the problems mentioned. An even broader scope might consider whether or not the insecticide even needs to be made and used; perhaps there are more acceptable alternatives to its use.

Life cycle assessment involves an **inventory analysis** to provide information about the consumption of material and release of wastes from the point that raw material is obtained to the ultimate fate of the product, an **impact analysis** to consider the environmental and other impacts of the product, and an **improvement analysis** to determine measures that can reduce impacts. A life-cycle assessment gives a high priority to the choice of materials in a way that minimizes wastes. It considers which materials and whole components can be used or recycled. And it considers alternate pathways for manufacturing processes or, in the case of chemical manufacture, alternate synthesis routes.

In doing life-cycle assessments it is useful to consider the three major categories of **products**, **processes**, and **facilities**, all of which have environmental and resource impacts. Products are obviously the things and commodities that consumers use. They are discussed further in Section 11.4. Processes refer to the ways in which products are made. Facilities consist of the infrastructural elements in which products are made and distributed.

Some of the greater environmental impacts from commerce result from the processes by which items are made. An example of this is paper manufacture. The environmental impact of the paper *product* tends to be relatively low. Even when paper is discarded improperly, it does eventually degrade without permanent effect. But the *process* of making paper, beginning with harvesting of wood and continuing through the chemically intensive pulping process and final fabrication has significant environmental impact. In addition to potential air and water pollution, paper making consumes energy and requires large amounts of water. Processes can be made much more environmentally friendly by application of the principles of industrial ecology enabling maximum recycling of materials with significant pollution potential.

The impact of facilities can vary over a wide range. A specialized facility such as a steel mill or petroleum refinery can have a significant environmental impact. Abandoned sites of these facilities can be blighted and difficult and expensive to restore for some other use. (The term "brownfields" is sometimes used to describe sites of abandoned industrial facilities, and restoration of blighted "brownfields" is often a major goal of urban renewal projects.) One of the more challenging kinds of facilities to decomission are sites of nuclear power reactors in which there is a significant amount of radioactivity to deal with in dismantling and disposing of some of the reactor components. The impact of facilities can be minimized by designing them with future use and eventual decommissioning in mind. Typically, well-designed commercial buildings may have a number of lives in which they are used by a sequence of enterprises. A key aspect of a building destined for multiple use is structure flexibility so that it can easily be rearranged for new uses.

Product Stewardship

The control of the life cycle of products that are sold is difficult. Attempts are made to exercise some degree of control by facilitating return of spent products. For example, laser printer cartridges commonly come with a return address and postage to return the used cartridge to the manufacturer. Another approach is to pay for the return of spent products. The price of a new automobile battery is commonly increased to cover a modest refund paid upon return of the spent battery. Automobile tires have a financial disincentive for recycling in that the customer is charged a disposal fee.

The success rate with these kinds of measures is mixed. Automobile tires are almost invariably mounted at a dealer's place of business, so customers pay the disposal fee in order to not be responsible for used tire disposal. Customers who install their own batteries usually turn in the old ones because they do not have a good disposal option. The return rate with printer cartridges is probably lower because people do not want to go to the trouble to mail them back.

The rate of return of items at the end of their life cycles is higher in companies and institutions where returns are more readily facilitated. The presence in a large office of a recycled paper bin or an "out box" where returned laser printer cartridges can be dropped for mailing certainly increases the recycle rate of these items. One key to recycling is in **product stewardship** where there are clear lines of accountability for items and materials. A good way to ensure product stewardship for things such as office equipment is through the leasing of equipment. In such a case, the concern holding the lease, rather than the user, is responsible for the ultimate fate of the item. The same approach can even be used with some kinds of materials, such as activated carbon used in a water treatment plant. Here the concern leasing the carbon retains ownership and is responsible for picking up the spent material for recycling. Another way to help ensure product stewardship through the consumer sector is by charging deposits on items and refunding the deposit upon return for recycling.

11.4. KINDS OF PRODUCTS

In considering life-cycle assessments, it is useful to divide products into three major categories. The first of these are **consumable products** which, by the nature of their use, are used up or dispersed to the environment with no possibility of recovery. Such materials include laundry detergents that are flushed down the drain with washwater or windshield washer fluid, which is squirted onto windshields, then wiped off. Another class of product consists of **recyclable commodities**. Engine antifreeze and motor oil are potentially recyclable commodities in that, in principle (though somewhat rarely in practice), they can be reclaimed after use, purified, reformulated, and sold again. **Service products** (sometimes called durable products) are usually devices that have multiple uses and last for a long time. The washing machine in which consumable laundry detergent is used is a typical service product.

Since consumable products are dispersed to the environment, it is important that they have environmentally friendly characteristics. They should first of all be **nontoxic** at the levels and manner in which organisms are exposed to them. In addition to not causing acute toxicity, they should not be carcinogenic or mutagenic and should not cause birth defects. Another characteristic that consumable products should have is that they should not be **bioaccumulative**. As discussed under the topic of "Biological Interaction with Environmental Chemicals" in Section 9.9, bioaccumulation is the term given to the uptake and concentration of xenobiotic materials by living organisms. Poorly biodegradable, lipid-soluble materials such as PCB compounds have a strong tendency to bioaccumulate, and such substances should be avoided in consumable products. Consumable products should also be **degradable**. The most common type of degradation is biodegradation, which occurs primarily through the action of microorganisms. The practice of green chemistry can aid in making biodegradable products by, for example, avoiding branched chain hydrocarbon structures in organic compounds and by attaching functional groups, such as the organic carboxylic acid group, $-CO_2H$, that are amenable to microbial attack.

Recyclable commodities should be designed with durability and recycling in mind. In order for them to last through a normal life cycle, such commodities should not be as degradable as consumables. An example of making a product more amenable to recycling is the use of bleachable and degradable inks on newsprint, which makes it easier to recycle the newsprint to a product that meets acceptable color standards.

Although service products are designed to last for relatively long times, they do reach a stage requiring disposal or recycling. A key factor in recycling is the need for channels through which such products can be recycled. Proposals have been made for "de-shopping" centers where items such as old computers and broken small appliances can be returned for recycling. Service products should be designed and constructed to facilitate disassembly so that various materials can be separated for recycling.

11.5. ATTRIBUTES REQUIRED BY AN INDUSTRIAL ECOSYSTEM

Under the heading of "Response of Life Systems to Stress" in section 9.6, is a discussion of the *inertia* of biological communities, which is their resistance to alteration and damage. The key factors involved in inertia were mentioned as *productivity* of basic food materials, *diversity* of species, *constancy* of numbers of various organisms and *resilience* in the ability of populations to recover from loss. Industrial ecosystems likewise have key attributes that are required for their welfare. These include **energy**, **materials**, and **diversity**. One big difference between biological and industrial ecosystems is the time scale involved. In the evolution of organisms, a time period of several thousand years is very brief without noticeable evolutionary changes, whereas in industrial systems several decades may be a very long time.

Energy

With enough energy, almost anything is possible. Therefore, the provision of adequate amounts of energy that can be used without doing excessive damage to the environment is essential for the proper function of industrial ecosystems. And the energy that is available has to be used as effectively and efficiently as possible. It was once believed that the world's vast coal resources would provide enough energy to meet human needs for several centuries. Now it is apparent that consuming most of these energy resources would cause unacceptable global warming effects. Solar energy and wind energy, which derives from solar energy, come about as close as any energy source to offering ideal renewable sources of energy. But there are major problems with the intermittent nature of these sources and the need that they present for short-term energy storage. Furthermore, they both require vast areas of land in order to provide a significant share of energy needs. Then there are unexpected problems, such as the one arising from the accumulation of dead insects on windmill blades, spoiling their finely tuned aerodynamic characteristics and reducing power output by about half in strong winds. Properly run nuclear power facilities can provide abundant energy for many decades, but this source comes with its own set of problems and is strongly opposed by many.

Cogeneration represents the most efficient energy use within an industry or within an industrial ecosystem. The two major reasons that an industrial plant uses energy are (1) for steam used in processing, such as heating chemical reaction mixtures to cause a reaction to go faster, and (2) to generate electricity. Traditionally, industrial operations, such as petroleum refineries, have bought electricity from external power plants to run pumps and compressors, for lighting, and other purposes that consume electricity. Steam, which can only be shipped economically for relatively short distances, is normally generated by burning fossil fuels in boilers on the site. Since a maximum of only approximately 40% of the heat generated in burning a fuel in a power plant can be converted to electricity, and because of losses in electrical transmission lines, obtaining electricity from an external source is a relatively inefficient means of getting power. Much greater efficiencies can be attained by burning fuels, such as natural gas, in large turbines connected to an electrical generator and using the hot exhaust from the turbine to raise steam. This approach can double the overall efficiency of energy utilization.

The ExxonMobil company has been very active in cogeneration at its various petroleum refining and petrochemical manufacturing facilities in which about 70% of the facilities' energy needs are now produced by cogeneration. As of 2001, the total electricity generating capacity by cogeneration at these facilities was 2,300 megawatts (a 1,000 megawatt power plant is a very large one). The company points out that this amount of electricity was about one-fourth of the world's total solar and wind power electrical production in 2001. The company also points to reductions of approximately 6 million tons per year of greenhouse gas carbon dioxide emissions. The total financial cost savings have also been substantial. Although these cogeneration facilities were established within a single company, transfers of steam and electricity between companies in an industrial ecosystem are also possible (see the Kalundborg example in Section 11.6).

Materials

There are several approaches to providing materials. These can be classified as **dematerialization** in which less material is used for a specific purpose, **substitution** of abundant materials for scarce ones, **recycling** materials, and **waste mining** in which needed materials are extracted from wastes.

Examples abound of areas in which the need for materials has been reduced in recent decades. Higher voltage electrical transmission carries more power over lighter copper or aluminum wires, the switch from 6-volt to 12-volt electrical systems in automobiles has enabled lighter wiring in automotive electrical systems, modern photographic film uses much less silver than in years past, and the switch from biased-ply to radial tires has greatly extended tire life, so that much less rubber is required. Dematerialization has been spectacular in the electronics area. The popular laptop computer has far more computing power than did the earliest vacuum-tube-equipped computers that each required an entire air conditioned building for housing.

Material substitution is an area in which green chemistry has made a significant contribution and will continue to do so at an accelerating pace in the future. The most spectacular advances have been made in electronics where material substitution, which enabled dematerialization to occur, has provided electronic circuits with many orders of magnitude more capability than the circuits that they replaced. The glowing, electricity-consuming vacuum tubes, capacitors, resistors, and transformers of the receiver circuit of a 1950s table-top radio have been replaced with a tiny circuit almost invisible to the human eye. The huge numbers of copper wires that carried telegraph and telephone messages in the 1940s have now been replaced by fiber optic signal conductors that carry unimaginably more information per unit mass of carrier. Polyvinylchloride (PVC) pipe has replaced copper and steel for water and wastewater transmissions. Toxic liquid sulfur dioxide and ammonia used in early refrigerator models were replaced by nontoxic, nonflammable chlorofluorocarbons (CFCs). When the CFCs were found to deplete stratospheric ozone, they were replaced with similar compounds containing at least one bound H atom per molecule (HCFCs) that break down in the troposphere before reaching stratospheric altitudes. Many more similar examples could be cited.

Recycling is of course one of the major objectives of a system of industrial ecology and one in which significant progress is being made. There are some consumable items that are not practical to recycle and for which the raw materials are abundant enough that recycling is not required. Household detergents fall into this category. Recyclables in a second group are those that are not particularly scarce, but for which recycling is feasible and desirable. Wood and paper fall into this category. A third category of recyclable materials consists of metals, particularly the more valuable and scarce ones, such as chromium, platinum, and palladium. These metals definitely should be recycled. A third category of recyclables consists of parts and apparatus that can be refurbished and reused.

Waste mining, the extraction of useful materials from wastes, provides more materials while benefiting the environment. One of several important examples of waste mining is the extraction of combustible methane gas, a low-polluting premium fossil fuel, from municipal refuse landfills in which the biodegradation of organic matter in the absence of oxygen generates the gas. Sulfur in sulfur dioxide extracted from the flue gases generated in burning coal that contains sulfur can be reclaimed and used to make sulfuric acid. Methods have been developed to extract aluminum from finely divided coal fly ash generated in coal combustion. In this case, the finely divided, homogeneous, dry nature of the fly ash is a definite advantage in processing it. It is anticipated that growing scarcity of resources combined with the need to dispose of a variety of wastes will lead to the development of additional waste mining processes in the future.

Diversity

Diversity in industrial ecosystems, tends to impart a **robust** character to them, which means that if one part of the system is diminished, other parts will take its place and keep the system functioning well. Many communities that have become dependent upon one or just a few major enterprises have suffered painful economic crises when a major employer leaves or cuts back. One area of vulnerability that has been manifested in 2001 in California, Brazil, and some other areas is in the energy sector where cutbacks in electricity production resulting especially from low precipitation required for hydroelectric power have forced painful economic adjustments. In many parts of the world, water supply from a single vulnerable source threatens diversity.

11.6. KALUNDBORG

Industrial ecosystems of various degrees of sophistication have existed ever since the first industries were established, although they were not called industrial ecosystems or even recognized as such until the latter 1980s. The system most commonly cited as a fully developed industrial ecosystem is the one that developed spontaneously in Kalundborg, Denmark. This industrial ecosystem system is centered around two very large energy enterprises. The first of these is the ASNAES electrical plant fired by coal and with a capacity of 1,500 megawatts. The second enterprise is the Statoil petroleum refinery, processing 4–5 million tons per year. The initial relationships in the Kalundborg system consisted of exchanges between these two enterprises. The power plant sold steam to the refinery to use in processing, and the refinery provided the power plant with fuel gas and cooling water. Both enterprises produce low-level energy in the form of steam that is used for district heating of homes and commercial buildings. The heat is also used in a large greenhouse operation, as well as in a fish-farming enterprise. Another big player in the Kalundborg industrial ecosystem is the Novo Nordisk pharmaceutical plant, which receives steam from the energy suppliers. This huge enterprise makes 40% of the world's supply of insulin as well as industrial enzymes. Fermentation processes in this plant generate excess yeast, which is used as protein supplement for swine. The plant produces large quantities of biological sludge, which, along with wastewater treatment sludge from the waste and wastewater treatment plant associated with the the fish farm, is used as fertilizer and soil conditioner in area farms.

The Kalundborg industrial ecosystem is often cited for the spontaneous way in which it developed, beginning in the 1960s with steam and electricity provided to the petroleum refinery from the power plant. Some of the enterprises were driven by required measures to lower pollution. As a result of the requirement for lime scrubbing of the stack gas from the power plant, large quantities of calcium sulfate were produced, which were used to manufacture gypsum wallboard for buildings. Air pollution control measures resulted in the substitution of clean burning hydrocarbons from the petroleum refinery in place of some of the coal to generate electricity in the power plant. The requirement to remove sulfur from petroleum led to the development of a sulfuric acid plant.

It is interesting to consider the conditions that lead to such a well developed industrial ecosystem at Kalundborg. It did not develop from directives from any centralized authority, but rather from agreements between various entities acting in their own corporate self-interests. A rather close social system that promoted contact between individuals was helpful. The relatively small geographic area involved has been helpful in enabling facile communication and the transfer of materials and energy among the various enterprises. This is especially so in that several of the main commodities involved — steam, water, waste treatment sludge — cannot be shipped economically for any great distances.

11.7. ENVIRONMENTAL IMPACTS OF INDUSTRIAL ECOSYSTEMS

The practice of industrial ecology in the anthrosphere certainly has profound potential effects upon the atmosphere, hydrosphere, geosphere, and biosphere. Anthrospheric influences may range from highly localized effects to global effects, such as greenhouse warming or stratospheric ozone depletion. The magnitude of the effects may be minor, or they may be catastrophic. Until recently, the effects of human activities on the surrounding environment were of relatively little concern, resulting in neglect that is the cause of many of the environmental problems that exist even today. However, the proper practice of industrial ecology requires that consideration be given to the various influences that anthrospheric activities have upon the surrounding environment.

Consider the kinds of effects that industrial activities may have upon the natural environmental spheres. One of the most obvious influences is upon the atmosphere because of the emission to the atmosphere of pollutant gases, vapors from volatile compounds, and particles. Released carbon dioxide and vapors such as those of fluorinated hydrocarbons have a high potential to cause greenhouse warming. Particles obscure visibility and cause adverse health effects in people who must breathe the air in which they are contained. Chlorofluorocarbons lead to stratospheric ozone depletion and hydrocarbons and nitrogen oxides released to the atmosphere can cause formation of photochemical smog.

Industrial activities often utilize large quantities of water for cooling and other purposes. Water may become polluted or warmed excessively when used for cooling (thermal pollution).

Many industries require large quantities of materials that are taken from the Earth by the extractive industries. This may result in disruption of the geosphere from mining, dredging, and pumping of petroleum. The other major effect upon the geosphere results from the need to dispose of wastes. Scarce land may be required for waste disposal dumps and the geosphere may become contaminated with pollutants from disposal of wastes.

The biosphere is most affected by industrial activity when toxic substances are released. Other effects upon the biosphere may be indirect as the result of adverse effects upon the atmosphere, hydrosphere, or geosphere.

Industrial systems are largely dependent upon the utilization of fossil fuels, so many environmental effects are due to fossil fuel extraction and combustion. Greenhouse-warming carbon dioxide emissions, acid gas emissions, smog-forming hydrocarbons and nitrogen oxides, and deterioration of atmospheric quality from particles released from fossil fuel combustion are all atmospheric effects associated with fossil fuel combustion. Coal mining activities have the potential to release acid mine water to the hydrosphere, petroleum production can release brines or result in ocean oil spills, acid precipitation may acidify isolated lakes, and water used as cooling water in power plants may become thermally polluted. The geosphere may be disrupted by fossil fuel extraction, especially in the surface mining of coal. Coal is extracted from some areas of West Virginia by cutting off entire mountain tops overlying coal seams and dumping the overburden into valleys below in order to get to the coal. Effects upon the biosphere from fossil fuel utilization may be direct (birds coated with tar from oil spills come to mind), but are more commonly indirect, such as acidified bodies of water from acid rain resulting from sulfur dioxide emissions from coal combustion.

Agricultural activities certainly have to be considered as parts of the anthrosphere, and modern agricultural practices are part of vast agriculturally based industrial systems. Large quantities of greenhouse-warming methane are released to the atmosphere from the action of anaeobic bacteria in rice paddies and in the intestines of ruminant animals. "Slash and burn" agricultural techniques practiced in some tropical countries release greenhouse gas carbon dioxide to the atmosphere and destroy the capacity of forests to sequester atmospheric carbon dioxide by photosynthesis. Enormous quantities of water are run through irrigation systems. Some of this water is evaporated and lost from the hydrosphere. The water that returns to the hydrosphere from irrigated fields picks up significant amounts of salt from the land and fertilizers applied to the land, so water salinity can become a problem. Underground aquifers become severely depleted by pumping large quantities of water for irrigation. The production of protein from livestock requires much more water overall than does the production of an equivalent amount of protein from grain. Animal wastes from huge livestock feedlots are notorious water polluters, adding oxygen-depleting biochemical oxygen demand (BOD, see Chapter 7, Section 7.4) and potentially toxic inorganic nitrogen compounds to water. The disturbance of the geosphere from crop cultivation is enormous. Raising livestock for food entails a much greater degree of land cultivation than does the cultivation of cereal grains. Agricultural production replaces entire, diverse biological ecosystems with artificial ecosystems, which causes a severe disturbance in the natural state of the biosphere. Another agricultural activity that affects the biosphere is the loss of species diversity in the raising of crops and livestock. In addition to the loss of entire species of organisms, the number of strains of organisms grown within species tends to become severely diminished in modern agricultural practice. Obviously, those varieties of crops and livestock that are most productive are the ones that will be used to produce grain, meat, and dairy products. However, if something happens, such as a particular variety becoming susceptible to a newly mutated virus, alternative resistant varieties may no longer be available. Finally, the raising of transgenic crops and livestock (see Section 10.12, "Agricultural Applications of Genetically Modified Organisms") promises profound and potentially unforeseen effects upon the biosphere.

Design of Industrial Ecosystems to Minimize Environmental Impact

From the discussion above it is obvious that industrial activity, broadly defined to include agriculture as well, has a high potential to adversely affect the atmosphere, hydrosphere, biosphere and geosphere. Inherent to the nature of industrial ecosystems, however, are measures and systems designed to minimize such impacts.

Several measures may be taken to minimize the effects of industrial ecosystems upon the geosphere. Since most of the raw materials required for manufacturing originally have to be extracted from the geosphere, the recycling of materials inherent to well designed industrial ecosystems minimizes impact upon the geosphere. The selection of materials can also be important. As an example, the mining of copper to make copper wire once widely used to carry communications signals involves digging large holes in the ground and exposing minerals that tend to release metals and acidic pollutants. The silica used in the fiber optic cables that now largely substitute for copper for communications is simply obtained from sand. The impacts of disturbing the geosphere for food and fiber production can be minimized by some of the conservation methods and agricultural practices discussed in Chapter 10.

Well designed industrial ecosystems emit much less harmful material to the atmosphere than do conventional industrial systems. Industrial atmospheric emissions have been decreasing markedly in recent years as the result of improved technology, more stringent regulation, and requirements to release information about atmospheric emissions. One of the main classes of industrial atmospheric pollutants has consisted of the vapors of volatile organic compounds (VOCs). These have been significantly reduced by modifying the conditions under which they are used to lower emissions and by measures such as activated carbon filters to trap the vapors. The practice of industrial ecology goes beyond these kinds of measures and attempts to find substitutes, such as water-based formulations, so that volatile organic compounds need not even be used.

Years of regulation have resulted in much lowered releases of water pollutants from industrial operations. These lowered levels have been due largely to sophisticated water treatment operations that are applied to water before it is released from a plant. Desirable as these "end-of-pipe" measures are, the practice of industrial ecology goes beyond such pollution control, minimizing the use of water and preventing its pollution in the first place. One way to ensure that water pollutants are not released from an industrial operation is to completely recycle water in the system — no water out, no water pollutants.

In past years, many hazardous solid and liquid wastes have been improperly disposed to sites in the geosphere, giving rise to a large number of "hazardous waste sites," the subject of Superfund activity in the United States. The practice of industrial ecology seeks to totally eliminate any such wastes that would require disposal. Ideally, such wastes simply represent material resources that are not properly utilized, a fact that can serve as a guideline for the prevention of such wastes.

The expenditure of energy entails the potential to cause environmental harm to the various spheres of the environment. A prime goal in the proper practice of industrial ecology is the most efficient use of the least polluting sources of energy possible. More efficient electric motors in industrial operations can significantly reduce electricity consumption. The proper design of buildings to reduce heating and cooling costs can also reduce energy consumption. Many industrial operations require heat (process heat in industrial parlance) and steam. Rather than generating these separately, they can be produced in combined power cycles along with the generation of electricity, thereby greatly increasing the overall efficiency of energy utilization.

11.8. GREEN CHEMISTRY IN THE SERVICE OF INDUSTRIAL ECOSYSTEMS

Green chemistry has an essential role to play in the development of successful industrial ecosystems, especially in making industrial metabolism as efficient, nonpolluting, and safe as possible. Recall from the discussion of "Yield and Atom Economy" in Section 1.6 that atom economy refers to the fraction of reactant material that actually ends up in final product in the course of a chemical synthesis. It is expressed by as a percentage by the equation

$$\text{Percent atom economy} = \frac{\text{Total mass of desired product}}{\text{Total mass of reactants}} \times 100 \qquad (11.8.1)$$

When all of the mass of reactants ends up in the desired products, the atom economy is 100%. The goal of green chemical synthesis is the achievement of 100% atom economy.

A major advantage of the practice of green chemistry to reduce environmental impact is that, ideally, it is inherently safe and clean. By using nontoxic chemicals and processes that do not threaten the environment, green chemistry avoids posing threats to the people who practice it and to the surrounding environment. Of course, these are ideals that can never be completely realized in practice, but by having these ideals as goals and making constant incremental improvements, the practice of green chemistry can become increasingly safe, environmentally friendly, and sustainable. This reduces dependence upon the command and control measures that require constant vigilance to maintain. Rather than depending upon regulations imposed from the outside to maintain its safe operation, green chemistry is much more self-regulating.

Green chemistry gives prime consideration to the chemical reactions and processes by which chemicals are manufactured. One approach to making chemical synthesis greener is to use existing chemical synthesis processes but make the process itself safer and less polluting while also making the reagents required for it by greener processes. An example of the former might be to substitute a less volatile, less toxic solvent as a reaction medium for a chemical synthesis reaction. In some cases, a reagent may be made more safely by using biological processes for its preparation in place of chemical processes. A second general approach to making chemical preparations greener is to use different reagents for the synthesis that are safer and less likely to pollute.

Hazard Reduction

The conventional approach to making chemical processes less dangerous to workers and less harmful to the environment has emphasized *exposure reduction*. In the arena of worker safety, this has involved measures such as wearing protective gear to prevent contact with hazardous chemicals. For the environment as a whole it has consisted largely of "end-of-pipe" measures to prevent release of pollutants once they are generated.

In contrast to exposure reduction, green chemistry relies upon **hazard reduction**. The first step in hazard reduction is to know what the hazards are and where they originate. Hazards may arise from the raw materials used, the media (solvents) in which chemical processes are carried out, catalysts that enable chemical reactions to occur, and byproducts. The direct hazards posed to workers in a chemical process fall into the two main categories of toxicity hazards and hazards associated with uncontrolled events such as fires and explosions.

Toxic substances are most logically classified according to their biochemical properties that lead to toxic responses. A useful means of relating toxic effects to the chemical nature of toxic substances is through **structure activity relationships**, which use computer programs to find correlations between features of chemical structure, such as groupings of functional groups, and the toxicity of the compounds. As an example, organic compounds containing the N-N=O functional group are N nitroso compounds, a family noted for members that cause cancer. Structural features that indicate a tendency to donate methyl (-CH_3) groups are also suspect because attachment of methyl groups to cellular DNA is a major mechanism in causing cancer. This hazard may be reduced by substituting alkyl groups with more carbons for the methyl group.

Three kinds of chemicals have a high priority in eliminating the toxicity hazards in green chemistry. The first such category consists of **heavy metals**, such as lead, mercury, and arsenic (a metalloid). These indestructible elements have a variety of toxic effects, such as impaired function of renal tubules in kidneys (cadmium), neurological damage (mercury), and inhibition of the production of ATP (see Section 9.2 and Figure 9.1). A second major category of toxic substance that should be avoided in the practice of green chemistry consists of **lipid-soluble organics** that are not readily degraded. These compounds often consist of relatively high molecular mass organohalides, such as PCBs, and accumulate in lipid (fat) tissue. Released into the environment, these materials can undergo

biomagnification in moving through a food chain. **Volatile organic compounds** (**VOCs**) constitute a third class of troublesome toxic substances. These are primarily hydrocarbons, such as heptane, and low-molecular-mass organohalides, such as trichloroethylene.

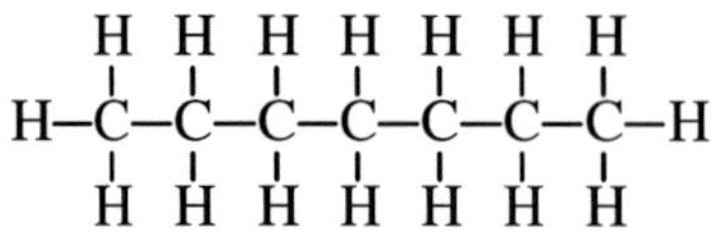

Heptane

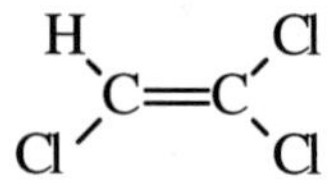

Trichloroethylene

VOCs have had many uses in industry for cleaning parts, as vehicles to carry organic-soluble paint and coatings, and as solvents for organic chemical reactions. When used as vehicles, VOCs have to evaporate to leave a coating behind, and the most straightforward way of removing excess VOC from washed parts is to let the solvent evaporate. Chemical products synthesized in VOC solution always contain residual solvents, which may be allowed to evaporate. This means that there is a strong tendency for VOCs to get into the atmosphere, including air in the workplace that employees have to breathe. There are a number of toxic effects of VOCs. Inhalation of some hydrocarbon solvents, for example, dissolves lipid coverings around nerve fibers, resulting in a condition called peripheral neuropathy. Another example is vinyl chloride (see Section 5.4 and Figure 5.5) known to have caused liver cancer in workers exposed to the vapor in polyvinylchloride polymer manufacture.

Chemicals that pose hazards because of their potential to undergo destructive chemical reactions fall into several often overlapping categories. **Combustible** or **flammable** substances are those that are liable to burn vigorously and destructively in air or other sources of oxygen. Hydrocarbon solvents may closely resemble gasoline in their characteristics and are highly flammable. Adding to the hazards of such materials is their volatility, which enables them to form explosive mixtures of vapor in air. Whereas combustible substances are chemical *reducers*, another category of hazardous chemical substances consists of chemical **oxidizers** that provide sources of oxygen for the reaction of reducers. One such compound is ammonium perchlorate, NH_4ClO_4, a potent oxidizer used in rocket fuel. Many chemical synthesis procedures involve oxidation steps and a variety of chemical oxidizers are used for these steps. Hazardous chemical substances in a fourth category are those that are **reactive**. Explosives are prime examples of reactive substances. One of the more treacherous explosives is nitroglycerin which undergoes the following reaction when it explodes:

$$4C_3H_5N_3O_9 \rightarrow 12CO_2 + 10H_2O + 6N_2 + O_2 \qquad (11.8.1)$$

This reaction shows that the nitroglycerin molecule actually contains an excess of oxygen because elemental oxygen is released when the nitroglycerin explodes. Some structural features of molecules are known to be associated with reactivity. One example is the close presence of oxygen and nitrogen on the same molecule, particularly where multiple nitrogen atoms are bonded together. A final category of hazardous chemical substances consists of those that are **corrosive**. In general, a corrosive substance is one that attacks materials, including even human flesh. A more specific definition is that a corrosive substance is one that produces high concentrations of either H^+ ion or OH^- ion — that is, a strong acid or a strong base.

Often hazardous chemicals belong to two or even more of the categories outlined above. An example of such a substance is concentrated nitric acid, HNO_3. In addition to its being a strong acid that is completely ionized in water, this material acts as a source of oxygen, represented as {O} in the reaction below:

$$2HNO_3 \xrightarrow{\text{Heat}} 3\{O\} + H_2O + 2NO \qquad (11.8.1)$$

Hazardous concentrated nitric acid is used in some chemical synthesis reactions in which an acidic oxidant is needed. Among the many other potentially hazardous strong oxidants used are permanganate (MnO_4^-) and oxygen-containing compounds such as potassium dichromate, $K_2Cr_2O_7$.

The ideal way to deal with hazardous substances in the practice of green chemistry is to totally avoid making or using them. A laudable goal in principle, it is often impossible to completely avoid hazardous materials in practice. In cases where dealing with hazardous substances is necessary, every effort must be made to prevent their release, exposure to humans, or circumstances in which their hazards may be manifested.

11.9. FEEDSTOCKS, REAGENTS, MEDIA, AND CATALYSTS

The main components of a chemical process can be divided into the four categories of feedstocks, reagents, media, and catalysts, all of which are important in the practice of green chemistry. These four aspects of chemical production are addressed in this section.

Feedstocks

Feedstocks are the main ingredients that go into the production of chemical products. As discussed below, feedstocks may be acted upon by reagents, and often there is some overlap between the two categories of materials. Feedstocks are addressed briefly here and in greater detail in Chapter 12.

There are three major components of the process by which raw materials from a source are obtained in a form that can be utilized in a chemical synthesis, then converted to a product. The first of these is the **source** of the feedstock, an aspect that has a number of environmental and sustainability considerations. This may consist of a depleting resource, such as petroleum, in which case the lifetime of the resource and the environmental implications of obtaining it must be considered. A preferable source consists of recycled materials, although not many recycled materials are suitable for chemical feedstocks. A third source that is very desirable consists of renewable resources, particularly from materials made by photosynthesis and biological processes.

The second major aspect of converting feedstocks to final products is separation and isolation of the desired substance. An example of this step is the isolation of specific organic compounds from crude oil to provide a feedstock for organic chemical synthesis. It may be necessary to process raw materials from a source to convert it to the specific material used as a feedstock for a chemical process. Often most of the environmental harm in providing feedstocks comes during the isolation process, in large part because of the relatively large amount of waste material that must be discarded in obtaining the needed feedstock.

Once a suitable chemical feedstock is obtained, it is subjected to chemical processes that give the final product. As addressed below, this may consist of reactions with various kinds of reagents in media such as organic solvents, often using catalysts.

Reagents

The term **reagents** is used here to describe the substances that act upon basic chemical feedstocks through chemical reactions that convert the feedstocks to new chemicals in synthetic processes. The kinds of reagents used have a very strong effect upon the acceptability of a chemical process with respect to green chemical aspects. Much of the work that has been done in developing and using green reagents has involved organic chemical processes, many of which are beyond the scope of this book. However, some of the general aspects of chemical reagents from a green chemical perspective are discussed here.

The most obvious characteristic required of a good chemical reagent is that it do what it is supposed to do, completely and at an acceptable rate. A reagent with a high **product selectivity** produces a high percentage of the desired product with a low percentage of undesired byproducts. Another desirable characteristic of a good reagent is high **product yield** meaning that most of the feedstocks are converted to product. The use of reagents that provide high selectivity and yield means that less unreacted feedstock and byproduct material have to be handled or disposed.

One of the most common measures taken in implementing green chemical processes is selection of alternative reagents. The criteria used in selecting a reagent include whether or not it is available, how efficient it is, and the effect that it has. Important considerations with the chemical transformation are whether it is stoichiometric or catalytic, the degree to which it is atom economical, and the quantities and characteristics of any wastes produced.

One of the main kinds of reactions for which reagents are used is **oxidation**, which usually consists of the addition of oxygen to a chemical compound or a functional group on a compound. (See Chapter 4, Section 4.7 for a discussion of oxidation and its accompanying phenomenon, reduction.) An example of an oxidation reaction is the conversion of ethanol to acetic acid,

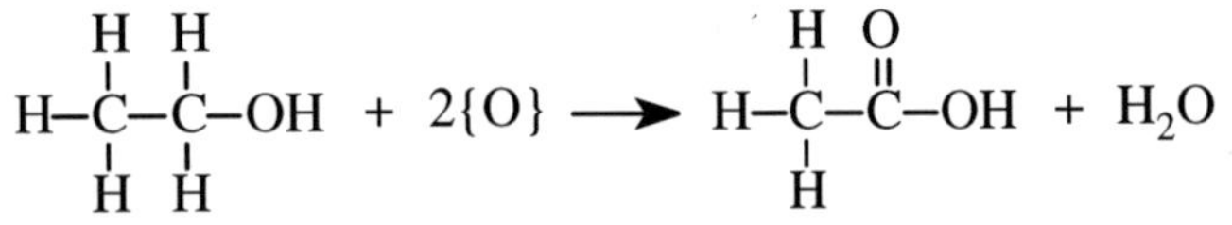

Ethanol Acetic acid (11.9.1)

where {O} is used to represent oxygen from some unspecified oxidant. Oxidation is one of the most common steps in chemical synthesis. A number of reagents are used as oxidants. Some of these reagents, such as potassium dichromate, $K_2Cr_2O_7$ are dangerous (dichromate salts are considered to be carcinogenic when inhaled for prolonged periods) and leave troublesome residues that require disposal.

Because of problems with oxidants that are commonly used, a major objective in the practice of green chemistry is to use more benign oxidants. Alternatives to the more traditional oxidant reagents include molecular oxygen (O_2), ozone (O_3), and hydrogen peroxide (H_2O_2), usually used with a suitable catalyst that enables the oxidation reaction to occur. Under the right conditions, hydrogen peroxide can be used as an alternative to elemental chlorine, Cl_2, a strong oxidant used in bleaching colored materials, such as paper pulp and cloth. Since chlorine is toxic (it was used as a poison gas in World War I) and has a tendency to react with organic compounds to produce undesirable chlorinated organic compounds, hydrogen peroxide is a much preferable bleaching agent.

In contrast to the usually harsh conditions under which chemical oxidations are carried out, organisms carry out biochemical oxidations under mild conditions. In so doing, they use monooxygenase and peroxidase enzymes that catalyze the oxidizing action of molecular oxygen or hydrogen peroxide. An area of significant interest in green chemistry is to perform such oxidations in biological systems or to attempt the use of catalysts that mimic the action of enzymes in catalyzing oxidations with molecular oxygen or hydrogen peroxide.

Reduction, which consists of loss of O, gain of H, or gain of electrons by a chemical species is also a common operation in chemical synthesis. As is the case with oxidants, the reagents used to accomplish reduction can pose hazards and produce undesirable byproducts. Such reductants include lithium aluminum hydride ($LiAlH_4$) and tributyltin hydride.

As an alternative to the potentially troublesome oxidation and reduction procedures using reagents, **electrochemistry** provides a reagentless means of doing oxidation and reduction. This is possible because an electrical current consists of moving electrons and oxidation consists of electron removal from a chemical species and reduction is addition of an electron. The passage of an electrical current between metal or carbon graphite electrodes through a solution resulting in oxidation and reduction reactions is called **electrolysis**. Consider the simplest possible case of electrolysis, that of water containing a non-reactive salt, such as Na_2SO_4, shown in Figure 11.3. At the **cathode**, where electrons (e^-) are pumped into the system and where reduction occurs, reduction of water occurs releasing H_2,

$$2H_2O + 2e^- \rightarrow H_2 + 2OH^- \qquad (11.9.1)$$

and at the **anode** where electrons are removed, O_2 is released as the water is oxidized:

$$2H_2O \rightarrow O_2 + 4H^+ + 4e^- \qquad (11.9.2)$$

In the setup shown, H^+ ion generated at the anode would eventually contact OH^- ion generated at the anode, and the two would react to produce water again. At the cathode, a dissolved chemical species could be reduced directly or the hydrogen generated could add to a species, reducing it. And at the anode another species could be oxidized directly by loss of electrons or the oxygen generated could add to a species, oxidizing it.

As another example of a commonly performed reaction that normally requires potentially troublesome reagents, consider **alkylation** with alkylating reagents in which an alkyl group, most frequently the $-CH_3$ (methyl) group, is added to an atom on an organic compound. The methylation reaction,

$$2R{-}N(H)_2 + H_3C{-}O{-}S(=O)_2{-}O{-}CH_3 + 2NaOH \longrightarrow 2R{-}N(CH_3)(H) + Na_2SO_4 + 2H_2O \qquad (11.9.3)$$

shows attachment of a methyl group to an amine group, $-NH_2$, that is part of an unspecified molecule represented "R." Methylation of nitrogen is used in a number of chemical syntheses including preparation of analgesics such as Ibuprofen. The dimethyl sulfate reagent used to accomplish the methylation poses toxicity problems in that it is a known primary carcinogen (a compound that does not require bioactivation to cause cancer). The reaction also produces a byproduct of Na_2SO_4, which if contaminated with dimethyl sulfate reagent may pose disposal problems.

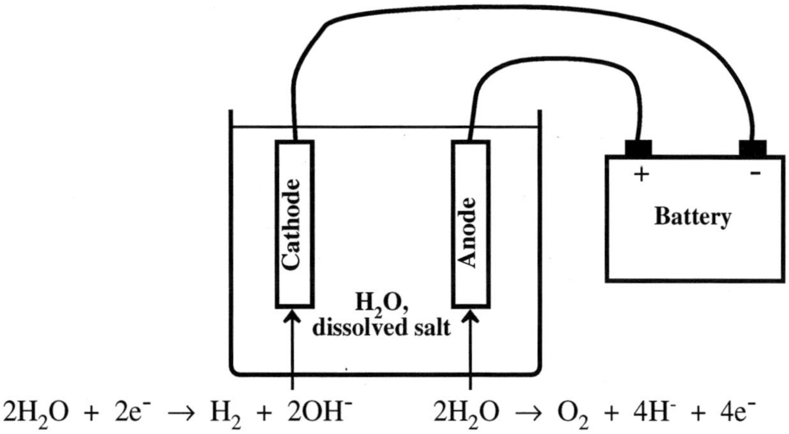

Figure 11.3. Apparatus for electrolysis in which a direct current of electricity is passed through a reaction medium, in this case water with a dissolved salt to make it electrically conducting. Reduction occurs when electrons are added to the medium at the cathode and oxidation when they are removed at the anode. Electrolysis is a reagentless way of doing oxidation and reduction.

Dimethyl carbonate prepared by reacting methanol, CH_3OH, with carbon monoxide, CO, in the presence of elemental oxygen and a copper salt catalyst has been developed as a green alternative to dimethyl sulfate as a methylating reagent. When dimethyl carbonate acts as a methylating agent,

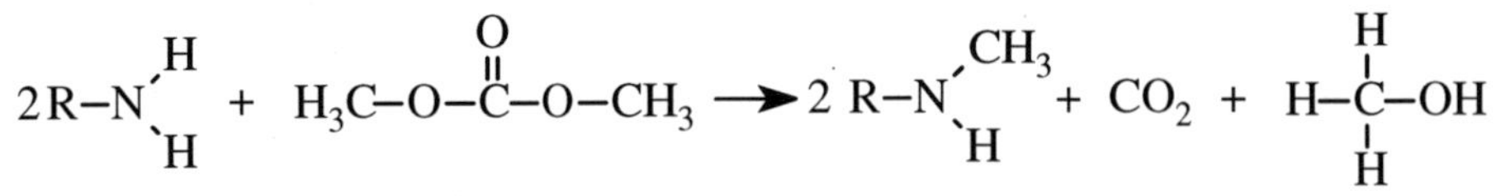

(11.9.4)

Dimethyl carbonate **Methanol**

methanol and innocuous carbon dioxide are generated as byproducts. The methanol can be recirculated through the system to generate addional dimethyl carbonate reagent.

Media

Chemical reactions are often carried out in **media**, usually organic **solvents** or water. The major role of a solvent in chemical reactions is to provide a medium in which feedstocks and reagents can dissolve and come into close, rapid contact at the molecular level. Substances dissolved in a solvent are **solvated** by binding of the solvent to the molecules or ions of the dissolved substance, the **solute**. Because of its polar nature and the ability to form hydrogen bonds (see Chapter 7, Section 7.1 and Figure 7.1), water is a particularly good solvent for ionic compounds — acids, bases, and salts. However, many organic feedstocks and reagents are not soluble in water or are decomposed by it, so organic solvents have to be used as reaction media.

Many environmental and health problems associated with making chemicals are the result of the use of organic solvent media. Hydrocarbon solvents will burn and their vapors are explosive. Although many hydrocarbon solvents are not particularly toxic, some can cause the condition of peripheral neuropathy mentioned in Section 11.8, and benzene is regarded as a carcinogen thought to cause leukemia. Released to the atmosphere, hydrocarbons can also participate in photochemical processes leading to the formation of photochemical smog (see Chapter 8, Section 8.10).

One approach to making chemical synthesis processes greener is to replace specific solvents with less hazardous ones. For this reason, toxic benzene solvent is replaced by toluene wherever possible. As shown by their structural formulas below, toluene has a methyl group, -CH_3, that benzene does not have. The methyl group in toluene can be acted upon by human metabolic systems to produce a harmless metabolite (hippuric acid) that is eliminated in the urine, whereas metabolic processes acting

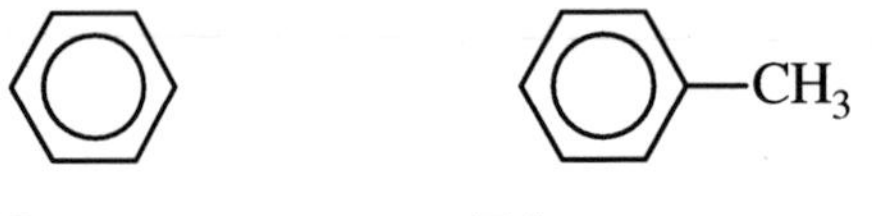

Benzene **Toluene**

upon benzene convert it to a toxic intermediate. As another example, *n*-hexane, which can cause peripheral neuropathy in exposed individuals, can be replaced with 2,5-dimethylhexane, which does not cause this condition, for reactions where the higher boiling temperature of the latter compound is not a problem.

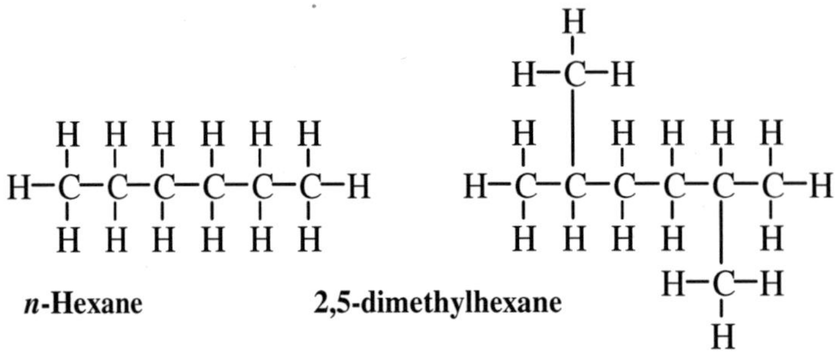

***n*-Hexane** **2,5-dimethylhexane**

Significant effort has been made to replace organic solvent reaction media with water, the greenest solvent Although water does not appreciably dissolve many nonpolar organic compounds, in some cases these may be suspended as very small colloidal particles in water, enabling close enough contact of organic materials to undergo reactions. Water is a good solvent for some of the biological materials, such as glucose, now favored as chemical feedstocks where they can be used.

At a high pressure above 73.8 atm (73.8 times normal atmospheric pressure at sea level) and a temperature exceeding 31.1° C, carbon dioxide becomes a **supercritical fluid**, a relatively dense state of matter in which there is no longer a distinction between liquid and gas. Supercritical carbon dioxide is a good solvent for organic compounds and can be used as a reaction medium for organic chemical reactions. An advantage of supercritical carbon dioxide in this application is that its viscosity is only about 1/30 that of common liquid organic solvents, which enables reactant species to migrate much faster through the fluid, thus speeding the reactions that they undergo. At temperatures and pressures below, but near those at which carbon dioxide becomes critical, it exists as separate gas and liquid phases while retaining many of the solvent properties of supercritical carbon dioxide. Under these conditions carbon dioxide is called a **dense phase fluid**, a term that also encompasses supercritical fluids.

Adjustment of the composition and conditions under which dense phase fluid carbon dioxide is maintained can provide significant variations in its solvent properties and adjustment of its ability to act as a reaction medium. In addition to variations in temperature and pressure, dense phase fluid carbon dioxide may be mixed with small quantities of other solvents, such as methanol, to further vary its solvent properties.

In addition to its solvent properties, dense phase fluid carbon dioxide offers the advantage of low toxicity and low potential for environmental harm (the small amounts of greenhouse gas carbon dioxide released from its application as a solvent are negligible compared to quantities released from combustion of fossil fuels). A big advantage of dense phase fluid carbon dioxide is its volatility, meaning that it separates readily from reaction products when pressure is released. Furthermore, carbon dioxide released from a reaction mixture can be captured and recycled for the same application. Carbon dioxide, itself, can be regarded as a green chemical because it can be obtained at low cost from biological fermentation processes.

Ionic liquids present another alternative to organic solvents for use as media for chemical synthesis. Inorganic salts consisting of ions, such as NaCl composed of Na^+ and Cl^- ions, are normally hard, high-boiling solids. However, when one or both of the ions are composed of large charged organic molecules, as shown by the cation in the example below,

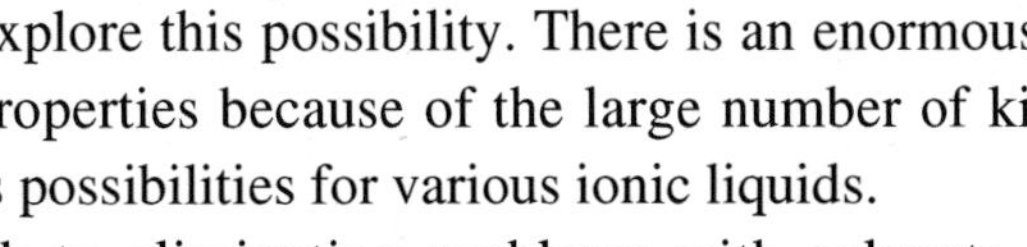

1-Butyl-3-methylimidazolium hexafluorophosphate

the salts can be liquids at room temperature and are called ionic liquids. These materials have the potential to act as suitable media in which substances can be dissolved and undergo reactions, and active research is underway to explore this possibility. There is an enormous variety of such ionic liquids with widely varying solvent properties because of the large number of kinds of ions that can be combined leading to almost limitless possibilities for various ionic liquids.

The ultimate approach to eliminating problems with solvents in chemical synthesis is to do reactions without solvents of any kind. Some reactions can be performed in which the reactants are simply mixed together or are held on solid supports, such as clays. Microwave heating of such reaction mixtures has proven effective in providing energy to enable reactions to occur rapidly.

Catalysts

Recall from Chapter 4, Section 4.5, that catalysts are materials that enable reactions to occur without themselves being changed. Large numbers of different catalysts are used in chemical processes and factors such as their potential toxicities, production of byproducts and contaminants, recycling, and disposal are matters of considerable importance in the chemical industry. Catalysts are divided into the two major categories of **heterogeneous catalysts** that are held upon some sort of support where they interact with reactants and **homogeneous catalysts** that are actually mixed with the reactants, often in solution in the media in which the reactions are carried out. Heterogeneous catalysts offer the advantage of being readily separated from reaction products, whereas homogeneous catalysts require measures such as distillation to remove them from the product. In many cases, however, homogeneous catalysts are much more effective in carrying out reactions. One of the objectives of green chemistry, therefore, has been to develop heterogeneous catalysts that equal homogeneous catalysts in their performance.

An important area of endeavor in the development of improved catalysts with respect to green chemistry is **selectivity enhancement**. Basically, this means developing a catalyst that is very selective in what it does, ideally making the right product and nothing else. A highly selective catalyst increases the percentage utilization of raw material and in so doing provides increased percent yield. Also, a highly selective catalyst decreases the amount of waste byproducts from undesired side reactions thereby providing increased atom economy.

Another important attribute of a good catalyst is related to the basic way in which a catalyst works, which is by lowering the activation energy that is required to make a reaction proceed at a significant rate. As a consequence, catalysts lower the total amount of energy that must be put into a chemical process to get it to occur. Lowered energy requirements are a basic part of the practice of green chemistry and in this respect good catalysts can be extremely beneficial in lowering costs and environmental impact.

Nature provides an abundance of catalysts that carry out a wide variety of chemical processes. These catalysts are the enzymes in organisms. Since life molecules are for the most part stable under conditions under which humans are comfortable, enzymes are expert in carrying out chemical processes efficiently under mild conditions near normal room temperature and atmospheric pressure. In consideration of this fact, a great deal of attention is being devoted to using organisms to carry out chemical processes. Bacteria have gotten particular attention because of their rapid growth, fast reproduction, and relative simplicity compared to multicelled organisms. By splicing desired genes for making specific enzymes into bacteria so that they will carry out desired reactions, genetic engineering has the potential of making an enormous contribution to the development of enzyme-catalyzed green chemical processes.

Chemists are trying to use enzymes as models for synthetic catalysts that have performance characteristics of enzymes, but which are much simpler and work under conditions that would destroy enzymatic catalysts. A promising area in which this might be accomplished is the use of iron-containing catalysts to oxidize alkene (C=C) groups in organic compounds using relatively mild hydrogen peroxide reagent, H_2O_2. Organisms accomplish this task using catalysts in which the Fe^{2+} ion is bonded by four N atoms in relatively large heme porphyrin molecules. The same oxidation has now been accomplished with a catalyst in which Fe^{2+} is bound by four N atoms by an organic molecule with the formidable name of N,N'-dimethyl-N,N'bis(2-pyridylmethyl)-ethylenediamine as shown in Figure 11.4. A big advantage of this catalyst that is shared with enzyme catalysts that enable peroxide oxidations is that it does not cause the decomposition of hydrogen peroxide as do a number of synthetic catalysts.

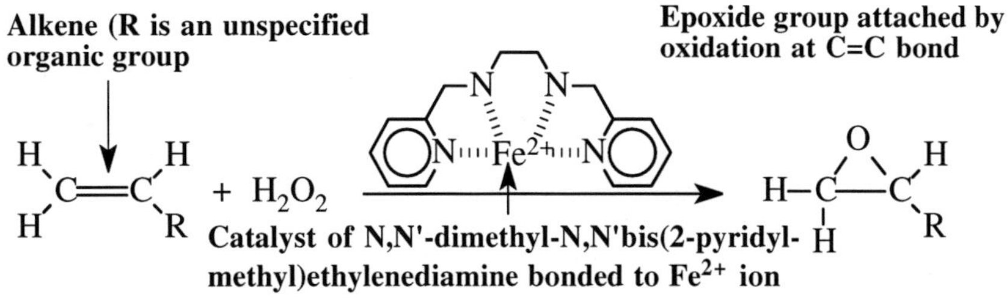

Figure 11.4. Action of an iron-containing compound modeled after iron-based enzymes as a catalyst to bring about the oxidation of an alkene hydrocarbon group with hydrogen peroxide.

SUPPLEMENTARY REFERENCES

McDonough, William, and Michael Braungart, *Cradle to Cradle: Remaking The Way We Make Things*, North Point Press, 2002.

QUESTIONS AND PROBLEMS

1. Define industrial ecology.
2. Define an industrial ecosystem.
3. Name four major parts of an industrial ecosystem.
4. Give the name of the processes to which materials and components are subjected in industrial ecosystems.
5. Give a definition of wastes in terms of natural resources.
6. What is the general pathway of materials through industrial systems as they currently operate?
7. What is meant by "level of recycling" and how is it related to embedded utility?
8. Name the three kinds of analyses and three categories considered in a life-cycle assessment.
9. What are the three kinds of products, classified in part on their amenability to recycling, that are normally considered in life-cycle assessments?
10. Give three important useful characteristics of consumable products related to their potential environmental effects.
11. Name three key attributes of industrial ecosystems that largely determine the well being of the systems.
12. Given that an abundant source of energy can make almost anything possible in an industrial ecosystem, in what respects do vast reserves of coal, wind power, and solar energy fall short of being ideal energy sources?
13. Explain cogeneration of energy. What are its advantages?
14. Name three approaches to providing materials other than from virgin sources.
15. Consumable items and products cannot be recycled on a practical basis. Name three other categories of goods or products that can be recycled.
16. What is Kalundborg, Denmark, noted for?

17. Name several characteristics that facilitated development of the Kalundborg industrial ecosystem.
18. Name an environmental effect of industrial activities in the anthrosphere upon each of the other four environmental spheres.
19. Name an environmental effect of agricultural activities in the anthrosphere upon each of the other four environmental spheres.
20. For each of the other four environmental spheres, name a measure that may be taken in anthrospheric activities to reduce the environmental impact of these activities.
21. What does atom economy have to do with reducing the environmental impact of industrial ecosystems?
22. Describe two general approaches to making chemical synthesis greener
23. What are structure activity relationships applied to toxic substances?
24. Why should the use of lipid-soluble organics be reduced in the practice of green chemistry?
25. What is a characteristic of VOCs that both makes them useful in industrial applications, but also increases their hazards?
26. Distinguish between reactive and corrosive substances. May a compound be both?
28. Distinguish among feedstocks, reagents, media, and catalysts.
29. Name two desirable characteristics of reagents insofar as their ability to generate products is concerned.
30. What are two oxidants used by organisms? What catalysts are used with these oxidants?
31. What is a nonchemical alternative to the use of oxidant and reductant reagents?
32. What are the two most common media used for chemical reactions?
33. Why is toluene preferred to benzene as a solvent medium for organic reactions?
34. What is a common toxicity problem with some lighter hydrocarbons, such as hexane?
35. Normally encountered as a gas, how can carbon dioxide be used as a medium for chemical reactions?
36. What are ionic liquids and what is their potential use in green chemistry?
37. Name two general categories of catalysts based upon physical form. Which is more desirable from the green chemistry standpoint?
38. What is a desirable characteristic of catalysts that chemists try to enhance?

12. FEEDSTOCKS: MAXIMUM UTILIZATION OF RENEWABLE AND BIOLOGICAL MATERIALS

12.1. SOURCES OF FEEDSTOCKS

Recall from Chapter 11, Section 11.9 that **feedstocks** are the main ingredients that go into the production of chemical products. Reagents act upon feedstocks and often the two are not readily distinguished. Feedstock selection largely dictates the reactions and conditions that will be employed in a chemical synthesis and is, therefore, of utmost importance in the practice of green chemistry. A feedstock should be as safe as possible. The source of a feedstock can largely determine its environmental impact, and the acquisition of the feedstock should not strain Earth's resources. The process of isolating and concentrating a feedstock can add to the potential harm of otherwise safe materials. This is true of some metal ores in which corrosive and toxic reagents (in the case of gold, cyanide) are used to isolate the desired material.

As a general rule, it is best if feedstocks come from renewable sources rather than depletable resources A biomass feedstock, for example, can be obtained as a renewable resource grown by plants on land, whereas a petroleum-based feedstock is obtained from depletable crude oil resources. However, the environmental tradeoffs between these two sources may be more complex than first appears in that the petroleum feedstock may simply be pumped from a few wells in Saudi Arabia, whereas the biomass may require large areas of land, significant quantities of fertilizer, and large volumes of irrigation water for its production. Another important decision is whether or not the feedstock should be made entirely from virgin materials or at least in part from recycled material.

For the manufacture of organic chemicals and the vast variety of products made from them, such as textiles, plastics, and rubber, the overwhelming choice of feedstock now is petroleum, amounting to all but about two percent of raw material used for organic chemical manufacture in United States. In many important respects petroleum is an ideal feedstock for this purpose; during the last 100 years it has been readily available and relatively inexpensive except during times of temporary supply disruption. There are, of course, disadvantages to the use of petroleum as a feedstock, not the least of which is the fact that eventually available supplies will become exhausted. The transportation and refining of petroleum consume large amounts of energy, amounting to more than 15 percent of total energy use in United States. Chemically, a consideration with the use of petroleum as a raw material is that the hydrocarbon molecules that compose petroleum are in a highly reduced chemical state. In order to be utilized as feedstocks, petroleum hydrocarbons often must be oxidized. The oxidation process (see Section 11.9) entails a net consumption of energy and often requires the use of severe and hazardous reagents. Although commonly used oxidation processes are remarkably well contained and safe, there is always the consideration of possible combustion and explosion hazards in the partial oxidation of petroleum.

Much of the challenge and potential environmental harm in obtaining feedstocks is in separating the feedstock from other materials. This is certainly true with petroleum, which consists of many different hydrocarbons, only one of which may be needed as the raw material for a particular kind of product. Cellulose from wood, which can be converted to paper and a variety of chemicals, is mixed intimately with lignin, from which it is separated only with difficulty. Some metals occur at levels of less than 1% in their ores, requiring energy-intensive means of separating out the metals from huge quantities of rock. The smelting of copper and lead ores releases significant quantities of impurity arsenic with the flue dust, which must be collected from the smelting operation. Indeed, this byproduct arsenic provides all the arsenic needed in commerce.

In evaluating the suitability of a feedstock, it is not sufficient to consider just the hazards attributable to the feedstock itself and its acquisition. That is because different feedstocks require different processing and synthetic operations downstream that may add to their hazards. If feedstock A requires use of a particularly hazardous material to convert it to product, whereas feedstock B can be processed by relatively benign processes, feedstock B should be chosen. This kind of consideration points to the importance of considering the whole life cycle of materials rather than just one aspect of them.

12.2. UTILIZATION OF FEEDSTOCKS

Before considering sources of feedstocks, it is useful to consider how those feedstocks can be used in the least polluting, most sustainable way possible. Feedstocks are modified by chemical processes to produce new chemical materials with commercial uses. The ideal feedstock is renewable and poses no hazards. And it can be converted to the desired product using few steps with 100% yield and 100% atom economy. This should be done with minimum quantities of reagent using only safe media in which the reaction occurs.

There are three major categories of reactions that are involved in chemical processing of feedstocks as shown in a general sense in Figure 12.1. In an **addition reaction**, all feedstock material becomes

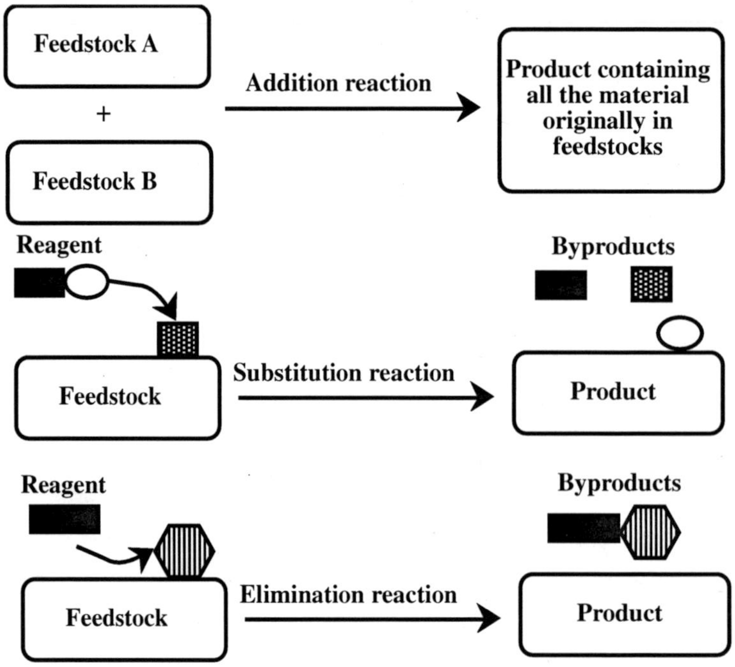

Figure 12.1. Illustration of three major categories of reaction processes by which feedstocks are acted upon by reagents to produce desired products.

part of the product and there are no byproducts. These are the best kinds of reactions from the viewpoint of green chemistry because, when they work ideally, there are no wastes. A **substitution reaction** uses a reagent to replace a functional group on the feedstock molecule. As its name implies, an **elimination reaction** removes a functional group from a feedstock molecule. Both of these latter kinds of reactions produce byproduct materials from the feedstock and from spent reagent. Their impacts can be reduced by reclaiming byproducts, if a use can be found for them, and by regenerating reagent, when that is possible. In some cases, elimination reactions can be carried out without use of a reagent, reducing the impact of this kind of reaction.

12.3. BIOLOGICAL FEEDSTOCKS

Organisms have provided a huge share of the materials used by humans throughout their existence. Trees have served as sources of wood for construction and firewood. Animals provided hides and fur to protect primitive humans from Ice Age cold. The American bison was the source of food, shelter, clothing, and a wide variety of other useful items to plains-dwelling Native Americans. Protein silk is obtained from the cocoons of insects, and protein wool from sheep and related animals.

Biomass, the plant material generated from photosynthesis is the leading candidate to replace petroleum as a feedstock for the organic chemicals industry. There are several major categories of biomass that can be used for feedstock:

1. Carbohydrate, which has the general formula of approximately CH_2O. Carbohydrate is the biomass that is produced initially as glucose sugar from water and carbon dioxide during photosynthesis. It is contained in the structural parts of plants as cellulose, a biopolymer.

2. Lignin, a biological polymer with a complex structure, which occurs with carbohydrate cellulose in woody parts of plants, binding fibers of cellulose together. Relatively few uses have been found for lignin, and it poses impurity problems in extracting cellulose for feedstock use.

3. Lipid oils extracted from seeds, including soybeans, sunflowers, and corn.

4. Hydrocarbon terpenes produced by rubber trees, pine trees, and some other kinds of plants.

5. Proteins, produced in relatively small quantities, but potentially valuable as nutrients and other uses.

Biological materials used as sources of feedstocks are usually complex mixtures, which makes separation of desired materials difficult. However, in some biological starting materials nature has done much of the synthesis. Most biomass materials are partially oxidized as is the case with carbohydrates, which contain approximately one oxygen atom per carbon atom (compared to petroleum hydrocarbons which have no oxygen). This can avoid expensive, sometimes difficult oxidation steps, which may involve potentially hazardous reagents and conditions. The complexity of biomass sources can make the separation and isolation of desired constituents relatively difficult.

There are several main pathways by which feedstocks can be obtained from biomass. The most straightforward of these is a simple physical separation of biological materials, such as squeezing oil from oil-bearing biomass or tapping latex from rubber trees. Only slightly more drastic treatment consists of extraction of oils by organic solvents. Physical and chemical processes can be employed to remove useful biomass from the structural materials of plants, which consist of lignocellulose composed of cellulose bound together by lignin "glue."

Carbohydrates are perhaps the most likely candidates as feedstocks for chemical processes. Carbohydrates come in several forms. Sucrose sugar, $C_{12}H_{22}O_{11}$, can be squeezed from sugar cane as sap and can be extracted from sugar beets and sugar cane with water. The exceptional photochemical productivity of sugar cane makes sucrose from this source an attractive option. Larger amounts of carbohydrates are available in starch, a polymer of glucose readily isolated from grains, such as corn, or from potatoes. An even greater source is found in cellulose, which occurs in woody parts of plants. It is relatively easy to break down starch molecules with the addition of water (hydrolysis) to give simple sugar glucose. Breaking down cellulose is more difficult, but can be accomplished by the action of cellulase enzymes.

Lipid oils are extracted from the seeds of some plants. Volatile solvents, most commonly the 6-carbon straight-chain alkane *n*-hexane, C_6H_{14}, are used to extract oils. In this process, the solvents are distilled off from the extract and recirculated through the process.

The hydrocarbon terpenes that occur in rubber trees can be tapped from the trees as a latex suspension in tree sap. Steam treatment and distillation can be employed to extract terpenes from sources such as pine or citrus tree biomass.

Grain seeds are rich sources of protein, almost always used for food, but potentially useful as chemical feedstocks for specialty applications. An exciting possibility just now coming to fruition in a practical sense is to transplant genes into plants so that they will make specialty proteins, such as medicinal agents.

12.4. FERMENTATION AND PLANT SOURCES OF CHEMICALS

In the provision of specialty and commodity chemicals and feedstocks, there are two main biological sources of materials. One of these consists of plants, which make huge quantities of cellulose and lesser quantities of other materials by photosynthesis. The other source is microorganisms, especially bacteria and yeasts.

Fermentation

Fermentation refers to the action of microorganisms on nutrients under controlled conditions to produce desired products. Fermentation for some products is anaerobic (absence of air) and for others aerobic fermentation is used. Fermentation processes have been used for thousands of years to produce alcoholic beverages, sauerkraut, vinegar, pickles, cheese, yogurt, and other foods. Ethanol, the alcohol in alcoholic beverages, is the most widely produced chemical made by fermentation. Lactic acid,

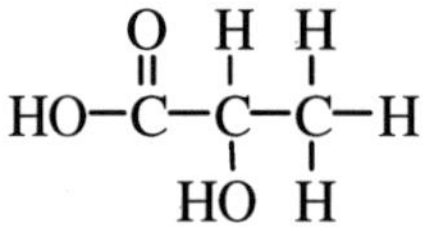

Lactic acid

has also been produced by fermentation processes for many years. More recently, fermentation has been applied to the production of a wide variety of organic acids, antibiotics, enzymes, and vitamins.

Starting in the 1940s, one of the major products of industrial fermentation has been penicillin, of which there are several forms. Figure 12.2 shows a simplified diagram of a facility for production of this life-saving antibiotic. Following penicillin, fermentation processes were developed for the production of several other significant antibiotics.

Selection of the appropriate microorganism is the most important consideration of a successful fermentation production process. The microorganisms have to have the proper nutrients, the choice of which can affect the kind and yield of the product. Sterile conditions must be maintained, and sterilization of equipment and media is accomplished by heating to 125–150° C for appropriate lengths of time. Air entering the fermenter must be filtered and sterilized. The temperature of fermentation is important, with fermentation rates increasing up to an optimum temperature, after which they decrease sharply with increased temperatures as the enzymes used by the microorganisms are destroyed (denatured). This kind of temperature relationship has increased interest in the use of thermophilic microorganisms that exist at boiling water temperatures in hot springs. If such organisms can be engineered to produce desired products, the rate of product generation may increase markedly. Both the levels of oxygen (which must be excluded from anaerobic processes) and pH must be controlled precisely. Modern fermentation processes use a variety of sensors to continuously monitor conditions in the fermentation tank and computerized control to accurately control all the parameters.

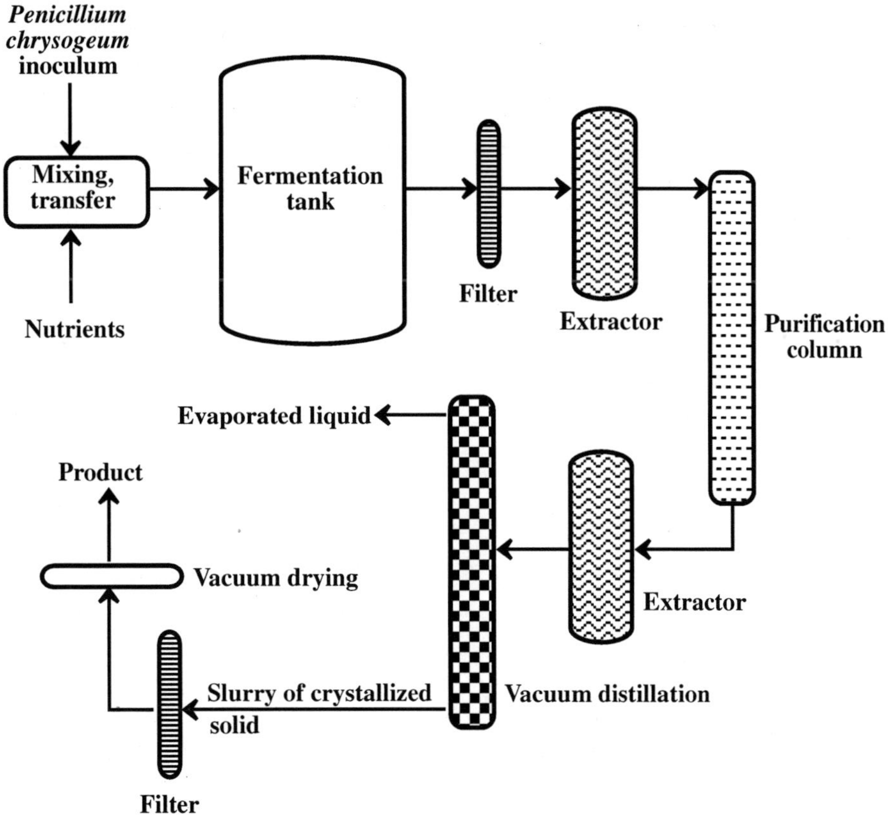

Figure 12.2. Simplified schematic diagram of the process for making penicillin by fermentation.

Fermentation is undergoing tremendous development with the use of transgenic microorganisms to which genes havc bccn transfcrrcd to make specific kinds of substances. The most common and valuable substances made by transgenic microorganisms consist of a variety of proteins. These include proteins and smaller molecule polypeptides that are used as pharmaceuticals. The best example of such a substance is human insulin, which is now produced in large quantities by transgenic microorganisms.

Until recently, fermentation has not been widely employed to make commodity chemicals used on a large scale. An exception is the large-scale production of ethanol from the fermentation of glucose sugar by yeasts. Now mandated as a gasoline additive in some parts of the U. S. by law, huge and growing quantities of ethanol are made by fermentation of glucose derived from corn and is an important market for corn. It is not clear that this is a truly green technology and some authorities believe that the energy consumed and the environmental damage from more intensive cultivation of corn outweigh the benefits of using this grain to produce ethanol fuel. Advances in transgenic microbiology have now raised the possibility of using fermentation for the production of a variety of chemicals and chemical feedstocks, several examples of which are discussed in this chapter.

Production of Materials by Plants

The uses of microorganisms operating in fermentation processes to generate commodity chemicals were discussed above. Plants are the other kind of organism that can be used for producing chemicals. Indeed, the nutrients used for fermentation processes come originally from plants. Fermentation is in a sense not a very efficient means of producing chemicals because of the consumption of nutrients to support the microorganisms and their reproduction and because of the generation of large quantities of byproducts. Plants, which generate their own biomass from atmospheric carbon dioxide and water are very efficient producers of materials. Wood and the cellulose extracted from it are prime examples of such materials.

In addition to their efficient production of biomass, plants offer distinct advantages in their production and harvesting. Genetics determine the materials that a plant makes, and once a crop is growing in a field, the products it is programmed for will be produced without fear of contamination by other organisms, which is always a consideration in fermentation. Plants can be grown by relatively untrained personnel using well known agricultural practices. Plant matter is relatively easy to harvest in the form of grains, stalks, and leaves, which can be taken to a specialized facility to extract needed materials.

The production of feedstocks and other chemical commodities from plants has been limited by the genetic restrictions inherent to plants. Now, however, transgenic plants can be bred to produce a variety of materials directed by genes transplanted from other kinds of organisms. For example, as discussed in Section 12.9, plants have even been developed to synthesize plastics. Another limitation of the production of materials by plants has been the mixture of these materials with other matter generated by plants. The intimate mixture of useful wood cellulose with relatively useless lignin is a prime example of this problem. Again, transgenic technology can be expected to be helpful in developing plants that produce a relatively pure product (such as the almost pure cellulose in cotton).

The potential of plants to produce useful products has been greatly increased by the development of hybrid plants with spectacular abilities to generate biomass by photosynthesis. Corn is one of the more productive field crops, and hybrid varieties produce large quantities of grain and plant biomass. Sugar cane is noted for its ability to produce biomass, some in the form of sugar, much more in the cane stalk biomass, which has relatively few uses, other than for fuel. One of the more exciting developments of productive hybrid plants is the hybrid poplar tree which, nourished by minimal amounts of fertilizer and watered by economical trickle irrigation systems, grows within a few years to a harvestable size for the production of wood pulp, and wood for plywood. The ability of these trees to generate cellulose that can be converted to glucose means that they may serve as the basis of an entire plant-based chemicals industry. The possibility exists that they can be genetically engineered to produce other chemicals as well.

12.5. GLUCOSE AS FEEDSTOCK

The glucose molecule,

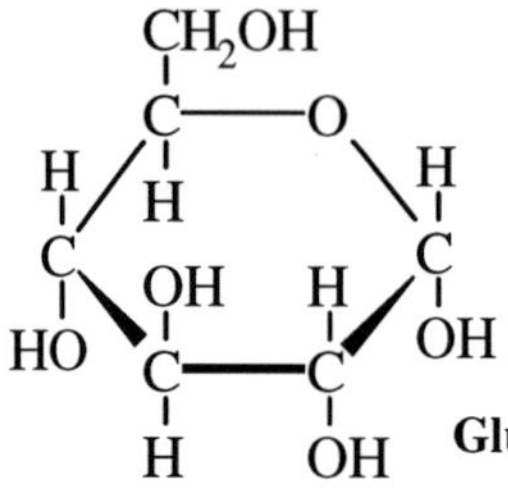

Glucose

provides a promising platform for a number of different organic syntheses. In addition to being produced in abundance by plants, glucose is a partially oxidized material, advantageous where a partially oxidized product is made. It also contains hydroxyl groups (-OH) around the molecule, which act as sites for the attachment of various functionalities. Glucose is metabolized by essentially all organisms, so it serves as an excellent starting point for biosynthesis reactions using enzymes, and it and many of its products are biodegradable, adding to their environmental acceptability.

Glucose can be obtained by enzyme-catalyzed processes from other sugars including sucrose and fructose. Virtually all of the glucose that is now used is obtained from the enzymatic hydrolysis of cornstarch. It is also possible to obtain glucose by the enzymatic hydrolysis of cellulose. The difficulty in doing this is compensated by the enormous quantities of cellulose available in wood and other biomass sources. The greatest use of glucose for synthesis is by fermentation with yeasts to produce ethanol,

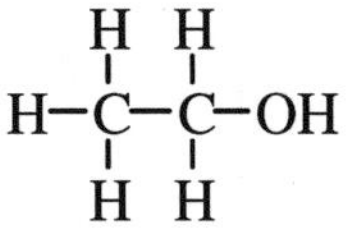
Ethanol

an alcohol widely used as a gasoline additive, solvent, and chemical feedstock. A byproduct of this fermentation process is carbon dioxide, the potential of which in green chemical applications as a supercritical fluid solvent are discussed in Section 11.9.

Glucose is widely used as a starting material for the biological synthesis of a number of different biochemical compounds. These include ascorbic acid, citric acid, and lactic acid. Several amino acids used as nutritional supplements, including lysine, phenylalanine, threonine, and tryptophan, are biochemically synthesized starting with glucose. The vitamins folic acid, ubiquinone, and enterochelin are also made biochemically from glucose.

In addition to the predominantly biochemical applications of glucose mentioned above, this sugar can be used to make feedstocks for chemical manufacture. The possibilities for so doing are now greatly increased by the availability of genetically engineered microorganisms that can be made to express genes for the biosynthesis of a number of products. Sophisticated genetic engineering is required to make chemical feedstocks because these are materials not ordinarily produced biologically. As an example of the potential of glucose for making important feedstocks, consider the synthesis from glucose of adipic acid,

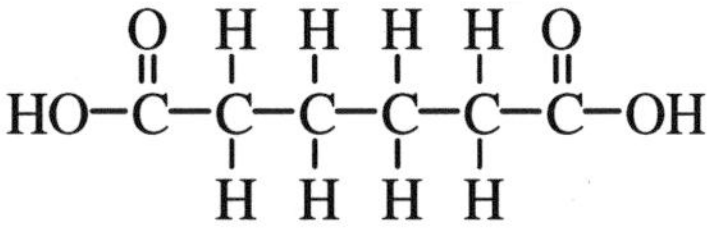
Adipic acid

a feedstock consumed in large quantities to make nylon. The conventional synthesis of this compound starts with benzene, a volatile, flammable hydrocarbon that is believed to cause leukemia in humans. The synthesis involves several steps using catalysts at high pressure and corrosive oxidant nitric acid, which releases air pollutant nitrous oxide, N_2O. The first step is the addition to benzene over a Ni/Al_2O_3 catalyst at a pressure 25 to 50 times atmospheric pressure of explosive hydrogen gas, H_2,

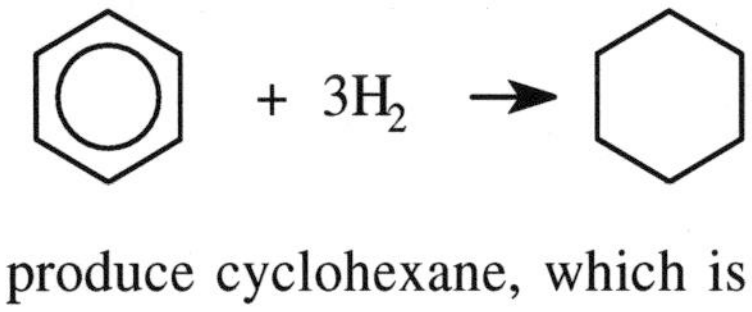
(12.5.1)

to produce cyclohexane, which is then subjected to oxidation in air at 9 atm pressure over a cobalt catalyst,

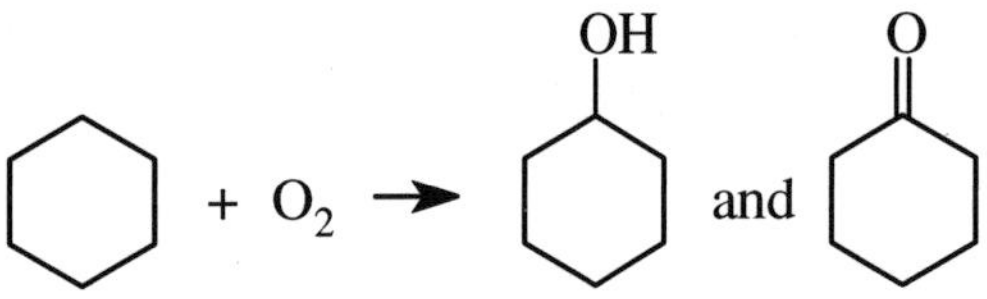

(12.5.2)

Cyclohexanol Cyclohexanone

to produce a mixture of cyclohexanol, a cyclic alcohol, and cyclohexanone, a cyclic ketone. This mixture is then reacted with oxidizing, corrosive, 60% nitric acid over a Ni/Al_2O_3 catalyst at 25–50 atm pressure to give the adipic acid feedstock:

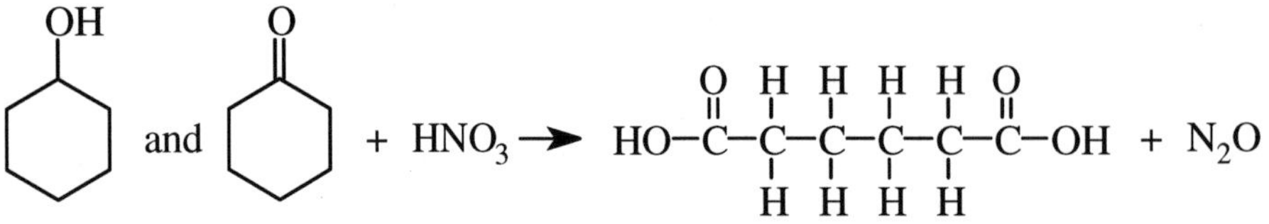

(12.5.3)

Throughout the synthesis process, elevated temperatures of approximately 250° C are employed. The N_2O released by the synthesis of adipic acid in the manufacture of nylon accounts for a significant fraction of worldwide N_2O releases. The potential dangers and environmental problems with this synthesis are obvious.

As an alternative to the chemical synthesis of adipic acid described above, a biological synthesis using genetically modified *Escherichia coli* bacteria and a simple hydrogenation reaction has been devised. The bacteria convert glucose to *cis,cis*-muconic acid:

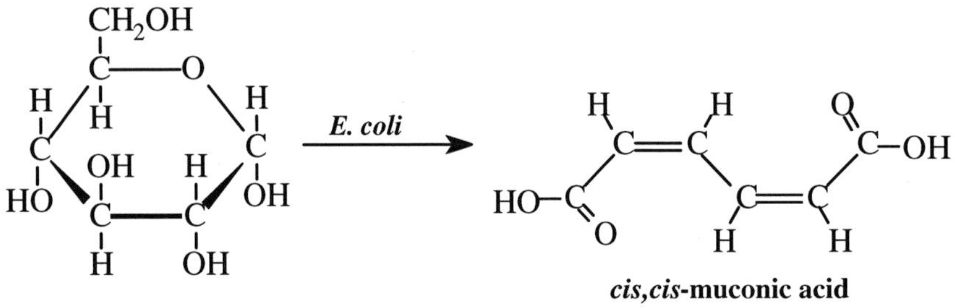

(12.5.4)

***cis,cis*-muconic acid**

The muconic acid is then treated under relatively mild conditions with H_2 under 3 atm pressure over a platinum catalyst to give adipic acid.

Another organic chemical that potentially can be produced by the action of transgenic microorganisms on glucose is catechol, used as a feedstock to make flavors, pharmaceuticals, carbofuran pesticide, and other chemicals. About 20 million kilograms per year worldwide of this compound are now manufactured chemically starting with propylene and carcinogenic benzene, both derived from depleting petroleum sources. Toxic phenol is generated as an intermediate, and it is oxidized to catechol with 70% hydrogen peroxide, which at this concentration is a violently reactive, hazardous oxidant. These steps require some rather severe conditions and stringent precautions in handling hydrogen peroxide reagent. *E. coli* bacteria of a genetically modified strain designated AB2834/pKD136/pKD9/069A, produce catechol from glucose and, if yields can be gotten to acceptable levels, biosynthesis could become a major source of this important chemical.

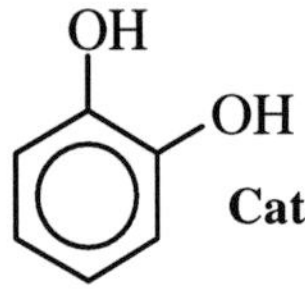

Catechol

Another potentially important organic feedstock that has now been synthesized from glucose using transgenic *E. coli* is 3-dehydroshikimic acid:

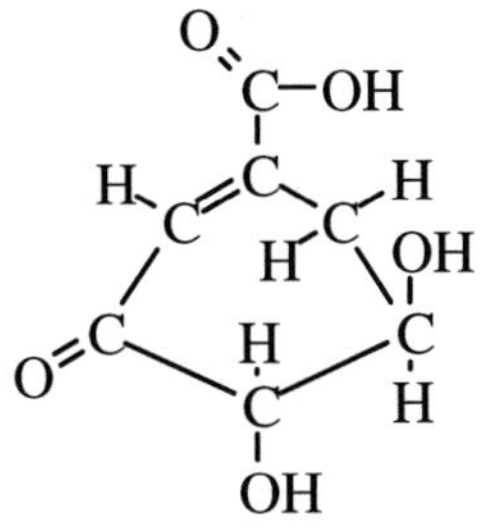

3-Dehydroshikimic acid

This compound is an important intermediate in the production of aromatic amino acids, gallic acid, vanillin, and other chemicals. It also has antioxidant properties. Antioxidants are organic compounds that react with oxygen-containing, reactive free radical species, such as hydroxyl radical, HO·. With their unpaired electrons (which make them free radicals), these species oxidize materials such as oils, fats, and lubricating oils and greases, causing deterioration in quality. By reacting with the free radicals, antioxidants stop their action. An abundant source of 3-dehydroshikimic acid could lead to its much wider application as an antioxidant.

12.6. CELLULOSE

The most abundant natural material produced by organisms is **cellulose** synthesized biologically by the joining of glucose molecules with the loss of 1 H_2O molecule for each bond formed (see Figure 12.3). This makes the chemical formula of cellulose $(C_6H_{10}O_5)_n$, where n ranges from about 1500 to 6000 or more. Most cellulose is made by plants, with total amounts exceeding 500 *billion* metric tons per year world-wide. Cellulose makes up the sturdy cell walls of plants. Wood is about 40% cellulose, leaf fibers about 70%, and cotton, one of the purest sources of cellulose, about 95%. Cellulose occurs in different forms and is always associated with hemicellulose (a material also composed of carbohydrate polymers) and lignin, a biopolymer of varied composition and bonding composed largely of aromatic units.

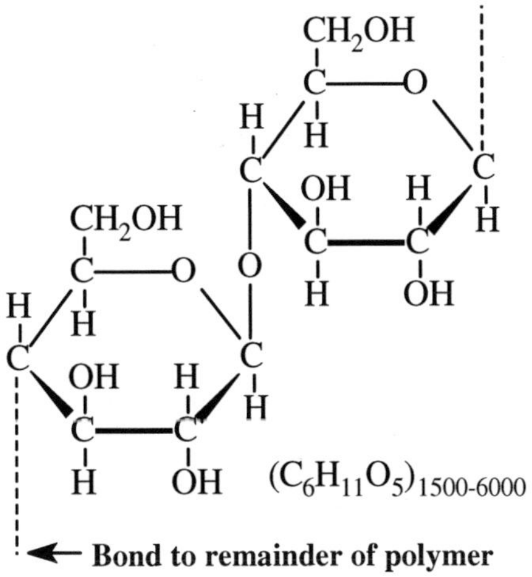

Figure 12.3. Segment of the cellulose molecule in which from 1500 to several thousand anhydroglucose units (glucose molecules less H_2O) are bonded together.

The first major step in cellulose utilization, such as extraction of cellulose fibers for making paper, consists of separating the cellulose from its matrix of lignocellulose (hemicellulose and lignin). This step has been the cause of many problems in utilizing cellulose because of the harsh chemical processing that has been employed. Lignin residues impart color to the cellulose, so wood pulp used in making paper has to be bleached with oxidants that alter the structure of the coloring agents. Bleaching used to be done almost entirely with elemental chlorine, Cl_2, and salts of hypochlorite ion, ClO^-, which produced chlorinated organic impurities and pollutants. Therefore, ozone and hydrogen peroxide are preferred bleaching agents.

A finely divided form of cellulose called **microcrystalline cellulose** is produced by appropriate physical and chemical processing of cellulose. This material has many uses in foods in which they impart smoothness, stability, and a quality of thickness and in pharmaceutical preparations and cosmetics. Added to food, indigestible cellulose contributes bulk and retains moisture.

Chemically modified cellulose is used to make a wide variety of materials. Like the glucose that comprises it, cellulose has an abundance of -OH groups to which various other groups can be bonded to impart a variety of properties. One of the oldest synthetic fabrics, rayon, is made by treating cellulose with base and carbon disulfide, CS_2, then extruding the product through fine holes to make thread. In a similar process, chemically treated cellulose is extruded through a long narrow slot to form a sheet of transparent film called cellophane.

As seen by the structure in Figure 12.3, each unit of the cellulose polymer has three -OH groups that are readily attached to other functional groups leading to chemically modified cellulose. One of the most common such products is cellulose acetate, an ester (see Section 5.4 and Reaction 5.4.1) used primarily for apparel and home furnishings fabrics in which most of the -OH groups on cellulose are replaced by acetate groups by reaction with acetic anhydride (see below):

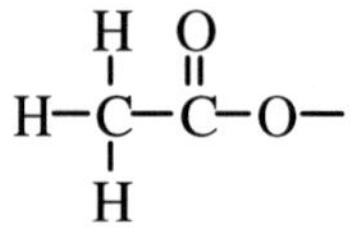

Acetate group

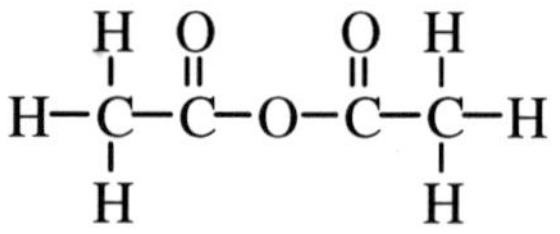

Acetic anhydride reagent

Although the cellulose feedstock for cellulose acetate synthesis is certainly a "green" material, acetic anhydride used to make the acetate is a corrosive, toxic chemical that produces poorly healing wounds on exposed flesh. Furthermore, potentially hazardous solvents, such as dichloromethane, are used in some processes for making cellulose acetate.

Another cellulose ester that has been widely manufactured is cellulose nitrate in which the -OH groups on cellulose are replaced by $-ONO_2$ groups by treating cellulose with a mixture of nitric acid (HNO_3) and sulfuric acid (H_2SO_4). Cellulose nitrate makes transparent film and was used in the early days of moving pictures for movie film. However, one of the other major uses of this material is as an explosive, so cellulose nitrate can burn violently giving off highly toxic fumes of NO_2 gas. In years past this characteristic has lead to several tragic fires involving human fatalities. Its use is now largely restricted to lacquer coatings, explosives and propellants. Although the cellulose raw material is green, neither the process for making cellulose nitrate involving strong acids, nor the flammable product would qualify as green.

From the discussion above, it is apparent that cellulose is an important raw material for the preparation of a number of materials. The reagents and conditions used to convert cellulose to other products are in some cases rather severe. It may be anticipated that advances in the science of transgenic organisms will result in alternative biological technologies that will enable conversion of cellulose to a variety of products under relatively mild conditions.

12.7. FEEDSTOCKS FROM CELLULOSE WASTES

Large quantities of cellulose-rich waste biomass are generated as byproducts of crop production in the form of straw remaining from grain harvest, bagasse residue from the extraction of sucrose from sugar cane, and other plant residues. This biomass represents a large amount of essentially free raw material that potentially can be converted to valuable chemical feedstocks. One way in which this can be done is by the use of enzyme systems to break the cellulose down into glucose sugar used directly as a feedstock (see Section 12.4) or fermented to produce ethanol. Direct conversion of cellulose wastes to feedstocks is another route. Fortunately, nature has provided efficient microorganisms for this purpose in the form of rumen bacteria that live in the stomachs of cattle and related ruminant animals. It has been found that these bacteria function well in large fermenters from which oxygen is excluded if the plant residues are first treated with lime ($Ca(OH)_2$ and $CaCO_3$), producing short-chain organic acids that exist as their calcium salts in the presence of lime.

The organic acids produced by rumen bacteria in animals are absorbed from the digestive systems of the animals and used as food. The acids produced in digesters are in the form of calcium salts, primarily calcium acetate, calcium propionate, and calcium butyrate. These materials can be processed to produce feedstocks for a variety of organic syntheses. Acidification of the salts yields the corresponding organic acids as shown by the structural formulas below:

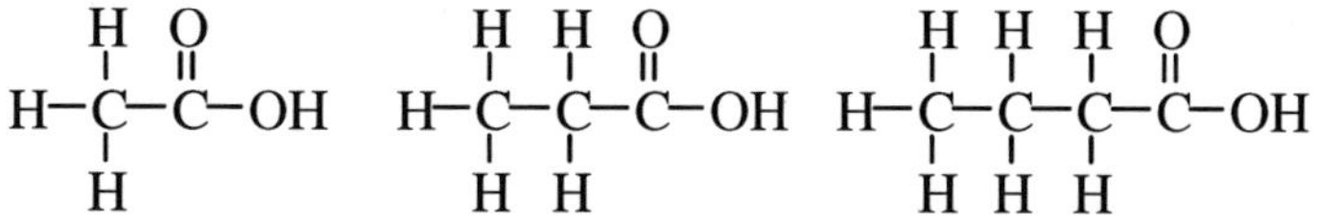

Acetic acid **Propionic acid** **Butyric acid**

Reaction of these acids with elemental hydrogen (hydrogenation) can be used to convert them to alcohols:

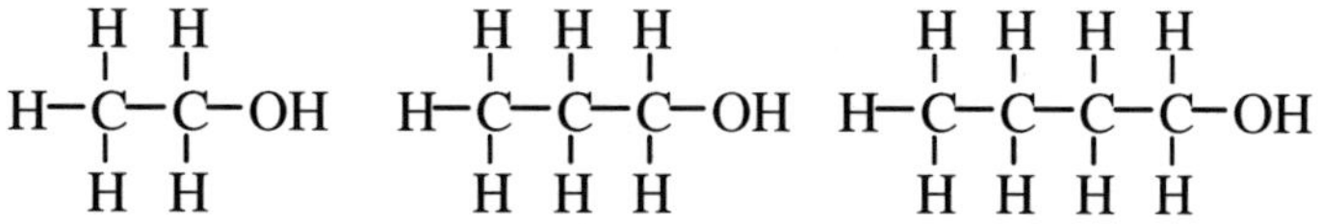

Ethanol **Propanol** **Butanol**

Heat treatment of the calcium salts of the organic acids at 450° C produces ketones, such as those shown below. These compounds are valuable feedstocks for a number of different chemical synthesis operations.

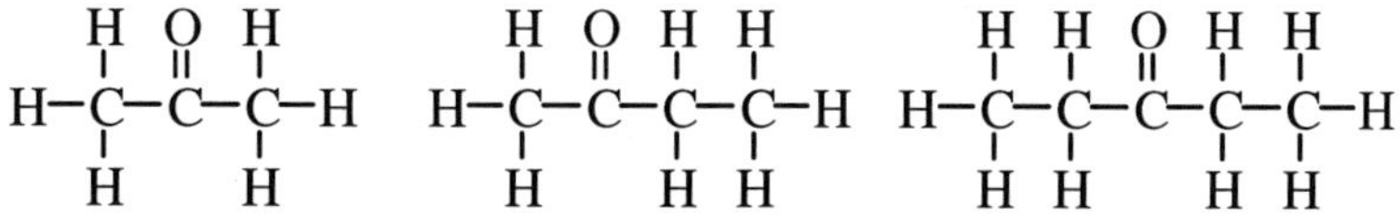

Acetone **Methylethyl ketone** **Diethyl ketone**

12.8. LIGNIN

Lignin, a chemically complex biopolymer that is associated with cellulose in plants and serves to bind cellulose in the plant structure, ranks second in abundance only to cellulose as a biomass material produced by plants. Lignin is normally regarded as a troublesome waste in the processing and utilization of cellulose. The characteristic that makes lignin so difficult to handle in chemical processing is its inconsistent, widely variable molecular structure as shown by the segment of lignin polymer in Figure 12.4. This structure shows that much of the carbon is present in aromatic rings that are bonded to oxygen-containing groups. Because of this characteristic, one of the potential uses of lignin is to produce phenolic compounds, which have the -OH group bonded to aromatic rings. The abundance of hydroxyl (-OH), methoxyl ($-OCH_3$), and carbonyl (C=O) groups in lignin also suggests potential chemical uses for the substance. A significant characteristic of lignin is its resistance to biological attack. This property, combined with lignin's highly heterogeneous nature makes it a difficult substrate to use for the enzyme-catalyzed reactions favored in the practice of green chemistry to give single pure products useful as chemical feedstocks.

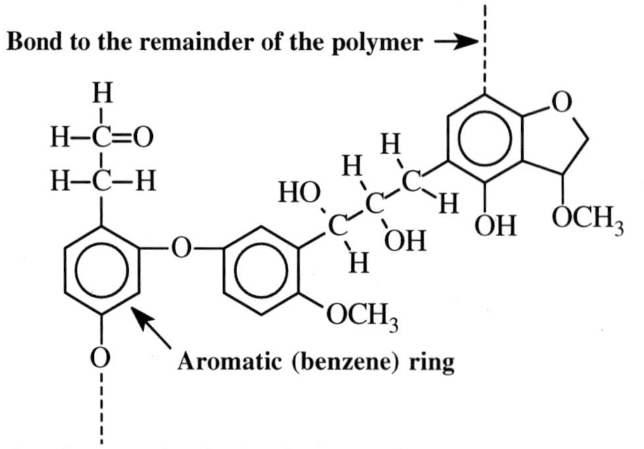

Figure 12.4. Segment of a lignin polymer molecule showing aromatic character and the disorganized, variable chemical structure that makes lignin a difficult material to use as a feedstock.

Lignin generated as a byproduct in the extraction of cellulose from wood is now largely burned for fuel. It has some uses for binders to hold materials together in coherent masses, fillers, resin extenders, and dispersants. There is also some potential to use lignin as a degradation-resistant structural material, such as in circuit boards.

12.9. DIRECT BIOSYNTHESIS OF POLYMERS

Cellulose in wood and cotton is only one example of the numerous significant polymers that are made biologically by organisms. Other important examples are wool and silk, which are protein polymers. A big advantage of these kinds of polymers from an environmental viewpoint is that polymers made biologically are also the ones that are most likely to be biodegradable. Attempts have been made to synthesize synthetic polymers that are biodegradable, These efforts have centered on those prepared from biodegradable monomers, such as lactic acid.

From the standpoint of green chemistry, it is ideal to have polymers that are made by organisms in a form that is essentially ready to use. Recently, interest has focussed on poly(hydroxyalkanoate) compounds, of which the most common are polymers of 3-hydroxybutyric acid:

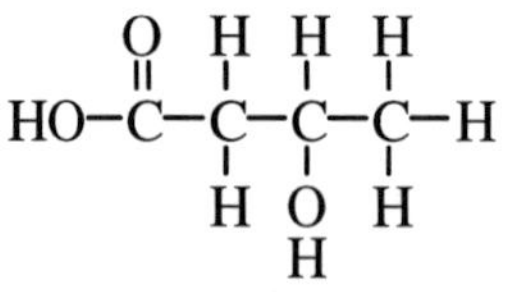

3-Hydroxybutyric acid

This compound and related ones have both a carboxylic acid (-CO_2H) and an alcohol (-OH) group. As discussed in Section 5.4 and shown in Reaction 5.4.1, a carboxylic acid can bond with an alcohol with the elimination of a molecule of H_2O forming an *ester linkage*. Since the hydroxyalkanoates have both functional groups, the molecules can bond with each other to form polymer chains:

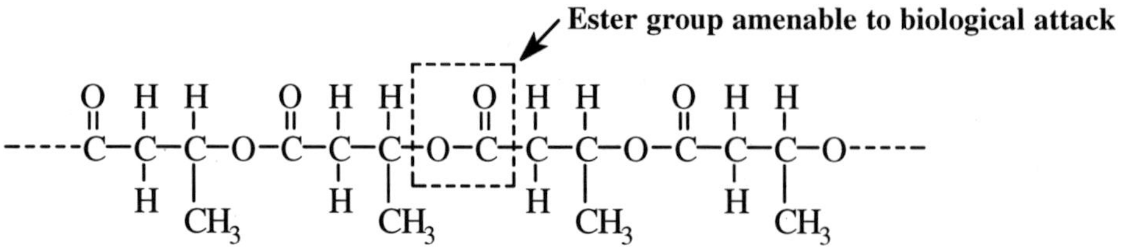

Segment of poly(3-hydroxybutyrate) polymer

Ester groups are among the most common in a variety of biological compounds, such as fats and oils, and organisms possess enzyme systems that readily attack ester linkages. Therefore, the poly(hydroxyalkanoate) compounds are amenable to biological attack. Aside from their biodegradability, polymers of 3-hydroxybutyric acid and related organic acids that have -OH groups on their hydrocarbon chains (alkanoates) can be engineered to have a variety of properties ranging from rubberlike to hard solid materials.

It was first shown in 1923 that some kinds of bacteria make and store poly(hydroxyalkanoate) ester polymers as a reserve of food and energy. In the early 1980s it was shown that these materials have thermoplastic properties, meaning that they melt when heated and resolidify when cooled. This kind of plastic can be very useful, and the thermoplastic property is rare in biological materials. One commercial operation was set up for the biological synthesis of a polymer in which 3-hydroxybutyrate groups alternate with 3-hydroxyvalerate groups, where valeric acid has a 5-carbon atom chain. This process uses a bacterium called *Ralstonia eutropia* fed glucose and the sodium salt of propionic acid (structure in Section 12.7) to make the polymer in fermentation vats. Although the process works, costs are high because of problems common to most microbial fermentation synthesis processes: The bacteria have to be provided with a source of food, yields are relatively low, and it is difficult to isolate the product from the fermentation mixture.

Developments in genetic engineering have raised the possibility of producing poly(hydroxyalkanoate) polymers in plants. The plant *Arabidopsis thaliana* has accepted genes from bacterial *Alcaligenes eutrophus* that have resulted in plant leaves containing as much as 14% poly(hydroxybutyric acid) on a dry weight basis. Transgenic *Arabidopsis thaliana* and *Brassica napus* (canola) have shown production of the copolymer of 3-hydroxybutyrate and 3-hydroxyvalerate. If yields can be raised to acceptable levels, plant-synthesized poly(hydroxyalkanoate) materials would represent a tremendous advance in biosynthesis of polymers because of the ability of photosynthesis to provide the raw materials used to make the polymers.

12.10. BIOCONVERSION PROCESSES FOR SYNTHETIC CHEMICALS

Most of the biochemical operations described so far in this chapter pertain to natural products which, by their nature, would be expected to be amenable to the action of enzymes. The mild conditions under which enzymes operate, the readily available, safe reagents that they employ, such as molecular O_2 for oxidations, and the high specificity of enzyme catalysts make biocatalyzed reactions attractive for carrying out chemical processes on synthetic chemicals, such as those from petroleum sources. This section discusses two examples of enzyme-catalyzed processes applied to chemical processes on synthetic chemicals that would otherwise have to be performed with chemical reagents under much more severe conditions.

p-Hydroxybenzoic Acid from Toluene

The potential for use of biosynthesis applied to synthetic chemicals can be illustrated by the synthesis of *p*-hydroxybenzoic acid,

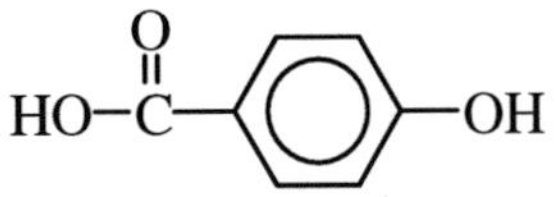

***p*-Hydroxybenzoic acid**

an important intermediate used in the synthesis of pharmaceuticals, pesticides, dyes, preservatives, and liquid crystal polymers.[1] It is currently made by reacting potassium phenolate,

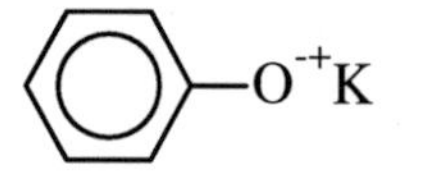

Potassium phenolate

with carbon dioxide under high pressure at 220° C, which converts slightly less than half of the potassium phenolate to the desired product and produces substantial impurities. The process dates back to the early 1860s almost 150 years ago, long before there were any considerations of pollutants and wastes. It requires severe conditions and produces metal and phenol wastes. Reactive alumina powder (Al_2O_3) used to catalyze the process has been implicated in a 1995 explosion at a facility to produce *p*-hydroxybenzoic acid that killed 4 workers.

A biosynthetic alternative to the synthesis described above has been attempted with *Pseudomonas putida* bacteria genetically engineered to carry out several steps in the synthesis of *p*-hydroxybenzoic acid starting with toluene. A key to the process is the attachment at the *para* position on toluene of a hydroxyl group by the action of toluene-4-monooxygenase (T4MO) enzyme system transferred to *Pseudomonas putida* from *Pseudomonas mendocina*:

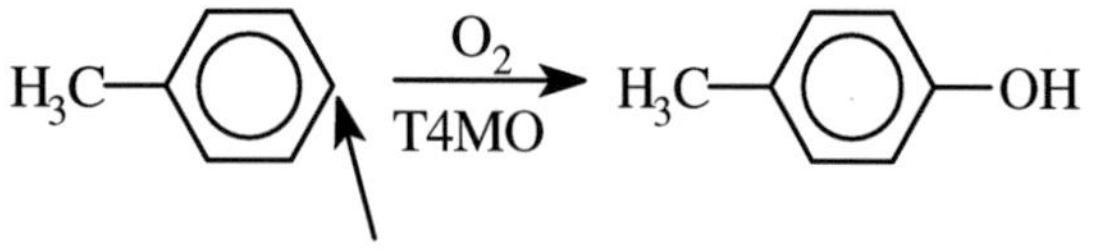

(12.10.1)

***Para* position on the aromatic ring**

The next step is carried out by *p*-cresol methylhydroxylase (PCMH) enzyme from a strain of *Pseudomonas putida* that yields *p*-hydroxybenzyl alcohol followed by conversion to *p*-hydroxybenzaldehyde:

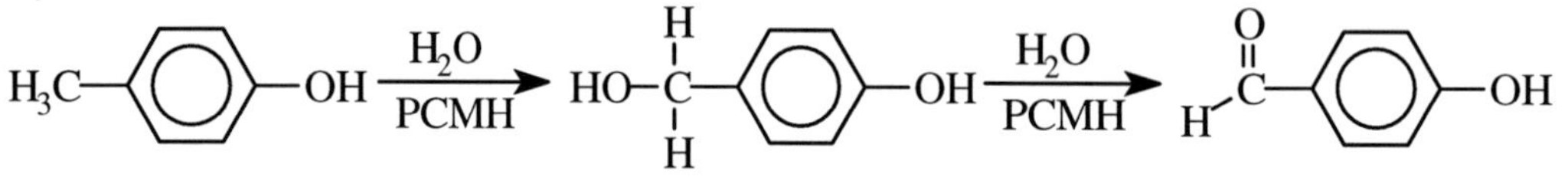

(12.10.2)

***p*-Hydroxybenzyl alcohol** **_p_-Hydroxybenzaldehyde**

The last step is carried out by an aromatic aldehyde dehydrogenase enzyme designated PHBZ also obtained from a strain of *Pseudomonas putida* and consists of the conversion of the aldehyde to the *p*-hydroxybenzoic acid product:

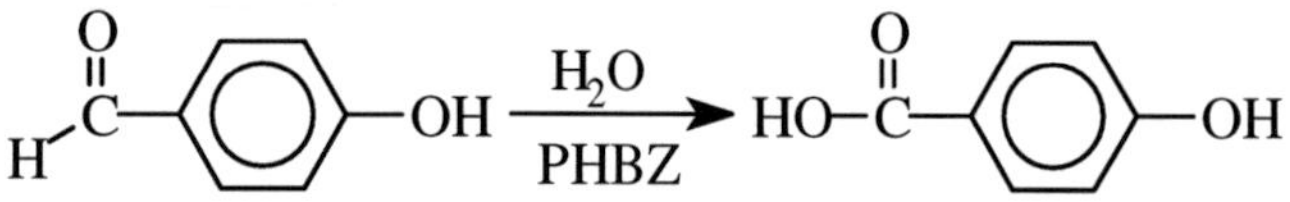

(12.10.3)

Through elegant genetic manipulation, the chemical processes described above were achieved leading to the desired product. In addition to providing the enzymes to carry out the desired steps, it was also crucial to block steps that would consume intermediates and give undesired byproducts that would consume raw material and require separation from the product. Although it is a long way from showing that the complex biochemical synthesis process actually gives the desired product to the final goal of having a practical process that can be used on a large scale, the results described above certainly show the promise of transgenic organisms in carrying out chemical syntheses.

Production of 5-Cyanovaleramide

The second biocatalyzed process to be considered is the conversion of adiponitrile to 5-cyanovaleramide. This conversion was required for the synthesis of a new chemical used for crop protection. This process can be carried out chemically with a stochichiometric mixture of adiponitrile with water and a manganese dioxide catalyst under pressure at 130° C as shown by the following reaction:

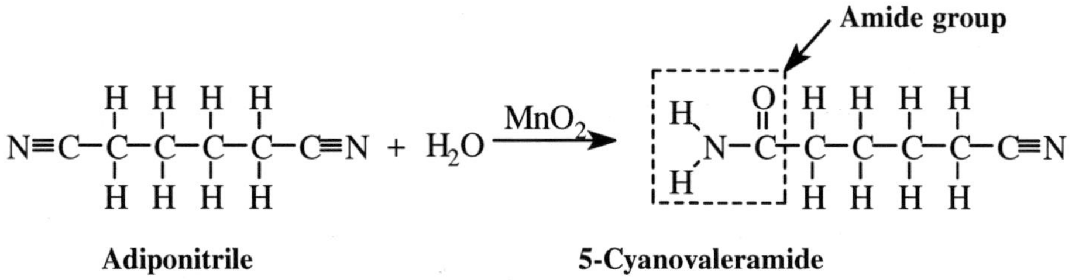

(12.10.3)

If the reaction is run to 25% completion, an 80% selectivity for the 5-cyanovaleramide is achieved, with the other fraction of the adiponitrile that reacts going to adipamide, in which the second C≡N functional group is converted to an amide group. Carrying the reaction beyond 25% completion resulted in unacceptable levels of conversion to byproduct adipamide.

The isolation of the 5-cyanovaleramide product from the chemical synthesis described above entails dissolving the hot reaction mixture in toluene solvent, which is then cooled to precipitate the product. The unreacted adiponitrile remains in toluene solution from which it is recovered to recycle back through the reaction. For each kilogram of 5-cyanovaleramide product isolated, approximately 1.25 kg of MnO_2 required disposal; this is definitely not a green chemical process!

As an alternative to the chemical synthesis described above, a biochemical synthesis was developed using organisms that had nitrile hydratase enzymes to convert the C≡N functional group to the amide group.[2] The microorganism chosen for this conversion was designated *Pseudomonas chloroaphis* B23. The cells of this organism were immobilized in beads of calcium alginate, the salt of alginic acid isolated from the cell walls of kelp. It was necessary to run the process at 5° C, above which temperature the enzyme lost its activity. With this restriction, multiple runs were performed to convert adiponitrile to 5-cyanovaleramide. During these runs, 97% of the adiponitrile was reacted, with only 4% of the reaction going to produce byproduct adipamide. The water-based reaction mixture was simply separated mechanically from the calcium alginate beads containing the microorganisms, which were then recycled for the next batch of reactant. The water was distilled off of the product to leave an oil, from which the 5-cyanovaleramide product was dissolved in methanol, leaving adipamide and other byproducts behind. In contrast to the enormous amount of waste catalyst produced in the chemical synthesis of 5-cyanovaleramide, only 0.006 kg of catalyst waste residue was produced per kg of product. And the waste microbial catalyst was 93% water, so its disposal was not a problem.

LITERATURE CITED

1. Miller, Edward S., and Steven W. Peretti, "Bioconversion of Toluene to *p*-Hydroxybenzoate Using a Recombinant *Pseudomonas Putida*," *Green Chemical Syntheses and Processes*, Oxford University Press, New York, 2000, pp. 126–142.

2. DiCosimo, Robert, "Biocatalytic Production of 5 Cyanovaleramide from Adiponitrile," *Green Chemical Syntheses and* Processes, Oxford University Press, New York, 2000, pp. 114–125.

QUESTIONS AND PROBLEMS

1. Discuss advantages that biological feedstocks have over petroleum. Are there disadvantages?

2. What is a fundamental chemical difference between petroleum and biological feedstocks?

3. Name some characteristics of an ideal feedstock.

4. Name three kinds of reactions used in processing feedstocks. Which is best from the viewpoint of green chemistry?

5. Name several categories of biomass that can be used for feedstocks. Which of these is the least useful?

6. How are oils extracted from plant sources?

7. Use chemical formulas to make the argument that carbohydrates are a more oxidized chemical feedstock than hydrocarbons.

8. What are the two main biological sources of materials?

9. Name some categories of chemicals routinely produced by fermentation.

10. Which pharmaceutical material has been produced by fermentation for many years?

11. What is the first, most important consideration in developing a fermentation process for production of a chemical?

12. What is the significance of temperature in fermentation processes? What happens if temperature is too high?

13. Which chemical is made in largest quantities by fermentation?

14. In which fundamental respect are plants more efficient producers of material than fermentation?

15. Which relatively recent advance in biotechnology has greatly increased the scope of materials potentially produced by plants?

16. Why are hybrid poplar trees particularly important in the production of raw materials?

17. Describe the structural characteristics of glucose and other carbohydrates that make them good platforms for chemical synthesis.

18. Give a disadvantage and an advantage of the use of cellulose as a source of glucose.

19. List some of the hazards associated with the chemical synthesis of adipic acid used to make nylon.

20. Give a major concern with the use of benzene as a feedstock.

21. What is a chemical characteristic of 3-dehydroshikimic acid that could lead to much greater uses for it?

22. Although the chemical formula of glucose is $C_6H_{12}O_6$, that of the cellulose polymer made from glucose is $(C_6H_{10}O_5)_n$ where n is a large number. Since cellulose is made from glucose, why is the cellulose formula not $(C_6H_{12}O_6)_n$?

23. Why is wood pulp consisting mostly of cellulose, treated with oxidants? Which oxidants are preferred, and which has lost favor?

24. Give some examples of useful chemically modified cellulose. Which of these has proven to be rather dangerous?

25. In ruminant animals that have bacteria in their stomachs that digest cellulose, the rumen bacteria and the organic acids they generate are passed on through the digestive tract where the bacterial biomass is dissolved, with the products and the organic acids previously generated absorbed by the animal as food. Suggest why basic limestone is used in the large batch processes that use rumen bacteria in digesters to produce organic acids from cellulose.

26. Why is it difficult to deal with lignin as a chemical feedstock?

27. What is the current main use of waste lignin?

28. Give the main advantage of biopolymers from an environmental viewpoint.

29. Which structural feature of hydroxyalkanoates enables them to make polymeric molecules?

30. What was the original source of poly(hydroxyalkanoate) polymers? How is it now proposed to produce them?

31. Although enzymes have not developed specifically to act upon synthetic compounds, they have some specific advantages that make them attractive for carrying out chemical processes on synthetic compounds. What are some of these advantages?

32. Name two chemicals for which it has been shown that enzymatic processes can actually convert synthetic raw materials to chemical products normally made by nonbiological chemical reactions.

13 TERRORISM, TOXICITY, AND VULNERABILITY: CHEMISTRY IN DEFENSE OF HUMAN WELFARE

13.1. VULNERABILITY TO TERRORIST ATTACK

Terrorist attacks upon human targets have become a constant fear in modern times. In the United States, vulnerability to such attacks were illustrated in horrifying detail by the suicide attacks by hijacked commercial aircraft on the New York World Trade Center on September 11, 2001. Other nations have long lived in the shadow of threats from groups that would cause them harm. Throughout the world, the possibility of deliberate attacks upon people, their support systems, and the anthrospheric infrastructure have come to be the greatest concern facing large numbers of people.

Chemicals and chemistry figure prominently in considerations of terrorist actions. The sudden release of a huge amount of chemical energy from a mixture of ammonium nitrate (a common agricultural fertilizer) and diesel fuel brought down the Alfred P. Murrah Oklahoma City Federal Building in 1995 with the loss of dozens of lives. Powerful explosives strapped to the bodies of suicide bombers have killed 20 or more people at a time in attacks in Israel. The extreme toxicity of military poison nerve gases is a constant concern in subways and other locations where large numbers of people are packed into small spaces. Biochemistry applied to recombinant DNA science may enable production of particularly virulent disease pathogens, such as vaccine-resistant smallpox. The accidental release of methyl isocyanate in an industrial chemical accident in Bhopal, India, in 1984 killed more people than even the 2001 attack on the World Trade Center. At least 243 people died from hydrogen sulfide contained in natural gas released from a pressurized deposit of this lethal mixture penetrated by a drilling operation in the Chuandongbei natural gas field of southwestern China in December, 2003. Hundreds of people were made ill and thousands were evacuated. A massive fire resulted when the escaping gas was ignited to convert the hydrogen sulfide (H_2S) to toxic, but much less lethal sulfur dioxide, SO_2.

Terrorist activities are not confined to direct attacks upon humans. The environment is susceptible to terrorist activities and may be severely damaged by them. For example, a major nuclear war — arguably the ultimate form of terrorism — could contaminate large areas of land and other parts of the environment with radioactive materials and, in the worst case scenario, could do substantial harm to the global climate resulting in a "nuclear winter."

So, what can green chemistry do to prevent terrorist attacks and mitigate their effects? Actually, green chemistry is a key discipline in such endeavors. For example, one of the basic tenets of green chemistry is to use the safest possible chemicals as safely as possible. When particularly dangerous chemicals are not made or used, they are not available to cause mischief. The practice of green chemical manufacturing calls for minimizing the accumulation of hazardous chemicals and seeks to eliminate hazardous chemical wastes. Safer materials made under the practice of green chemical technology minimize hazards from more dangerous substances. Highly sensitive analytical techniques developed by chemical science can be used to detect miniscule quantities of explosives or toxic substances slated for use in terrorist attacks. Biochemistry and recombinant DNA science have the potential to enable the development of better vaccines against pathogenic biological warfare agents and antidotes to chemical and biological toxins. More subtly, the use of green chemistry and chemical technology to produce effective substitute materials can reduce potential for "resource blackmail" that can lead to vulnerability to terrorist activity. A prime example is the substitution of biomass alternatives for petroleum feedstocks that to a certain extent many nations must obtain from other nations that are not necessarily friendly.

This chapter addresses potential terrorist threats with emphasis upon those that employ chemical and biological agents. Having identified threats that may occur, it then discusses ways in which chemistry, especially the proper practice of green chemistry, can minimize such threats.

13.2. PROTECTING THE ANTHROSPHERE

The anthrosphere constructed with a high degree of human ingenuity has provided a generally safe and comfortable environment for human beings. The underpinning of this entire support system is the **infrastructure**, which includes systems to purify and deliver water, electricity generation and distribution systems, communications, fuel distribution networks, highways, and railroads. The sophistication and complexity of the infrastructure is what makes it work so smoothly (for the most part), but also makes it vulnerable to attack. A key aspect of this susceptibility to attack is **vulnerability due to interconnectivity**, which arises from the high degree to which various parts of the infrastructure are interconnected and mutually dependent.[1] No part of the infrastructure illustrates vulnerability due to interconnectivity more so than modern electrical grids, which can stretch across vast geographical regions and across national borders. A reminder of this vulnerability occurred on August 14, 2003, when a failure of the electrical grid in the northeastern U.S. and southeastern Canada resulted in a power outage for tens of millions of people in New York City, Detroit, Cleveland, and Toronto. In this incident a total of 68,100 megawatts of generating capacity — equivalent to 68 very large, modern power generating facilities — was lost as dozens of high-voltage transmission lines shut down. This occurred within about 5 minutes, and the event that caused it probably took only about 10 seconds. The precipitating event was not terrorism — although it well could have been — and probably resulted from nothing more dramatic than tree limbs interfering with transmission lines.

The electrical power failure described above illustrates a phenomenon called **cascading failures** on complex networks.[2] An electrical power distribution system is a complex network of hundreds of electrical power plants interconnected by electrical transmission lines. Such a system enables very efficient generation, distribution, and utilization of electrical power in that a surge in demand on one part of the system can be compensated by redistribution of power from the network to that segment of the system. Therefore, local generating facilities do not have to have the capacity to meet temporary demand, resulting in high efficiency and much lower costs. Other parts of the infrastructure have similar networks. Routers on the Internet are designed to route Internet traffic around bottlenecks or other routers that are temporarily out of commission. Modern manufacturing operations using "just-in-time" deliveries of components make it unnecessary to stockpile large quantities of parts, thereby minimizing the need for production capacity and maximizing efficiency. The downside is that these systems operate "close to the edge" so that a relatively small failure, such as one brought about by human mischief, can rapidly cascade into a major failure.

Chemistry can be applied to infrastructure protection. One area in which this is true is the production of materials that resist heat and flame. Such materials used in buildings can provide substantial protection from fire. Processes that are consistent with the practice of green chemistry also provide protection from attack. For example, green chemistry attempts to reduce the production and use of hazardous materials. Sophisticated analytical chemistry and analytical instrumentation can be used to detect agents of attack before damage is done.

13.3. SUBSTANCES THAT EXPLODE, BURN, OR REACT VIOLENTLY

Substances that react violently and rapidly enough to cause fires, explosions, or corrosive destruction of materials are those that have been used most commonly in terrorist attacks and that have the most potential for future use. Reactive substances, such as explosives used to quarry rock, have many commercial uses and are therefore widespread and potentially readily available, both to legitimate users and terrorists.[3]

Explosives are the most common materials used in terrorist attacks. The devastating effects of explosives have been illustrated by a number of incidents including the 1995 bombing of the Murrah Federal Building in Oklahoma City, the 2003 bombing of the British consulate in Turkey, and frequent bombings of various sites in Iraq in 2003/2004. Explosives can be made from readily available materials; the Murrah Federal Building was brought down by a mixture of ammonium nitrate fertilizer and diesel fuel. Gunpowder has long been prepared from charcoal, sulfur, and nitrate salts. Nitroglycerin, the explosive in dynamite, is made from glycerin, a byproduct of soap preparation, reacted with nitric and sulfuric acids (most amateurs who attempt this synthesis succeed in blowing their heads off). More sophisticated explosives consist of organic compounds containing nitrogen and oxygen, such as 2,4,6-trinitrotoluene (TNT), 1,3,5-trinitro-1,3,5-triazacyclohexane (RDX), and pentaerythritol tetranitrate (PETN). Figure 13.1 gives structural formulas of several explosives that could be used by terrorists.

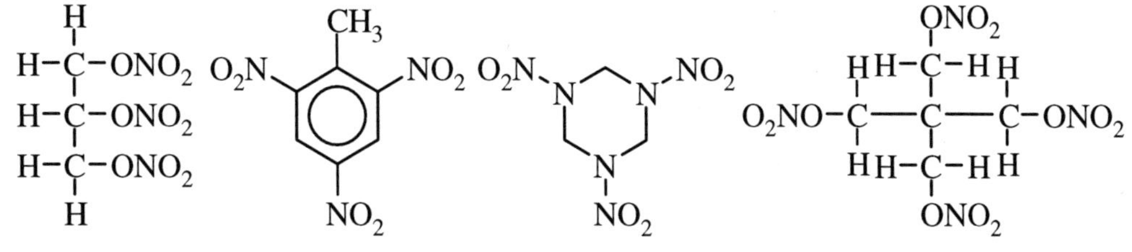

Figure 13.1. Structural formulas of four explosive compounds that have the potential for use in terrorist attacks. Note that all contain an abundance of oxygen bound with nitrogen

The potential of flammable substances to cause death and destruction was shown by the jet-fuel-fed fires that brought down both towers of the New York World Trade Center in the 9/11/01 attack. Fires set on ruptured petroleum pipelines have caused significant destruction in Iraq in 2003/2004. When mixed with air, vapors of flammable liquids can cause massive explosions. Adding to the hazard of flammable substances are oxidants that release oxygen to react with combustible materials. In 1997, oxidant sodium chlorate, $NaClO_3$, an ingredient of emergency oxygen generators in aircraft that were being transported contrary to regulations in the cargo hold of a Valujet airplane, caused a devastating fire of aircraft tires that brought the aircraft down in the Florida Everglades.

By their nature, fuels are flammable substances, as are some common industrial solvents. Flammable fuels and solvents are used throughout almost all societies and are readily available to those who would use them for malevolent ends. Adding to their availability is their transport by truck, rail, and pipeline. The hazard of flammable substances is increased by the ease by which they and their vapors can be distributed through sewers, elevator shafts, subway tunnels, and other conduits.

Corrosive substances that destroy materials and flesh have been used in attacks on equipment and people. Concentrated sulfuric acid, which dehydrates and destroys flesh, has been used by criminals to blind people. Corrosive materials can be used to damage relays and other devices in communications equipment as a means of sabotage.

A major concern with respect to flammable, reactive, and explosive substances is their widespread industrial use. Actually, such materials are relatively safe inside of manufacturing plants and properly secured storage areas. The greater threat comes from their transport. This is illustrated by very frequent transportation accidents involving rail cars, trucks, barges, and pipelines that result in explosions, fires, and release of corrosive materials. Hijacking of trucks transporting hazardous materials and even trucks driven by terrorists are a particular concern. The practice of industrial ecology and green chemistry can help minimize such threats by, for example, promoting the production of hazardous substances in minimal quantities where needed and as needed. "Just-in-time" production minimizes storage of hazardous substances.

13.4. TOXIC SUBSTANCES AND TOXICOLOGY

One of the greater concerns that the general public has with chemistry is the potential toxic effects of various substances including those that could be used for terrorist attacks. **Poisons**, or **toxicants**, are substances that can adversely affect biological tissue leading to harmful responses including, in the severest cases, even death. The study of such substances and their effects is the science of **toxicology**. The science that relates the chemical properties of toxic substances to their toxic effects is **toxicological chemistry**. Because poisons are among the leading terrorist threats, it is appropriate to consider toxic substances and toxicological chemistry here.

Toxic Responses

Any kind of tissue and all organs can be the subject of attack by toxic substances. The major human organ systems that are potentially adversely affected by toxic substances are given in Table 13.1.

Toxicities

The toxicities of substances vary over a wide range, and those that are toxic at lowest doses are of most concern for deliberate poisoning. This is illustrated in Figure 13.1 which gives the toxicities of several substances. It is important to note that the dosage scale in this figure is logarithmic; that is for each division *decrease* on the scale, a substance is *ten times* as toxic. The two circles in Figure 13.1 illustrate the enormous differences between toxicities of different substances. If the area of the large circle represents the size of a fatal dose of parathion, a once widely used insecticide that has killed a number of people and has now been banned because of its toxicity, a fatal dose of military poison nerve gas Sarin is represented by the miniscule dot below the circle! Toxicities are normally expressed as LD_{50} values, the dose in units of mass of poison per unit mass of test subject. Rats are usually used for tests, and toxicities to humans are inferred from these test values.

Metabolism of Toxic Substances

Toxic substances that enter the body and that are foreign to it, commonly called **xenobiotic substances**, are subject to metabolic processes that may activate them or make them less toxic (detoxification). The metabolism of toxic substances may be divided into two phases. Phase I reactions normally consist of attachment of a functional group, usually accompanied by oxidation. For example, benzene, C_6H_6, (see Chapter 5, Section 5.2) is oxidized in the body by the action of the cytochrome P-450 enzyme system as shown in Reaction 13.2.1:

Table 13.1. Major Target Systems of Toxic Substances

Target system	Typical toxic responses
Respiratory system	Emphysema from cigarette smoke, pulmonary edema (fluid accumulation), lung cancer from asbestos
Skin responses	Allergic contact dermatitis, such as from exposure to dichromate; chloracne from exposure to 2,3,7,8-tetrachlorodibenzo-*p*-dioxin ("dioxin"); skin cancer from exposure to coal tar constituents
Hepatotoxicity (toxic effects to the liver)	Steatosis (fatty liver), such as from exposure to carbon tetrachloride; cirrhosis (deposition and buildup of fibrous collagen tissue) from excessive ingestion of ethanol; haemangiosarcoma, a type of liver cancer observed in workers heavily exposed to vinyl chloride in polyvinylchloride plastic manufacture
Blood	Carboxyhemoglobin formation from binding of carbon monoxide to blood hemoglobin, methemoglobinemia consisting of conversion of iron(II) to iron(III) in hemoglobin from exposure to substances such as aniline or nitrobenzene, aplastic anemia from exposure to benzene
Immune system effects	Immunosuppression from exposure to radiation, hypersensitivity from exposure to beryllium
Endocrine system effects	Disruption of endocrine function by endocrine disruptors such as bisphenol-A
Nervous system	Encephelopathy (brain disorder), such as from exposure to lead; peripheral neuropathy from exposure to organic solvents; inhibition of acetylcholinesterase enzyme in nerve function by exposure to organophosphate military poisons
Reproductive system	Interference with sperm development by some industrial chemicals, interference with cells involved with egg formation by chemicals such as cyclophosphamide
Kidney and urinary tract system	Nephrotoxicity to the kidney by heavy metal cadmium

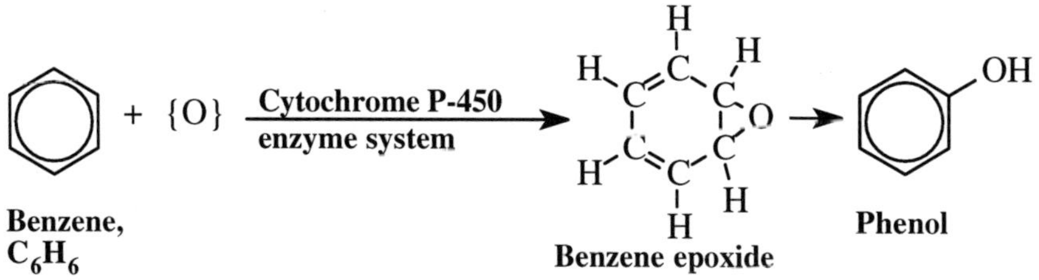

(13.2.1)

The Phase I oxidation product is phenol, a toxic substance. A reactive intermediate in the process is benzene epoxide, which interacts with biomolecules to cause toxic effects. The phenol Phase I oxidation product of benzene may undergo a second reaction, a Phase II reaction in which it is bound with a conjugating agent that is endogenous to (produced naturally by) the body, such as glucuronide:

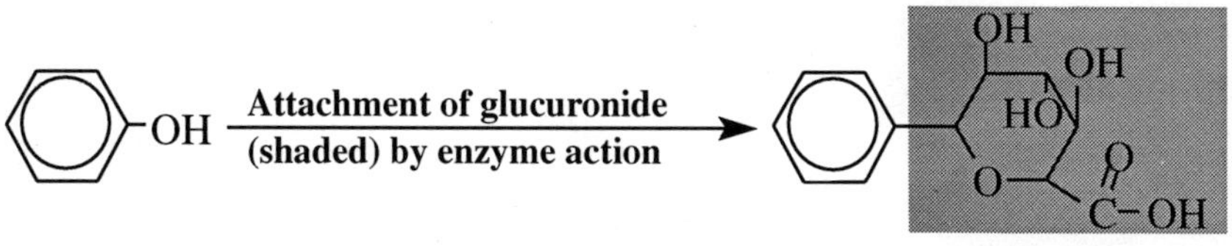

(13.2.2)

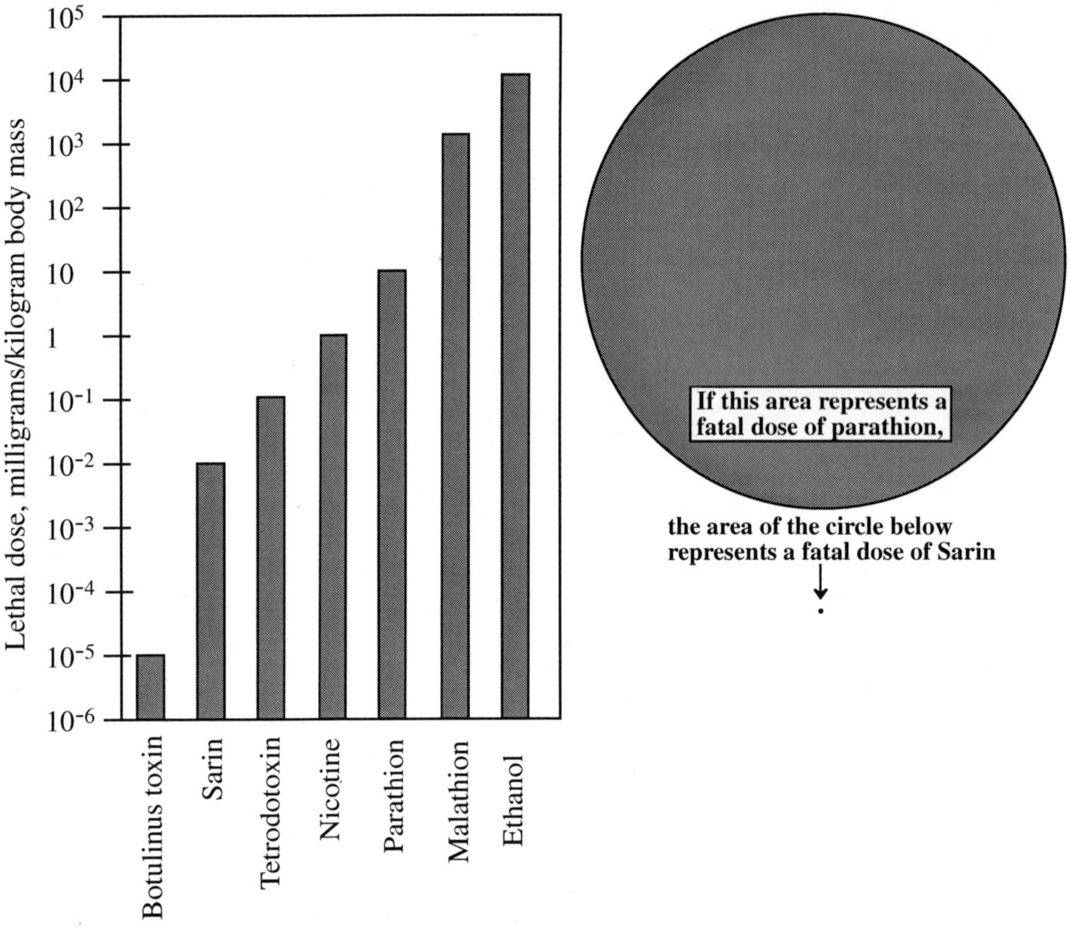

Figure 13.2. Relative toxicities of various substances in units of milligrams dose per kilogram of body mass required to kill 50% of test subjects (LD_{50}). Also shown are the relative toxicities of parathion, a once commonly used insecticide that has been discontinued because of its toxicity to humans, and Sarin, an extremely toxic military poison.

Although Phase I and Phase II reactions generally act to make xenobiotic substances more water soluble, more readily eliminated from the body, and less toxic, in some cases, the opposite occurs and metabolic processes make substances more toxic. Most known human carcinogens (cancer-causing agents) are actually produced by biochemical processes in the body from noncarcinogenic precursor substances.

The Action of Toxic Substances

Toxic substances, which, as noted above, are often produced by metabolic processes from nontoxic precursors, produce a toxic response by acting upon a **receptor** in the body. Typically, a receptor is an enzyme that is essential for some function in the body. As a consequence of the binding of the receptor to the toxicant there is a **biochemical effect**. A common example of a biochemical effect occurs when a toxicant binds to an enzyme such that the bound enzyme may be inhibited from carrying out its normal function. As a result of a biochemical effect, there is a response, such as a behavioral or physiological response, which constitutes the actual observed toxic effect. Acetylcholinesterase enzyme inhibited by binding to nerve gas Sarin may fail to stop nerve impulses in breathing processes, leading to asphyxiation. The phenomena just described occur in the **dynamic phase** of toxicant action as summarized in Figure 13.3.

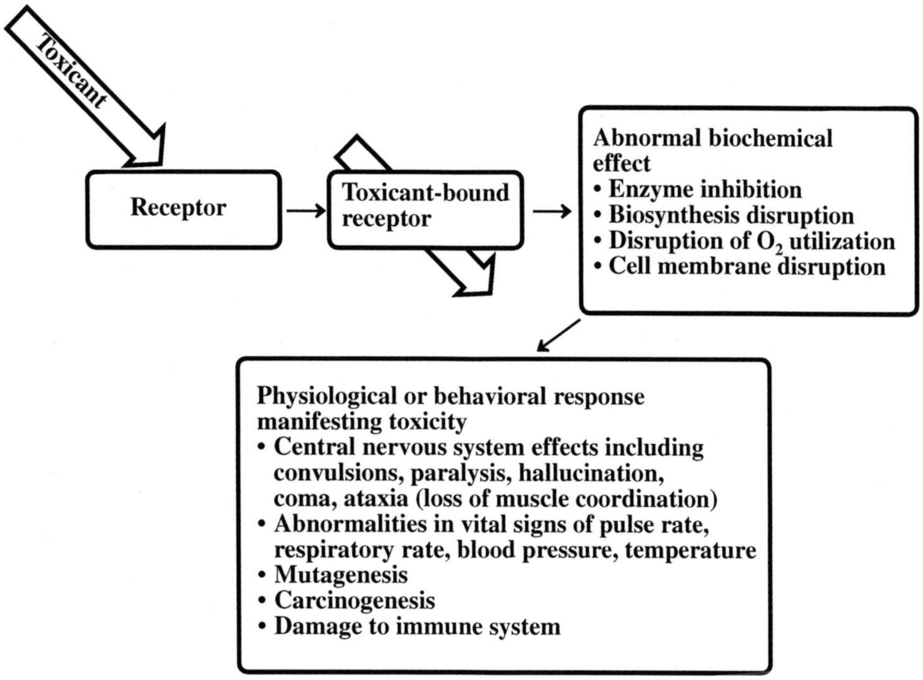

Figure 13.3. The dynamic phase of toxicity in which a toxicant interacts with a receptor in the body to cause a biochemical alteration that is manifested in a toxic response.

13.5. TOXIC CHEMICAL ATTACK

Bhopal

The tragedy that arguably illustrates most vividly the potential of chemicals to cause major loss of life was the accidental release of methyl isocyanate from a chemical manufacturing operation in Bhopal, India. This disaster occurred during the night of December 2/3, 1984, exposing many victims as they slept in their homes. A total of about 40 tons of this chemical was released during the incident exposing thousands of residents in the surrounding area. Of those exposed, more than 3000 died, primarily from pulmonary edema (fluid accumulation in the lung). Immunological, neurological, ophthalmic (eye), and hematological effects were also observed. Because of its high vapor pressure and toxicity to multiple organs,

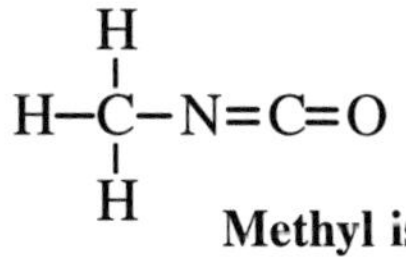

Methyl isocyanate

methyl isocyanate is the most toxic of the isocyanates. An interesting aspect of methyl isocyanate toxicity is its ability to cross cell membranes and reach organs far from the site of exposure, despite its very high reactivity.

It is unlikely that methyl isocyanate or of any any other substance with similar toxicity could be obtained and delivered to potential victims in large enough quantities and with sufficiently effective delivery to cause widespread death and illness in a deliberate attack. Nevertheless, the magnitude of the Bhopal catastrophe vividly illustrates the potential of toxic substances to cause harm and the incident serves as a grim reminder of the potential of toxic chemicals, especially those that may be airborne, to be used in attacks. And the Bhopal incident, which occurred as the result of a tragic accident, serves as a grim warning of the potential of terrorist attacks upon chemical plants to cause widespread harm.

Potential Chemical Agents

Although an unlikely "weapon of mass destruction," **carbon monoxide**, **CO**, can certainly cause fatal poisonings. It has killed thousands of people accidentally and through suicide. Given the fact that carbon monoxide is odorless and provides essentially no warning of exposure, it should be regarded as a potential weapon.

To understand the action of carbon monoxide, it is important to realize that blood is carried from the lungs to tissue by **hemoglobin**, a large-molecule protein in red blood cells that contains iron(II) (the Fe^{2+} ion) bound with nitrogen. This functionality exchanges oxygen,

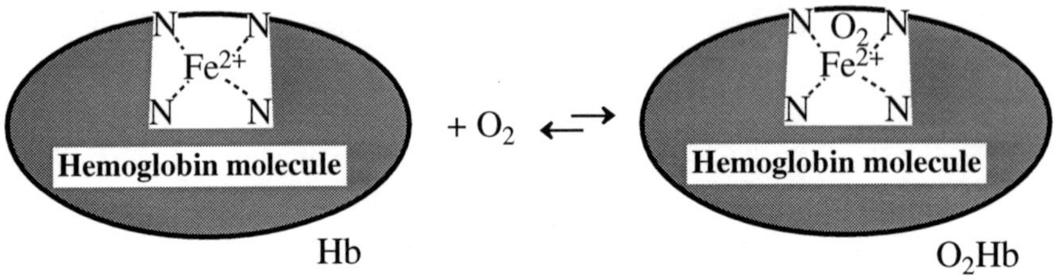

converting hemoglobin, Hb, into oxyhemoglobin, O_2Hb, which carries oxygen to tissues where it is released to be used for metabolic processes. When carbon monoxide is present in inhaled air, the oxygen bound to hemoglobin is displaced,

$$O_2Hb + CO \leftrightarrow COHb + O_2 \tag{13.5.1}$$

producing **carboxyhemoglobin**, COHb. This species is not only useless for carrying oxygen, it is much more stable than oxyhemoglobin, O_2Hb, so that a relatively low concentration of carbon monoxide will convert enough of the hemoglobin to carboxyhemoglobin to cause a serious oxygen deficiency. Rapid death ensues from inhalation of air containing 1,000 parts per million (ppm) carbon monoxide, and unconsciousness results from inhaling 250 ppm CO. Dizziness, headache, and fatigue result from inhalation of 100 ppm CO and levels as low as 10 ppm can impair visual perception and judgment.

Chlorine (Cl_2) gas could potentially be used in terrorist attacks because of its wide availability for water disinfection and other uses. Illustrative of this potential is the fact that chlorine was the first substance used as a military "poison gas" in World War I. Chlorine is a strong oxidizer that reacts with water, including water in tissue, to produce an oxidizing, acidic solution that is especially damaging to respiratory (lung) tissue. Air containing only 10–20 ppm chlorine causes acute discomfort to the respiratory tract and brief exposure to 1,000 ppm of Cl_2 can be fatal.

Hydrogen cyanide, HCN, is a potentially devastating gaseous pollutant. It has been used to carry out death sentences in gas chambers, causing death very rapidly when inhaled. Another toxic form of cyanide is **cyanide ion**, CN^-, in salts such as KCN. Only about 60 mg of KCN will kill a human. Glass ampules containing KCN were used by some doomed Nazi leaders to commit suicide near the end of World War II. There is concern that potassium cyanide or other soluble cyanide salts may be put into water supplies as toxic agents.

The metabolic action of cyanide depends upon its strong binding for iron in the +3 oxidation state. In the essential utilization of molecular oxygen in the body (the respiration process called oxidative phosphorylation) iron cycles between iron(III) in ferricytochrome oxidase enzyme and iron(II) in the chemically reduced form, ferrouscytochrome oxidase. By stabilizing ferricytochrome oxidase, cyanide stops this cycle, preventing utilization of oxygen and causing metabolic processes to cease. It is interesting to note that an antidote to cyanide poisoning — in those rare instances where the victim survives long enough for antidotes to be administered — is to have the victim inhale a volatile nitrite ester. The nitrite converts a fraction of the iron(II) in hemoglobin to iron(III), generating methemoglobin. This form of hemoglobin cannot carry oxygen, but it can bind with cyanide, preventing it from tying up ferricytochrome oxidase enzyme.

Hydrogen sulfide, H_2S, is a colorless gas with a foul, rotten-egg odor that is as toxic as hydrogen cyanide and may kill even more rapidly. Inhalation of 1000 ppm hydrogen sulfide causes rapid death from respiratory system paralysis. Nonfatal doses of this gas can cause excitement due to damage to the central nervous system; headache and dizziness may be symptoms of exposure.

Military Poisons and Nerve Gas Agents

Beginning with the use of toxic chlorine gas in World War I, nations have developed a variety of diabolical toxic agents to disable and kill opposing troops in war. One such agent is **mustard oil**, chemical name bis(2-chloroethyl)sulfide:

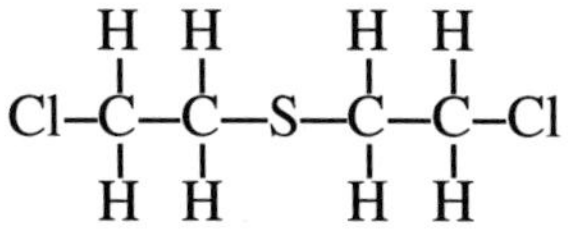
Mustard oil

The vapors of this substance penetrate rapidly and deeply into tissue causing tissue damage and destruction well below the point of entry. Because of its penetrating ability, efforts to remove mustard oil from exposed tissue are futile after about 30 minutes. Mustard oil is classified as a "blistering gas" producing severely inflamed lesions that are susceptible to infection. Such lesions in the lungs are likely to be fatal. Mustard oil is a mutagen that causes mutations and is thought to be a primary carcinogen that does not require metabolic activation to produce cancer.

The chemical agents of greatest concern for their potential use in terrorist attacks are the organophosphorus "nerve gases." The first of these deadly agents was reported in 1937 by Gerhard Schrader of the German concern Farbenfabriken Bayer AG. Work continued on these compounds in Germany during World War II and by other nations after the war and during the cold war between Western and Communist bloc nations until around 1990. The possibility that Iraq possessed large stores of military nerve gas "weapons of mass destruction" was part of the rationale for the U.S./Iraq war in 2003. Among the common nerve gases are Sarin, Soman, Tabun, CMPF, VX, and diisopropylphosphofluoridate (fluorodiisopropyl phosphate) Structural formulas of three of these compounds are shown in Figure 13.4.

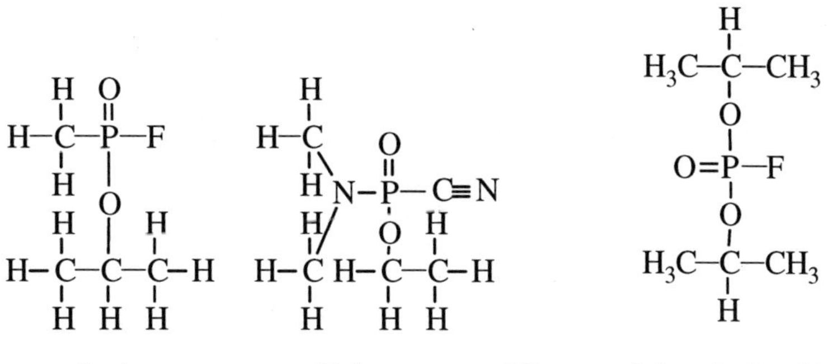

Sarin **Tabun** **Diisopropylphosphofluoridate**

Figure 13.4. Three common nerve gas organophosphate military poisons.

Sarin is perhaps the best known organophosphorus military poison because of its use in a 1995 attack by a terrorist group on the Tokyo subway system that killed several people and caused illness in a number of others. It is estimated that a dose of only about 0.01 milligrams of Sarin per kilogram of body mass is fatal; absorption of a single drop of liquid Sarin through the skin can kill a human. Sarin and other organophosphate military poisons act on the nervous system by binding with and inhibiting acetylcholinesterase enzyme, which is required to hydrolyze acetylcholine and stop nerve impulses once their function has been completed. The failure of this hydrolysis typically causes failure of the respiratory system. The binding of diisopropylphosphofluoridate to a serine side-chain on the active site of acetylcholinesterase enzyme is shown by the reaction below:

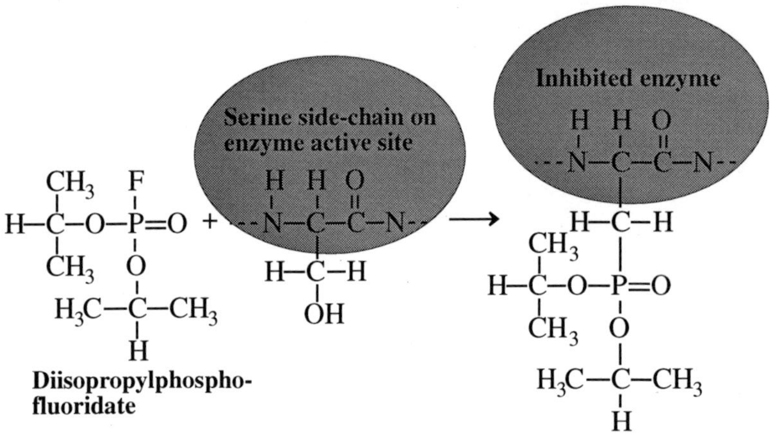

(13.5.2)

Biotoxins

Some of the most toxic substances known are produced by organisms and some of these have been used to attack humans. *Clostridium botulinum* bacteria growing in the absence of oxygen produce **botulinum** toxin, which has killed many people who have eaten improperly canned food (heating to 80–100° C for a sufficient time deactivates the toxin). Consisting of several proteins, botulinum toxin binds irreversibly to nerve terminals preventing the release of acetylcholine, an enzyme required for proper nerve function. Neurologic symptoms are followed by paralysis of the respiratory muscles and death. From the LD_{50} of about 1×10^{-5} mg/kg for botulinum toxin shown in Figure 13.1, it may be inferred that a 70 kg person could die from absorbing 70 kg $\times$ 1×10^{-5} mg/kg = 0.0007 mg of botulinum toxin, or only about 1 microgram of this extraordinarily deadly substance. A simple calculation would show that literally millions of people could be killed by the amount of this toxin that could be carried in a terrorist's pocket if some efficient means could be found to deliver it.

Another highly toxic natural product is **ricin**, a very stable proteinaceous material extracted from castor beans (*Ricinus communis*). Only about $^1/_2$ milligram of ricin (about the size of a pinhead) can be fatal when injected resulting in failure of kidneys, liver, and spleen along with massive blood loss from the digestive tract. Ricin gained notoriety in the 1978 assassination in London of the Bulgarian writer and journalist Georgi Markov injected with ricin from the tip of an umbrella. Although it is mentioned as a potential terrorist tool, ricin has its greatest toxicity by injection, which tends to limit its use as a weapon.

13.6. PROTECTING WATER, FOOD, AND AIR

Water to drink, food to eat, and air to breathe — the most basic of human needs — are likely targets for terrorist attack. The finding of a single cow with mad cow disease in Washington State in December, 2003, caused a major upheaval in cattle markets, loss of exports, and a final cost of perhaps hundreds of millions of dollars. Earlier outbreaks of mad cow disease and hoof and mouth disease in England had caused staggering economic loss. These disasters were the result of accident and poor agricultural and food production practices, not terrorism, but they illustrate the vulnerabilities of the food supply to potential terrorist attack.

A chemical attack on food supplies, though plausible, would be very difficult to carry out on a scale that would cause great damage. Spraying of food crops with toxic substances before harvest could cause some adverse effects and great anxiety, but would be relatively easy to detect and probably would not cause widespread harm.

The food supply is more susceptible to biological attack that could be carried out by terrorists than it is to chemical attack. Farmers are in a constant struggle with pests including insects and fungi. Introduction of exotic insects could cause severe crop devastation. Microorganisms that cause livestock disease can be spread by terrorists. The most likely such agent is the bacteria that cause anthrax, which afflicts both livestock and people. Viral diseases and fungal diseases of plants are also possibilities.

Direct contamination of food with disease-causing agents is a possible terrorist action. An incident has been described in which 12 laboratory staff were infected by acute diarrheal illness that hospitalized 4 due to ingestion of *Shigella dysenteriae* bacteria taken from a culture in the laboratory and deliberately placed on doughnuts or muffins in the facility break room.[4] Most people have experienced the misery of "food poisoning" caused by *Salmonella* bacteria in contaminated food. Although usually not fatal, *Salmonella* on food have the potential to cause temporary disability.

Because of their central distribution to large numbers of people, water supplies are susceptible to both chemical and biological attack. There are reports of terrorist groups trying to obtain deadly cyanide salts with the objective of contaminating water supplies. The tragic contamination of well water in Bangladesh described in Section 7.11 is still killing thousands and is a reminder of the potential for ill effects from chemical contamination of drinking water. Astoundingly toxic botulinus toxin from *Botulinus* bacteria (see Figure 13.2) is a potential chemical agent that could be put in water supplies. Though possible, it would be rather difficult to deliberately contaminate a municipal water supply with toxic chemicals.

Throughout history, drinking water contaminated by microorganisms that cause cholera, typhoid, and other maladies have killed millions and still cause disease and fatalities. In 1993, more than 400,000 people in Milwaukee were sickened and over 50 died from waterborne protozoal *Cryptosporidium parvum*. In May, 2000, approximately 3000 people were made ill and seven died in Walkerton, Ontario, Canada, from drinking water contaminated with *Escherichia coli* bacteria. Although usually *E. coli* bacteria are harmless and normal residents of human intestinal systems, they may develop strains with DNA transferred from *Shigella dysenteriae* bacteria that produce shiga toxin, the cause of potentially fatal dysentery, which is what happened in the Ontario incident. Bacteria that could be added deliberately to drinking water include *Shigella dysenteriae*, *Vibrio cholerae*, and *Yersinia pestis*.

Air is a possible medium for both chemical and biological attack. A means is required to deliver agents through the air, which makes it difficult to expose people through this medium. Although a low-flying crop-spraying plane would be an effective means of spreading either chemical or biological agents through the air, it would rapidly alert authorities leading to corrective action. Spores of bacteria that cause anthrax, *Bacillus anthracis*, are of particular concern for bioterror attack through air. Other microbial agents of concern for their potential for airborne attack include *Variola major*, which causes smallpox; *Francisella tularensis*, which causes tularemia, and viruses that cause viral hemorrhagic fevers, including Ebola, Marburg, Lassa, and Machupo.

Historically, the disease that has caused devastation most closely resembling the harm that could result from a massive attack by bioagents is "plague," which killed tens of millions of people in Europe during the Middle Ages. This malady is caused by infection with *Yersinia pestis* bacteria. This disease takes several forms, the most common of which is bubonic plague characterized by swollen, tender lymph glands called buboes. Readily cured by antibiotics in its early stages, it is transferred from infected rodents to humans by fleas, and several cases are reported each year in the Southwestern U.S. Pneumonic plague is readily spread through air between humans and is the form of most concern for terrorist attack. Initial symptoms similar to those of influenza progress to a fatal form of pneumonia.

13.7. DETECTING HAZARDS

A key to protection from terrorist threats is their detection before damage can be done. The detection of explosives immediately comes to mind. Another priority area is detection of disease-causing pathogenic organisms.

Hazardous substances are not readily detected by standard metal detectors and X-ray imagers used to find weapons and bombs on air travelers and in their luggage. Residues of TNT, RDX, and PETN explosives (see Figure 13.1) can be detected by sophisticated instruments including ion mobility spectrometers and chemiluminescence sensors. Such instruments normally detect residues of explosives on swabs from swabbing luggage; they can be circumvented by careful cleaning of luggage. Nuclear quadrupole resonance (NQR) may develop as an especially promising detection technique for explosives. One reason for this is its specificity for nitrogen, which is abundantly present in all common explosives. Secondly, NQR has the potential to detect explosives in containers and even land mines. It works by generating a pulse of radiofrequency radiation which excites nitrogen atoms to higher quantized energy levels. By following the signals given off as the atoms return to their ground energy levels the kinds and abundances of nitrogen functional groups in explosives molecules can be determined.

"Canine olfactory detection," which uses the sniffing abilities of dogs, is widely used to detect explosives, illicit drugs, and other potentially hazardous materials. A dog has approximately 220 million mucus-coated olfactory receptors, about 40 times the number possessed by a human, making the canine nose an extraordinarily sensitive detector. In order for canine olfactory detection to work, a rewards system must be used. This can lead to unpredictable, temperamental behavior in dogs of the type commonly attributed to humans and computers. As a result, dogs are not completely reliable and, according to an authority on the subject, "Dogs lie. We know they do."[5]

13.8. GREEN CHEMISTRY TO COMBAT TERRORISM

As *safe* and *sustainable* chemistry, green chemistry has an important role to play in the war on terrorism. When safe chemistry is practiced, hazards and hazardous substances that might be stolen or diverted for use in attacks are not made or used in large quantities. The successful practice of green chemistry means that chemical products do what they are supposed to do and are used in minimum quantities. With green chemistry, materials and processes that are likely to result in violent reactions, fires, high pressures, and other extreme conditions are avoided. The use of potentially hazardous auxiliary substances and flammable materials is avoided.

Green chemistry minimizes energy consumption, thereby reducing energetic, high-temperature processes that might be susceptible to sabotage. With its emphasis on biological processes, where applicable, green chemical processes are carried out under the mild, low temperature, toxic-substance-free, inherently less hazardous conditions conducive to biochemical reactions. By reducing demand for energy and raw materials, green chemistry reduces reliance on uncertain sources controlled by potentially hostile populations that are inherently subject to disruption and blackmail.

The practice of green chemistry requires exacting process control combined with real-time, in-process monitoring techniques. These are conditions that resist sabotage. In addition, green chemistry uses passive safety systems that function by default in the event of failure of or deliberate damage to intricate control systems. An example of such a system is gravity-fed cooling water for nuclear reactors that will continue to flow even if cooling system pumps fail.

The chemical industry and related enterprises continue to implement green chemistry practices to reduce hazards and vulnerabilities to attack. A DuPont chemical facility in Texas now makes methyl isocyanate, the agent of the catastrophic 1984 Bhopal, India, disaster, on site so that it does not have to store large quantities of this dangerous chemical. Some water treatment plants have started using relatively safe solid sodium hypochlorite in place of toxic, reactive liquid chlorine formerly stored in large pressurized tanks on site.

13.9. GREEN CHEMISTRY FOR SUSTAINABLE PROSPERITY AND A SAFER WORLD

Poverty, human misery, and hopelessness are conditions that feed terrorism. Although eliminating these conditions would not guarantee a safe world, reducing them would go far toward ensuring safe societies. People with satisfied material needs able to lead comfortable and fulfilling lives are relatively less likely to commit violent acts. To the extent that the practice of green chemistry fulfills human needs and makes life more comfortable, it can play a significant role in reducing terrorism.

Prosperity, narrowly defined, has resulted in consumption of increasingly scarce resources and environmental degradation. But, as stated by Elsa Reichmanis, President of the American Chemical Society, "We are past the days when we can trade environmental contamination for economic prosperity; that is only a temporary bargain, and the cost of pollution both economically and on human health is too high."[6] Green chemistry and the practice of industrial ecology can go far in providing high living standards without ruining the environment or recklessly exploiting resources.

The key to material prosperity consists of sources of abundant, inexpensive energy that can be tapped sustainably without major environmental harm; with such energy sources, all else is possible. Energy sources tend to be contentious and competition for them has precipitated past wars. The most abundant producers of petroleum, currently the key energy source for industrialized nations, are regions that are breeding grounds for terrorists. The provision of adequate energy independent of such sources would substantially reduce terrorist threats.

Figure 13.5 shows how abundant, sustainable energy is the foundation of the kind of prosperity that can lead to less terror-prone societies. Abundant energy can be used to produce food through synthesis of fertilizers (particularly by synthetic fixation of atmospheric nitrogen) and for irrigation, cultivation,

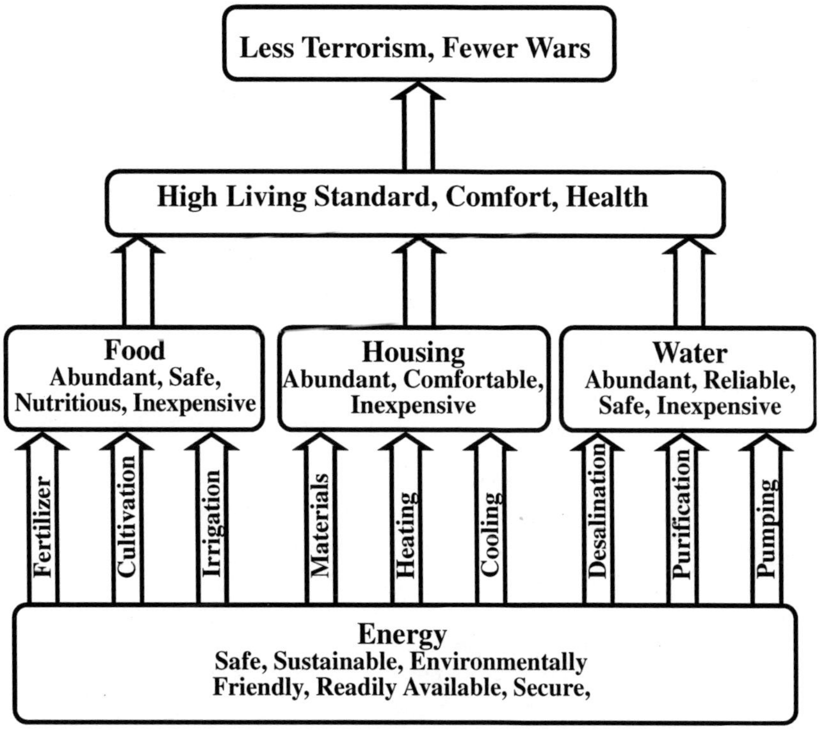

Figure 13.5. Abundant, sustainable energy is the base of a pyramid through which greater human well-being combined with suitable political and social systems leads to less conflict and terrorism.

and reclamation of farmland. Energy can be used to fabricate materials for housing and to provide the heating, cooling, and lighting required to make dwellings comfortable. Energy is required to pump water, in some cases over great distances from abundant sources to more arid regions. Energy can be used to purify marginal sources of water and to reclaim water after use. With an abundant source of energy, seawater can be desalinated for domestic, industrial, and agricultural use. Safe, comfortable, non-polluting transportation systems require an abundance of energy. These and other amenities based upon abundant, sustainable energy can go far toward building peaceful, productive societies with high living standards. They are not a guarantee because reasonable social systems, functional democratic governments, and sensible religions are needed as well, but material well being based upon a foundation of abundant, sustainable energy can go far in eliminating conditions that breed terrorism.

The provision of abundant, sustainable energy in the future requires the best practice of green chemistry, green engineering, and industrial ecology. Increased efficiency of energy utilization is a key aspect of providing more usable energy. Solar, wind, and biomass energy are leading contenders for renewable energy sources. Another essentially inexhaustible energy source is thermonuclear fusion, the stuff of hydrogen bombs and the sun's energy, but despite significant investments, a practical controlled system of energy production from this source has proven elusive. Fossil fuels will play an interim role, especially if sequestration of greenhouse gas carbon dioxide byproduct can be achieved. Despite its bad reputation in some quarters, nuclear fission with uranium fuel can provide abundant energy safely with new-generation nuclear reactors and reprocessing of nuclear fuel.

A key challenge in providing abundant renewable energy is its storage and transport. Wind and solar sources are by nature intermittent and dispersed, and they often produce electricity in locations far from where it is used, so the energy that they generate must be moved over long distances and stored for later use. For example, solar collectors function only in sunlight and, aside from rooftop installations, are often located in remote desert locations. Wind-powered electrical generators, which require at least some wind, are not usually welcome in urban areas where the energy is required, and some of the prime locations for them are the remote plains of Kansas or Texas or offshore. Superconductor or quantum conductor power cables are candidates for transport of electrical energy from source to use. Various means are available for energy storage, such as pumped water hydroelectric storage or high-speed flywheels coupled with electric motor/generators. In the future, elemental hydrogen, H_2, will be widely used for energy storage and transport as well as for fuel. Hydrogen can be produced by electrolysis of water and direct photoconversion of water to hydrogen and oxygen may eventually become practical. Hydrogen can be moved by pipeline and used to produce electricity directly in fuel cells.

LITERATURE CITED

1. Barabási, Albert-László, "We're all on the Grid Together," *New York Times*, August 16, 2003, p. A25.

2. Barabási, Albert-László, *Linked: The New Science of Networks*, Perseus Publishing, Boulder, CO, 2002.

3. "Toxic *Chemicals* and Explosive Materials," Chapter 4 in *Making the Nation Safer: The Role of Science and Technology in Countering Terrorism*, Bruce Alberts, William A. Wulf, and Kenneth I. Shine, Eds., The National Academies Press, Washington, DC, 2002, pp. 107–134.

4. Kolavic, Shellie A., Akiko Kimura, Shauna L. Simons, Laurence Slutsker, Suzanne Barth, and Charles E. *Haley*, "An Outbreak of *Shigella dysenteriae* Type 2 Among Laboratory Workers Due to Intentional Food Contamination," *Journal of the American Medical Association*, **278**, 396–8 (1997).

5. Derr, Mark, "With Dog Detectives, Mistakes Can Happen," *New York Times*, Dec. 24, 2002, p. D1.

6. As quoted in "Green Solutions to Global *Problems*," Stephen K. Ritter, *Chemical and Engineering News*, **81**, 31–33 (2003).

QUESTIONS AND PROBLEMS

1. At the beginning of this chapter, it was noted that the Alfred P. Murrah Oklahoma City Federal Building was destroyed in 1995 by the explosion of a mixture of ammonium nitrate (chemical formula NH_4NO_3) and diesel fuel ($C_{16}H_{32}$). An explosion occurs when chemicals react very rapidly to produce much more stable chemical species, gases, and heat. Consider that H_2O, CO_2, and N_2 are very stable forms of the elements present in a mixture of ammonium nitrate and diesel fuel, that CO_2, and N_2 are gases, and that at the temperature of an explosion H_2O is a gas as well. Attempt to write a chemical reaction that occurs when a mixture of ammonium nitrate and diesel fuel explodes.

2. Nitroglycerin explodes by itself without having to react with any other chemical. Look up its formula and, from the information given in the preceding question, give a possible chemical reaction for a nitroglycerin explosion.

3. Consider the infrastructure of your home. Suggest how it might be vulnerable because of the interconnectivity in it. Suggest how cascading failures might do great damage to your home life.

4. For a 75-kg person, estimate the lethal dose in grams, milligrams, or micrograms of each of the toxic substances shown in Figure 13.2.

5. Exposure of a person to toxic benzene can be estimated by measuring phenol in blood. Explain the rationale for such an analysis. Why is benzene epoxide not commonly determined to estimate benzene exposure?

6. Consider the toxicity of inhaled carbon monoxide in the context of Figure 13.3. Identify for carbon monoxide the receptor, the abnormal biochemical effect, and the physiological response manifesting toxicity.

7. Compare carbon monoxide to chlorine as agents of terrorist attack. What characteristic of carbon monoxide might make it a favored weapon. Why might chlorine be favored by terrorists? What characteristic would make it less attractive than carbon monoxide to terrorists?

8. Match each toxic substance and potential terrorist agent from the list on the left below with its characteristic from the list on the right.

A. Botulinus toxin	1. A disease-causing agent (pathogen)
B. Sarin	2. Inhibits acetylcholinesterase enzyme
C. Methyl isocyanate	3. Most toxic of those listed
D. *Shigella dysenteriae*	4. Toxic in gaseous, solid, or solution form
E. Cyanide	5. An industrial chemical that has killed thousands in a single incident

9. Explain how following the basic principles of green chemistry given in Section 1.10 reduces terrorist threats.

10. Figure 13.5 illustrating how abundant, sustainable sources of energy can lead to a high living standard emphasizes food, housing, and water. Suggest areas other than these three that depend upon abundant, sustainable energy and that can lead to a high living standard.

INDEX